AF552470

UNDERSTANDING GEOLOGY

UNDERSTANDING GEOLOGY

By

Dr. Veena

Dept. of Zoology

M.M.H. College

Ghaziabad (U.P.)

(India)

DISCOVERY PUBLISHING HOUSE PVT. LTD.

NEW DELHI-110 002

Published by:
Namit Wasan

DISCOVERY PUBLISHING HOUSE PVT. LTD.
4383/4B, Ansari Road, Darya Ganj
New Delhi-110 002 (India)
Phone : +91-11-23279245; 23253475; 43596065
E-mail : discoverybooksindia@gmail.com
discoverypublishinghouse@gmail.com
namitwasan9@gmail.com
web : www.discoverypublishinggroup.com

***Edition:* 2020**

ISBN: 978-81-8356-461-8

Understanding Geology

Printed at:
Infinity Imaging Systems
Delhi

Preface

The present title "Understanding Geology" has been written for those students interested in careers in diverse fields of biological sciences. It provides a structured approach to learning by covering all the important topics in a uniform, systematic format. The book has been comprehensively designed incorporating recent advances in this fast moving field. It also provides accessible information on geology in compact form for undergraduate students in biology and related life sciences. It is intelligible to the educated layman, though it deals with some complex ideas. It is an adequate text for all the requirements of students in this area. In addition, busy lecturers who require a quick reference compendium will find it useful, particularly for tuitional planning. Simple, yet hopefully clear figures and tables are provided throughout the book.

The over-riding goal of this book, and indeed of the whole *Understanding series,* is to present the essential information concerning geology in a compact, readily accessible form which leads itself to student learning and revision. The convergence of various approaches has generated a rich panorama of detail, the significance of which we are still attempting to unraval. The present text has been written as an introduction to this rapidly growing field.

To make the work more comprehensive and informative, the author has consulted many authoritative books, research journals, abstracts, monographs etc., so there can be no claim to originality except in the manner of treatment.

The author expresses his thanks to his friends and colleagues whose continue inspirations have initiated him to bring out this book.

The author expresses his gratitude to Mr. Wasan and staff of M/s Discovery Publishing House Pvt. Ltd. for their whole hearted cooperation in the publication of this book.

In the mean time, the author will remain sincerely responsible for any shortcomings of the book and be grateful to the readers for their suggestions and constructive criticism for the continuous betterment of the book. He takes this opportunity to appeal to the readers to send their suggestions straightaway to his Publisher.

Author

Contents

1

INTRODUCTION

In a dim corner of eastern Africa about 3 million years ago, an ancestral member of the human family stooped to 'pick up a sharp edged rock and discovered that it could be used as a tool. This simple act was the beginning of humanity's utilization of the many and varied resources of the earth. Unlike the antelope and bison on the horizon, these early members of ourown lineage were not content to passively-accept what the earth had to offer.

They, and their descendents, chose rather to exploit the environment and shape it to their needs. The resulting interaction between Homo sapiens and the earth has produced both benefits and problems. To understand the benefits, to gain knowledge useful, in solving the problems, and to enhance our aesthetic appreciation of this extraordinary planet, are sufficient reasons for undertaking the study of geology.

GEOLOGY: STUDY OF THE EARTH AND EARTHLIKE PLANETS

In a deep tunnel of a South African diamond mine, miners glance inquisitively at a young woman as she uses a geology pick to break loose a piece of hard, black, kimberlite rock. She seems not to notice the miners as she slowly turns the piece of rock in the light of her head lamp. A slight smile crosses her face as she recognizes nodules of iron-rich minerals that confirm a theory that the rock had been carried upward as a melt from depths far beyond the reach of the deepest drill hole.

On a quiet evening in the Mediterranean Sea, two scientists clad in shorts and tennis shoes shift their gaze slowly along the length of a cylindric core of sediment that had been brought up from the sea floor only moments before. Suddenly, one exclaims, "There it is!" "It" is a layer of gypsum that provides the two researchers with evidence that 6 million years ago the blue Mediterranean was a dry, hot, desert basin.

Tiny particles of ice decorate the beard of a man taking a sample of rock cuttings that flow from a pipe near the base of a thundering rotary drilling rig located in Alaska. He carries his sample into a trailer beside the huge derrick, washes the rock chips carefully, and examines them with the aid of a microscope. On the well record, next to the appropriate depth, he enters the words "sandstone, well sorted, fine-grained, *oil stained.*"

He holds a chip of rock to his nose and sniffs the faint odor of petroleum. Another sample of the rock is placed in a box containing a fluorescent lamp. The observer notes the vivid bluish hues that indicate the presence of crude oil. "Maybe," he thinks, "just maybe, this one will be an oil well." The researcher in the diamond mine, the investigators aboard the oceanographic vessel, and the explorer for petroleum are all engaged in the practice of a science devoted to the study of the earth-its origin, history, composition, properties, and resources. That science is called *geology,* a word derived from the Greek geo, meaning earth, and *logos* meaning discourse.

Today, a new breed of earth scientists has expanded the definition of geology to include geologic investigations of other planets in our solar system. This is not inappropriate, since geologic knowledge is indeed employed in such extraterrestrial tasks as interpreting photographs of the Martian surface, in estimating the power of volcanoes on Jupiter's moons, and in identifying the minerals that have crystallized to form the rocks of the lunar surface. Largely for convenience of study, the body of knowledge comprising geology can be divided into physical and historical aspects.

Although certain geologic topics may be included in either area, *physical geology* emphasizes the study of earth materials, examines the causes and results of crustal movements, and follows the manner in which water, ice, and wind modify planetary surfaces. Physical geology is thus *process* oriented. The *physical geologist* must be well trained in all the physical sciences, for he works as a chemist, physicist, mathematician, and engineer to formulate and test explanations regarding earth processes. *Historical geology* surveys the progress of change on earth as revealed by both rocks and fossils. It deals with ancient climates, with the evolution of life on earth, with the origin and historical development of continents and ocean basins, and with shifting patterns of lands and seas over millions of years. *Time is* the unifying theme in historical geology. Geology draws heavily on all other sciences and frequently merges into them. Many geologists consider themselves chemists or physicists who concentrate on the earth. Like chemists and physicists, geologists often set up carefully controlled experiments in the laboratory. Others collect data in the field. Their laboratory is an entire planet affected by a multitude of complex interacting processes.

Geologists are required to reach conclusions based on studies of the observed *results* of events whose *causes* may be traced back billions of years. An understanding of how earth materials originate, and the interpretation of the forces that produce mountain chains and volcanoes, must often be inferred from limited evidence or by analogy with events that can still he observed today. In arriving at an understanding of the earth, geologists employ the same mode of study used by scientists in other disciplines. That mode or procedure for investigating the earth is called the *scientific method.*

THE SCIENTIFIC METHOD IN GEOLOGY

Although there is no rigid pattern of thought used by geologists in solving problems, there is a general progression that begins with the *formulation of questions,* proceeds to the *collection of observations or data,* and is followed by the *development of an explanation or hypothesis* that is then rigorously *tested.* As an example of this scientific method, consider an investigation by Anita Fishman Harris, Jack Epstein, and Leonard Harris of the United States Geological Survey. These geologists were intrigued by the fact that a group of fossils called conodonts exhibited a range of colors from pale yellow to black.

Conodonts are the microscopic hard parts of organisms that lived on earth from about 600 to about 200 million years ago. They are composed of apatite—the same mineral of which bone is made and they are abundant at many localities in all kinds of sedimentary rocks. The earth scientists at the U.S. Geological Survey asked the question: "Why did conodonts of the same age, but from different parts of a region, have different colors?"

They then set about collecting the laboratory and field data needed to provide an answer. For their laboratory observations they selected pale yellow conodonts that came from a limestone formation that had never been deeply buried. The conodonts from the limestone were heated in an electric furnace. The investigators increased the temperature from 300°C to 600°C in 50° steps over a period of 10 to 50 days. To their delight, they observed that the conodonts changed in color through five phases from pale yellow to black. On further heating, so as to achieve conditions that change sedimentary rocks into metamorphic rocks, black conodonts changed sequentially to gray, opaque white, and finally crystal clear. The scientists then compared the color changes produced experimentally with the colors of untreated conodonts collected in the Appalachian Mountains from rocks that had been subjected to various depths and durations of burial.

When plotted on maps, their data dramatically revealed that light-colored conodonts are found in the western Appalachians where burial was least, and dark colored conodonts occur along the eastern side of the Appalachians where burial was greatest. The original question, "Why did conodonts of the same age, but from different parts of a region, differ in color?" could now be answered with a hypothesis stating that depth of burial and attendant increase in temperature are the dominant factors in conodont color alteration. Further, the duration of burial is an additional factor, and conodont color can be used as an index to assess the heating history of ancient rocks.

In the course of their investigation, another interesting fact was noted. It became apparent that rocks containing black conodonts, having experienced high temperatures, are less likely to yield commercial quantities of oil and gas. This index of alteration is now used routinely by many oil companies in their search for petroleum.

The observation has since been found to hold true for many other localities outside the original study area. It therefore has considerable economic importance as a predictor of success in finding new oil fields. It explains the absence of oil in some localities in which all other geologic conditions favor oil entrapment.

AN OVERVIEW OF THE THIRD PLANET

Movements

Imagine that you have taken a journey by spacecraft to a location high above our solar system. From your vantage point in space, and provided you had suitable tools, you might look downward and study the entire solar system. You would see at its center a rather ordinary star, our sun. Circling the sun would be the nine planets. On close examination you would discern that the four planets nearest the sun are relatively small, and that the next four are bigger and more widely spaced apart. Finally, the tiny planet Pluto would be seen at a distance from the sun of over 40 times that which separates the earth from the sun.

From your distant observation point, you would see little about the earth that is distinctive. It is the third planet from the sun and the largest of the inner group of rocky or terrestrial planets. Its moon is larger relative to the size of the earth than the moon of any other planet. Otherwise it would appear only as one planet among many, below average in size, and a mere speck when compared with the dimensions of the sun.

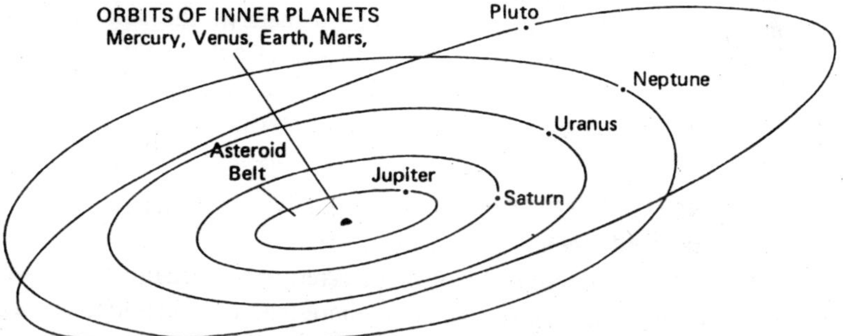

Figure 1.1: Orbits of the nine planets in the solar system.

With patience and mathematical skills, those aboard the spacecraft might next begin to work out the paths of the planets. It would soon be evident that the orbital routes of all the planets lie approximately in the same plane. Thus, the solar system is like a disc with the sun at its center. Looking down on the solar system from a point high above the North Pole, you would observe an interesting similarity of movements.

The revolution of the earth about the sun, of the moon about the earth, and the rotation of all three bodies on their individual axes would be similar, that is, in a *counterclockwise* direction. All of these movements play a role in our daily lives.

The sun, besides being our source of energy and light, is a reference object for the keeping of time. A calendar year, for example, is defined as the time required for the earth to make one entire orbit around the sun. During this time, the earth rotates 365.25 times about an imaginary line known as its axis. The axis of rotation makes an angle of 66.5° with the orbital plane (or 23.5° from a line perpendicular to that plane), and this is the reason why the earth experiences seasonal changes. Tides are caused largely by the gravitational attraction of the moon and sun for the earth, and tidal patterns of occurrence around the earth are directly related to rotational and orbital movement of the earth relative to the moon and sun.

Earth Dimensions and Density

From classical times it has been known that the earth is roughly spherical in shape. Actually, the planet is shaped more like a slightly flattened ball whose polar radius is about 21 km shorter than its equatorial radius. The average radius is 6371 km. The earth's overall density is about 5.5 grams per cubic centimeter (5.5 g/cm^3).

While the earth's density is greater than that of any other planet in the solar system, it is not appreciably different from that of Mercury, Venus, or Mars, the other three inner planets of the solar system. Because the average density of surface rocks is only about 2.7 g/cm^3, the material deep within the earth must have a density well in excess of the 5.5 g/cm^3 average. Very likely the material at the earth's center has a density of over 13.0 g/cm^3. The way this information is derived and its importance is discussed in other chapter.

Air, Water, and Rock

The splendid photographs of earth taken from space by the Apollo astronauts remind us that our planet is more than a rocky globe orbiting the sun. Wispy patterns of white clouds above the azure blue color tell us of the presence of an atmosphere and hydrosphere. Here and there one can even discern patches of tan that mark the locations of continents. Greenish hues provide evidence of the planet's most remarkable feature-life.

The Atmosphere

We live beneath a thin but vital envelope of gases called the atmosphere. We refer to these gases as *air.* "Pure air" is composed mainly of nitrogen (78.03 per cent by volume) and oxygen (20.99 per cent). The remaining 0.98 per cent of air is made of argon, carbon dioxide, and very small quantities of other gases.

One of these other components found mostly in the upper atmosphere is a form of oxygen called ozone. Ozone absorbs much of the sun's lethal ultraviolet radiation and is thus of critical importance to organisms on the surface of the earth. Air also contains 0.1 to 5.0 per cent water vapor. However, because this moisture content is so variable, it is not usually included in the list of atmosphere components.

Each day the earth receives radiation from the sun. At the surface of the earth, this energy is radiated back into the atmosphere as heat and absorbed by water vapor, carbon dioxide, and ozone. This returned energy heats the atmosphere and drives the winds. It is a major contributor to weather and climate, both of which influence the nature of weathering and erosion at the earth's surface.

The Hydrosphere

The discontinuous envelope of water that covers 71 per cent of the earth's surface is called the hydrosphere. It not only includes the ocean, but the water contained in streams and lakes, water frozen in glaciers, water that occurs underground in pores and cavities of rocks, and water vapor in the air. If surface irregularities such as continents and deep oceanic basins and trenches were smoothed out, water would completely cover the earth to a depth of more than 2 km.

Water is an exceedingly important geologic agent. Glaciers composed of water in its solid form alter the shape of the land by scouring, transporting, and depositing rock debris. Because water has the property of dissolving many natural compounds, it contributes significantly to the decomposition of rocks and, therefore, to the development of soils on which we depend for food. Water moving relentlessly downhill as sheetwash, in rills, and in streams loosens and carries away the particles of rock to lower elevations where they are deposited as layers of sediment. Clearly, the process of sculpturing our landscapes is primarily dependent on water.

The ocean basins hold most of the water in the hydrosphere. These basins are of enormous interest to geologists, who have discovered that they are not permanent and immobile as once believed, but rather are dynamic and ever changing. The sea floors move, and these movements have a direct relation to the formation of mountains, chains of volcanoes, deep sea trenches, and mid-oceanic ridges. The layers of sediment deposited on the ocean floor contain clues that allow geologists to decipher the earth's history. Here also one finds mineral resources and clues to the location of ore deposits elsewhere on the planet. The ocean provides part of our food supply and has a far-reaching influence on the climate we experience.

The lithosphere and the "spheres" beneath

Somewhat like the concentric shells of an onion, the earth is composed of a series of layers. As will be described in a later chapter, the existence of these layers has been deduced from the study of earthquake waves that have passed through the earth. At the surface is the thin outer shell known as the crust. The base of the crust is defined by a plane below which the velocity of certain earthquake waves is significantly greater than in the rocks above. A plane of this kind is called a seismic discontinuity, and the seismic discontinuity at the bottom of the crust has been named the Mohorovick discontinuity after its discoverer.

The crust of the earth really consists of two kinds of rock, each with its own distinctive general composition, thickness, and density. The continental crust has a composition somewhat similar to granite, has a relatively low density, and ranges in thickness between 35 and 60 km. The crust comprising the ocean basins is somewhat denser, rarely exceeds 5 km in thickness, and has an average composition of basalt.

The layer beneath the earth's crust is called the mantle. The mantle has not yet been penetrated by drilling, but earthquake data indicate that it extends from the base of the crust to a depth of about 2900 km. It comprises nearly 83 per cent of the earth's volume. The mantle has two upper parts that have the same composition but different physical characteristics.

The uppermost of these two parts lies immediately below the crust to a depth of about 120 km. It is solid and, together with the crust, is referred to as the *lithosphere*. The lithosphere rides over a plastic second layer of the mantle known as the *asthenosphere*, which extends to a depth of about 250 km. The velocity of earthquake waves in the asthenosphere is distinctly slower. Geophysicists believe that seismic waves are slowed in this zone because it is composed of weak material that is possibly able to flow plastically.

At the base of the mantle is yet another discontinuity that serves as the boundary between the mantle and the core. Geophysicists have recognized two parts of the core: a liquid outer zone and a solid inner core. Both parts are believed to be composed mainly of iron and nickel. As we have noted earlier, the specific gravity of the earth as a whole is greater than that of the common rocks making up its outer parts. Thus, we may conclude that the material of the core is indeed heavy.

Major features of the continents and ocean

It requires only a quick glance at a map of the world to become convinced that the most conspicuous elements of the earth's surface are continents and ocean basins. Both continents and ocean basins contain distinctive geologic features that have developed in response to particular geologic processes. For example, the major features of continents are stable regions and orogenic belts.

As is suggested by the name, stable regions are parts of the continents that are no longer disturbed by the kind of geologic forces that tend to distort rock layers and raise mountains. Plains and plateaus are characteristic of stable regions, although beneath their predominantly flat-lying strata may lie deformed rocks of former orogenic belts. Orogeny means mountainbuilding, and thus orogenic belts are zones in which great thicknesses of layered rocks have been strongly compressed, altered, and raised into lofty mountain chains.

A century ago, the ocean floors were considered to be rather featureless plains. Actually, they exhibit a variety of major features. Around the edges of the oceans are the submerged margins of the continents called continental shelves. The shelves are of enormous importance because they contain many off shore oil traps, as well as deposits of sand, gravel, oyster shells, and diamonds.

They are bounded seaward by the steeper continental slopes, which in turn drop off into less steep continental rises and eventually into the abyssal plains. Rising above the floors of the abyssal plains in the Atlantic, Pacific, and Indian oceans are perhaps the most impressive features of the ocean basins. These features, called mid-oceanic ridges, tower over 3500 meters above the sea floor. In contrast to the ridges, the ocean floor is cleaved by long narrow earthquake-prone *deep-sea trenches*.

The Aleutian trench extends 3500 km from Yakutat Bay in Alaska to the western tip of the Aleutian Islands. It has a maximal depth of more than 7500 km. Even greater depths are attained by the Marianas trench off the Phillipine Islands in the western Pacific. Here the ocean floor descends to the awesome depth of over 11,000 meters.

Plate tectonics—a preliminary overview

The world-encircling mid-oceanic ridges, the deepsea trenches, and even our great mountain ranges are neither haphazardly formed nor randomly located. These features are manifestations of a dynamic process called plate tectonics.

The central idea of plate tectonics is that the lithosphere of the earth is composed of seven or eight large "*plates*" and several smaller ones. The plates, which are 75 to 120 km thick, consist of both crust and part of the uppermost mantle. Lithospheric plates rest on a weak plastic layer of the mantle that we defined as the asthenosphere. Two characteristics of plates are that they move relative to one another and that volcanic eruptions and earthquakes frequently occur along their borders. Geologists are uncertain about what causes them to move. Most believe the plates are conveyed rather like the scum on boiling jam by the drag of slowly moving convection currents in the underlying mantle.

If one visualizes a globe sheathed in moving lithospheric plates, it becomes apparent that the boundaries between plates will differ according to the directions of plate movement. For example, zones along which plates move away from one another are called divergent boundaries.

Along these boundaries, black lavas pour from volcanoes and fill the area between the diverging plates.

The result is the great linear chains of mostly subsea volcanoes that constitute the mid-oceanic ridges. Iceland and the Azores represent small segments of the midAtlantic Ridge that have become emergent. On a sphere such as the earth, it is not possible for lithospheric plates to move apart from one another in some places, unless they converge or slide past one another along other boundaries. *Convergent boundaries* are those along which plates move toward one another.

They are also places where part of the crust is consumed in order to make room for new crust produced at divergent boundaries. This consumption may occur when one of the converging plates (usually the denser oceanic plate) sinks beneath the other plate to be melted and recycled in the underlying mantle. The process is called *subduction,* and the tops of the subjection zones are recognized by the presence of deep sea trenches.

Convergent boundaries are also zones of intense volcanic and earthquake activity. The third possibility for movement along a plate border is neither convergent nor divergent, but rather lateral motion as two plates slide past each other. Such zones of horizontal movement are called shear boundaries. They are located where crustal material is neither being added nor consumed. Shear boundaries are marked by great faults. Movements along these faults generate frequent and often intense earthquakes. The San Andreas fault zone in California is thought to mark a shear boundary.

A discussion of the historical development of plate tectonic theory and an explanation of the many lines of supporting evidence is presented in other chapter of this book.

A delicate balance

In speculating about the origin of such features as deep-sea trenches, mid-oceanic ridges, and mountain ranges, one becomes quickly aware of the mobility of the lithosphere. The old notion embodied in the phrase "good old terra firma" is no longer valid. Crustal movements are continuous and may vary from those that are imperceptibly slow to those that are rapid and violently destructive.

They may affect only small local areas or disturb a large region of a continent or ocean basin. The earth's instability is apparent to anyone who has been jolted by an earthquake or who has witnessed a volcano erupt. To the geologist, the continuum of unrest can be traced through folded strata, layers of lava, and great displacements of crustal rocks. These features are manifestations of powerful inner forces capable of uplifting entire continents and changing the dimensions of the sea floors.

However, just as there are powerful *constructional* forces on the earth, there are less dramatic, but equally significant, destructive forces. Geologic agents such as streams, glaciers, and wind erode the lands and carry the products of erosion to the ocean. One might surmise that the floors of the ocean would be the final resting place for these landderived sediments, but this is not likely.

As suggested earlier, the sea floors are not static but move along slowly like gargantuan conveyor belts. Sediment accumulating in the ocean basins is carried along on the moving floor to so-called subjection zones, and there it may be pulled downward into the asthenosphere and melted. Thus, the earth's surface is rather like a constantly changing battleground where internal forces raise and produce new land areas, where gravity and geologic agents operating near the surface work to wear the lands away, and where earth materials are continuously being changed and recycled.

THE RECYCLING OF EARTH MATERIALS

We have only to look about us to see that the crust of the earth is composed of rocks and minerals. Minerals are naturally formed chemical elements or compounds having a definite range in chemical composition and usually a regular crystal structure. Rocks are aggregates of one or more minerals that constitute an appreciable part of the earth's crust.

Rocks can be identified according to their physical properties and mineralogic composition. The properties that provide clues to the origin of the rock are of greatest importance. By selecting properties having the greatest genetic significance, geologists have been able to group rocks into three great families according to the way in which they formed. *Igneous rocks* are those that have cooled from a molten state. *Sedimentary rocks* are formed by the accumulation and consolidation of sediment. *Metamorphic rocks* are those that have formed in the solid state from preexisting rocks of any kind in response to heat, pressure, and associated chemical activity.

In thinking about the three major groupings of rocks, it is important to remember that rocks and minerals are not immune to change. The earth's crust is dynamic and ever-changing. Any sedimentary or metamorphic rock may be melted to produce igneous rocks, and a previously existing rock of any category can be affected by the pressure and heat accompanying mountain-building so as to produce metamorphic rocks. The weathered and eroded residue of any family of rocks can be observed today being transported by rivers to the sea for deposition and eventual conversion into sedimentary rocks. These changes can be incorporated into a schematic diagram that is designated the "rock cycle".

TIME AND GEOLOGY

The single most prevalent theme in geology is time. Time is involved in every change that occurs on and within the earth. The time during which geologic agencies act upon the crust influences the characteristics of continents and the configurations of land-scapes. Time is involved in the cooling of molten masses of the rocks and has a bearing on the characteristics of rocks that ultimately form. Time is an ingredient in organic evolution and provides the framework for the history of animals and plants. Indeed, every view of the earth contains the tangible imprint of the passage of time.

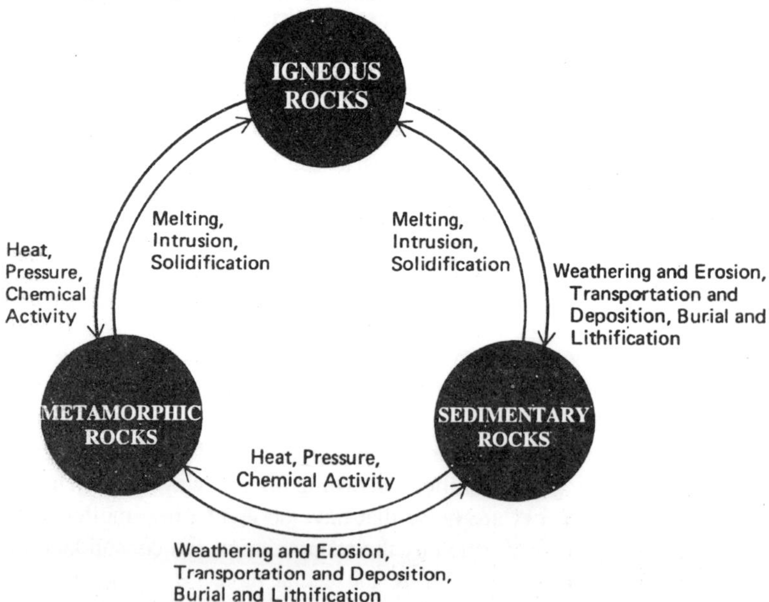

Figure 1.2: The rock cycle.

According to recent estimates, the earth is approximately 4700 million years old. The conception of so vast a span of years helps us to comprehend the fact that mountains are really only transitory forms and that processes that appear to be working imperceptibly slowly have nevertheless had ample time to change the surface of the earth. With geologic knowledge, each of us can expand his time window to better understand the earth. We need not be limited by what we perceive only in the span of our lives. In other chapter of this book, we will describe how geologists deal with time. We will learn how they determine the age of rocks and minerals and are thereby able to discern when momentous events occurred in the past.

2

MATTER AND ENERGY

The real world with which geology is concerned consists of two components: matter and energy. Physics and chemistry are the basic sciences that deal with the nature of matter and energy and the formulation of laws that govern their behavior. Modern physical geology depends so heavily upon the use of physics and chemistry that two major areas of geology-geophysics and geochemistry-overlap between geology and physics and between geology and chemistry, respectively. In this crucial overlap zone have taken place many of the great recent advances in geology in the realm of plate tectonics. One could make a similar statement about the role that physics and chemistry have played in modern biology.

In this chapter we review some important principles of physics and chemistry essential to an understanding of modern geology. If you are already familiar with this material, pass over it lightly, making sure, however, that you understand the topics. For those of you who have not studied physical science, we hope this presentation will be helpful, even though it is brief.

MATTER AND ENERGY

To define the terms *matter* and *energy* is not an easy task because they represent concepts that include everything of which we have any comprehension in the real world. To start, we can use the word "substance" for the word "matter," but this only postpones the problem. A substance occupies space; it is "tangible" because it can be seen, felt, tasted, measured, weighed, stored, and so forth. Matter has the mysterious property of *gravitation*, the mutual attraction between any two aggregates (lumps, groups, pieces) of matter. Rather than attempt a simple definition of matter, we shall proceed a little later to investigate the nature of matter in some depth. Energy, often defined in terms of its effects, is "the ability to do work." Energy is somehow always involved in the motion of matter, but energy can be stored in matter that does not appear outwardly to have motion.

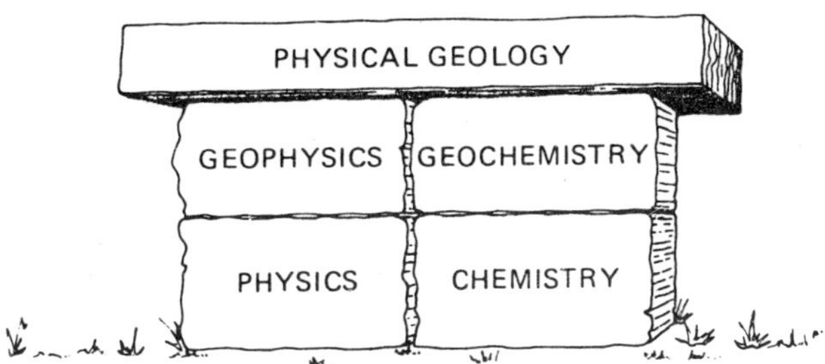

Figure 2.1: Physical geology as a capstone supported by the physical sciences-physics and chemistry—which are the foundation blocks. In the intermediate position are geophysics and geochemistry, which apply principles of physics and chemistry to the solving of problems in physical geology.

A brick poised on the brink of a high parapet contains a store of energy even though the brick is at rest. That stored energy will become very obvious if the brick is nudged slightly and swiftly falls to the sidewalk below. Whatever energy is, then, it can move or flow from place to place and it can also be held in storage in various ways. We frequently refer to energy as something "*expendable.*" For example, someone may say, "I used up a lot of energy playing those two sets of tennis." Actually, energy cannot be destroyed-that is, be removed from existence—by being "used"; it can only change from one form to another and move from one place to another. The same statement applies to matter, which cannot be destroyed, but only changed or moved from place to place.

LAWS OF CONSERVATION OF MATTER AND ENERGY

Two basic laws that emerged early in the history of science are the *law of conservation of matter* and the *law of conservation of energy*. The first of these laws simply states that within the universe as a whole, matter cannot be created or destroyed. The second law says the same thing with respect to energy. Not long after Marie and Pierre Curie in 1898 isolated two radioactive elements, polonium and radium, scientists were forced to conclude that matter can be transformed into energy, and thus it became necessary to combine the two yaws into a single yaw of conservation of matter and energy: The sum total of all matter and energy in the universe must remain constant.

The conversion of matter to energy through radioactivity within mineral substances of the solid earth has been of enormous importance to geologic processes over the entire 4-billion-year history of the planet. At the appropriate point we shall investigate this subject in some detail.

CLASSES OF MATTER

Let us now investigate the various kinds or classes of matter in the universe. A system devised by Professor John R. Holum, a chemist, collectively describes all matter by the word "substances." Next, *pure substances* are distinguished from *mixtures*, which consist of two or more pure substances mixed together in indefinite proportions. Pure substances are either *elements* or *compounds*. Examples of elements are several metals in the pure state: for example, iron, gold, copper, or silver.

Aythough never found in a condition of absolute purity, nuggets (lumps) or crystals of each of these *native elements* occur naturally in rock. Take native copper: In 1857 a single mass of native copper weighing 380,000 kg (about 420 U.S. tons) was discovered on the Keweenaw Peninsula in northern Michigan. Long before Europeans came to North America, Indians were using pieces of native copper as a malleable metal out of which to make various tools and ornaments. In addition, we are all familiar with flakes and nuggets of gold prospectors seek in the beds of streams. A large gold nugget is shown in Figure elsewhere in this chapter.

Compounds are also familiar: pure water, carbon dioxide gas, or crystals of amethyst quartz, for example. As we have described them, both elements and compounds exist "in the realm of things where samples can be seen and weighed directly," to use Professor Holum's words. The lower box in Figure elsewhere in this chapter takes us into "the realm where individual particles cannot be directly seen and weighed." The smallest representative samples of pure substances are of two classes: *atoms* and *compound particles*. The latter consist of groups of atoms, occurring either as molecules or as ions. We shall describe atoms, molecules, and ions in detail later in this chapter.

ATOMS AND COMPOUNDS

The notion that atoms represent the smallest units of matter was expressed as early as 500 a.c. by Greek philosophers. (The Greek word, *atomos*, means "indivisible," or "not cuttable.") The idea of the atom was revived by John Dayton, an English chemist and physicist, who proposed the theory that all matter consists of indestructible atoms. A given element, he reasoned, consists of only one variety of atom, having a given weight and other distinctive properties. Atoms of another element must have a different weight and a different set of properties. Atoms of two or more different elements can combine to form *compounds*. An important point in Dayton's theory is that the atoms of the component elements of a given compound must always combine in a definite ratio by weight, and must always involve whole numbers of atoms.

Mixtures of substances, referred to in the upper part of Figure elsewhere in this chapter, usually consist of different compounds mixed together physically in varying proportions. For example, most rocks exposed at the earth's surface are mixtures of several compounds; seawater is a mixture of water and several different dissolved salts, each of which is a distinctive type of compound.

Much later than Dayton's time came the discovery that atoms themselves are made up of various kinds of parts, which we can refer to as *subatomic particles*. If a single atom were to be enormously expanded, it would be found to consist of a dense central nucleus orbited at high speed by a large number of extremely minute particles called electrons.

Even the nucleus of the atom consists, in turn, of different kinds of parts, called *nuclear particles*. The only two classes of nuclear particles we will refer to are neutrons and protons, but you should be aware that there are many others.

MASS AND DENSITY

This is a good point at which to think in a very precise way about two of the most important terms relating to matter-mass and density. The measure of quantity of matter is called *mass*. The mass of a particular isolated solid object, such as a pebble or a billiard ball, is the quantity of matter contained within that object.

A practical method of determining the mass of an object at the earth's surface is to weigh it. However, when we place the object on an ordinary scale and read the weight in grams or kilograms, we are actually measuring the force with which the earth's gravitational force pulls down upon the object.

Performing this same operation on the moon, where the pulling force is much less, would give a much different weight on the same scale, but the mass of the object would be exactly the same as on earth, or as anywhere else in the solar system. A practical solution to the problem of establishing a unit of mass is to designate some particular quantity of matter as the standard mass (unit mass). As the French Revolution drew to a close, the French National Assembly organised the metric system and defined its units.

As originally planned, the unit of mass was to be the *gram* (*g*), a mass represented by one cubic centimeter (1 cc) of perfectly pure water at a temperature of 4°C. Later, for practical reasons, the standard of mass was established as the quantity of matter in a cylindrical metal block of platinum-iridium alloy equivalent in weight to 1000 g of pure water. This standard unit of mass is the *kilogram* (*kg*); it rests in a vault at Sevres, France. *A metric ton* (*MT*) is 1000 kg.

A second observation we can all make is that the quantity of matter (mass) in an object of a given volume varies a great deal, depending upon the substance of which the object is composed. Given two cubes of exactly the same dimensions, one made of rock crystal (quartz) and the other of iron, you need only lift each cube in turn to realize that the mass of the iron cube is much greater than that of the quartz cube.

The term *density* describes the quantity of matter (mass) in a unit volume of space. Using the gram as the unit of mass, the density of a substance is defined as the mass in grams per cubic centimeter (g/cc). Because the meter (m) and kilogram (kg) are now designated as the official units of length and mass in the SI (System Internationale), it is considered proper to define density in terms of 1000 kg (10^3 kg) per cubic meter (m^3). The numbers come out the same in any case.

Using grams and centimeters as units, the density of the mineral quartz is approximately 2.6 g/cc, that of iron is about 8.0 g/cc. Figure elsewhere in this chapter shows how four similar cubes, each of a different substance, cause a coil spring to be extended to greater lengths in proportion to increased density.

The concept of density, applied to minerals and rocks, is of great importance in understanding the classification of rocks and the structure of the earth in terms of a system of layers, or shells.

GRAVITATION AND MATTER

We mentioned earlier in this chapter that the nature of matter is closely tied in with the phenomenon of gravitation, the attraction that every particle of matter in the universe exerts upon every other particle. We can, in fact, define mass as the property of being susceptible to this attraction. Probably all of you have memorised at one time or another the *law of gravitation* formulated by Sir Isaac Newton.

It goes as follows: Any two bodies attract one another with a force that is directly proportional to the product of their masses and inversely proportional to the square of the distance separating them.

Written formally as an equation, the law of gravitation reads:

$$F_g = \frac{G\, M_1\, M_2}{D^2}$$

where

M_1 and M_2	are the two masses
D	is the distance separating the masses

and

F_g	is the gravitational force between them

(The letter *G* is a number known as the universal constant of gravitation; its numerical value is not important here.)

So far as geologic processes at or near the earth's solid surface are concerned, what counts is the gravitational attraction of the earth—a truly enormous mass—exerted upon very tiny masses, such as molecules of gas or liquid, and particles of soil, rock, or organic matter. (The infinitesimally small force with which the particle attracts the entire earth can be written off as inconsequential.)

The earth's gravitational attraction for very small objects within its surface region is called *gravity*. Because the earth is very nearly spherical, gravity has an almost constant value over the entire surface. In other words, the earth's attraction for a given quantity of matter (such as a mass of 1 kg) will be the same over the entire globe, subject only to minor corrections that are not particularly important in understanding most geologic processes. If a particle of matter is placed in a true vacuum (absence of substance) at the earth's surface, it will fall faster and faster toward the earth's center of mass.

This uniform increase in the velocity with time represents an *acceleration*, and is a constant value. By careful measurement, we know that the velocity of fall will increase by 9.8 m per second for each second of fall. The *acceleration of gravity* is, then, 9.8 m/sec^2. The acceleration of gravity multiplied by the quantity of mass upon which it acts constitutes the *force of gravity*. Stated in a simple equation,

$$F_g = M \times g$$

where

F_g	is the force of gravity
M	is the quantity of mass

and

g is the acceleration of gravity

Because the acceleration of gravity is nearly constant everywhere at sea level, the force of gravity varies according to the quantity of mass. We can therefore measure the quantity of mass with reasonable accuracy by the force registered upon a scale of the type that measures the amount of stretching of a coil spring. The standard unit of force is the *newton* (*N*). For one kilogram mass, the force of gravity is equal to 9.8 N.

STATES OF MATTER

The physical condition, or state, in which we find matter at a given place and time is a subject of great importance in geology.

The three common states of matter—solid, liquid, and gaseous-apply both to pure substances (elements, compounds) and to mixtures. Using only the simplest concepts of atoms and compounds, as Dalton visualised them, we can describe the three states of matter in terms of observable behavior. For this purpose, the atoms or molecules that comprise matter can be visualised as uniform spheres, all physically alike.

Certain states of matter are well known to everyone. Every day we drink water in its *liquid state*. We may also have close by a supply of ice cubes-water in the *solid state*. The *gaseous* state of water, called water vapor, cannot be seen, but is easily sensed by the human skin in summer when the relative humidity of the air is high.

To define these states further, *a gas* is a substance that expands easily and rapidly to fill any small, empty container. Atoms or molecules of the gas are in highspeed motion, and empty space between them is vast in comparison with their dimensions. The particles move in random directions and frequently collide. They rebound like perfect spheres at each impact, changing direction abruptly. They also strike and rebound off the walls of the container. A gas is usually very much less dense than a liquid or solid consisting of the same chemical substance. For example, the gaseous water vapor in warm, moist air has a density only about 1/100,000 that of liquid water.

A liquid is a substance that flows freely in response to unbalanced forces but maintains a free upper surface so long as it does not fill the container or cavity in which it is held. The molecules of a liquid compound-water, for example-move more or less freely past one another as individuals or small groups. For many practical purposes, liquids can be considered to be incompressible (not capable of being compressed). Under rather strong confining pressures (such as would exist at the bottom of the deep ocean), liquids are compressed only slightly into a smaller volume.

Both gases and liquids are classed as *fluids* because both substances freely flow. Put simply, both these substances flow downhill under the force of gravity wherever possible. Fluids of different density tend to come to rest in layers with the fluid of greatest density at the bottom and that of least density at the top.

A solid is a substance that resists changes of shape and volume. Solids are typically capable of withstanding large unbalanced forces (i.e., strong stresses) without yielding permanently, although they undergo a small amount of elastic bending. When yielding does occur, it is usually by sudden breakage, or rupture.

Because most rock of the outermost layer of the earth is in the solid state, these principles have important applications in many areas of geology, including the behaviour of lithospheric plates and the occurrence of earthquakes.

Changes of state are accompanied by either an energy input into or an energy output from the substance undergoing change. We are all familiar with this principle in the preparation of food. To boil water (change liquid to a vapor state), a great deal of heat must be applied; conversely, to freeze water, a great deal of heat must be removed. Our next step will be to examine the nature and forms of energy that can be easily observed to cause changes in pure substances and mixtures behaving as gases, liquids, and solids.

KINDS OF ENERGY

We stated earlier that energy is commonly defined as the ability to do work. In mathematical terms of physics, energy is the product of force and distance:

$$E = F \times d$$

where

E is energy

F is the force applied

and

d is the distance through which the force acts

Thus, energy is the ability to move an object (exert a force) for a certain distance. Energy is stored and transported in a variety of ways. Some of the recognised forms of energy are: mechanical energy, heat energy, energy transmitted by radiation through space (electromagnetic energy), chemical energy, electrical energy, and nuclear energy.

Mechanical Energy

Mechanical energy-energy associated with the motion of matter-has two forms: kinetic energy and potential energy. *Kinetic energy is* the ability of a mass in motion to do work. Thus, an automobile traveling down a highway possesses kinetic energy because it is a mass in motion.

Should this mass strike a telephone pole, its ability to do work upon its own body and upon the telephone pole will be quite obvious. The energy it will release in collision will increase with the weight (mass) of the car, and it will also increase with the square of the auto's speed. Kinetic energy, then, is proportional to the quantity of mass in motion multiplied by the square of its velocity. Stated as a simple equation,

$$E_k = M \times v^2$$

where

E_k is kinetic energy

M is quantity of mass in motion

and

v is velocity in a straight line

Kinetic energy is obvious in many forms of geologic processes acting at the earth's surface—a rolling boulder, a flowing stream, the pounding surf. Lithospheric plates in motion display kinetic energy. Although they seem to barely creep along, their mass is so enormous that the total kinetic energy in a single moving plate is very large indeed.

Potential energy, or energy of position, is equal to the kinetic energy an object would attain if it were allowed to fall under the influence of gravity. Imagine a brick balanced on the edge of a tabletop, then falling. The kinetic energy the brick possesses at the moment it hits the floor, as we have seen, is proportional to the mass of the brick multiplied by the square of its velocity.

Because the brick is being accelerated by gravity at a constant 9.8 m/sec^2, we could find its velocity at impact by measuring the distance it will fall and using the acceleration formula from physics. If the brick is again lifted to the tabletop, the work done in lifting gives the block a renewed quantity of potential energy. This energy will be released when the brick is again allowed to fall.

It should be obvious at this point that the floor is merely a convenient stopping place for both the brick and our discussion; if we sawed a hole in the floor and allowed the brick to fall further, it would possess even more kinetic energy at its impact on the floor below. Therefore, with respect to the lower floor level, the brick would have a greater potential energy when we return it to the tabletop. Thus, the three factors which determine the magnitude of potential energy are the mass of the object, the vertical distance it will fall, and the acceleration of gravity.

Looking around us outdoors, we can spot many instances of the existence of potential energy in a landscape, for example, a boulder poised at the top of a steep mountain face. In fact, the entire mountain represents a large reservoir of potential energy judged in reference to the level of the floor of an adjacent valley. Going even further, an entire continent possesses potential energy with reference to sea level or to any lower level we might choose to use as a reference.

The potential for any mass of rock, no matter how small or how large, to move toward the earth's center is always present. Geologists sometimes call this potential energy source "gravitational energy" and use it to explain a number of important earth phenomena. We should be careful at this point to note that gravitational force and gravity are not, in themselves, forms of energy.

Mechanical energy can be transmitted from one place to another in the form of *wave motion*, in which kinetic energy is carried through matter by an impulse passed along from one particle to the next. A sound wave is one example—a push on air molecules at one point will be transmitted outward in all directions. Geologists are particularly interested in the phenomenon of earthquake waves, which carry large amounts of energy for great distances, not only over the ground surface, but also in paths deep within the solid earth; the familiar Richter scale of earthquake intensity measures the quantity of energy released by an earthquake. In all mechanical forms of wave motion matter is displaced (up and down, sidewise, or forward and backward) in a rhythmic manner, but as frictional resistance within the moving substance withdraws energy, the waves gradually die out as they travel farther from the source.

Heat Energy

Sensible heat is another form of energy of paramount importance in geologic processes. Kinetic energy can readily be converted into sensible heat energy through the mechanism of friction. Braking a moving automobile is a familiar example. As the automobile slows to a stop (losing kinetic energy), the brake drums become intensely hot. What is probably not apparent, however, is just how much heat is related to how much motion. Laws of physics tell us that "a little goes a long way." The energy required to heat one cubic centimeter of water one degree Celsius is about the same as that contained in a mass of nearly 100 metric tons moving at a speed of one centimeter per second.

Sensible heat represents kinetic energy, but it is of an internal form, rather than the external form we see in moving masses. Thus, a cupful of water resting completely motionless on the table has internal energy because of constant motion of the water molecules on a scale too small to be visible.

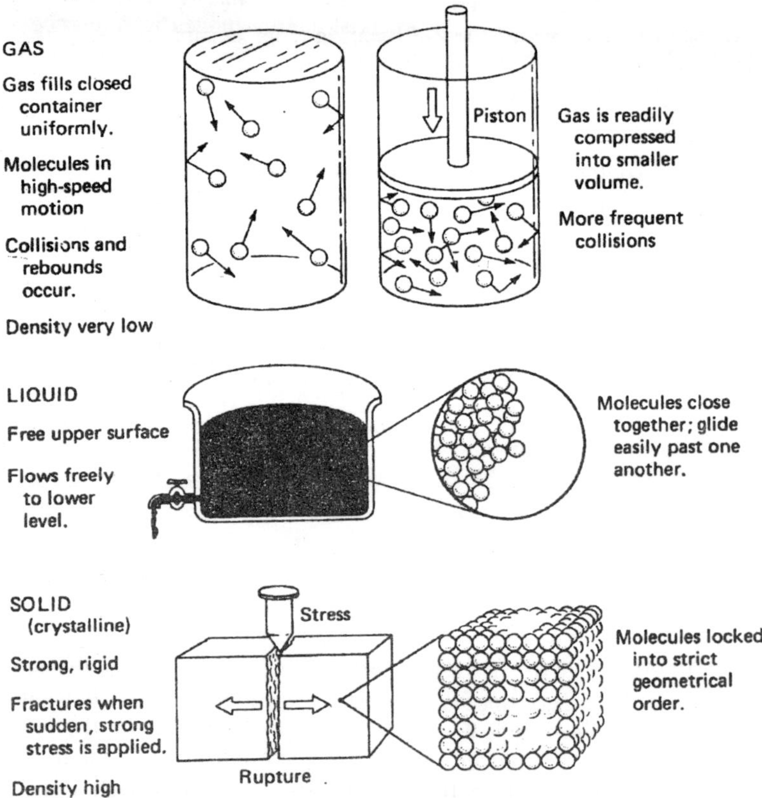

Figure 2.2: Some properties of gases, liquids, and solids.

This internal motion is the sensible heat of the substance, and its level of intensity is measured by the thermometer. For gases, the internal motion is in the form of high-speed travel of free molecules in space but with frequent collisions with other molecules. The energy level in the gaseous state of water (water vapor) is thus higher than within the liquid water. Ice, on the other hand, represents a lower energy level than liquid water, for here the molecules are locked into place in a fixed geometrical arrangement. For these molecules in the solid state, the motion is one of vibration without relative motion.

When ice melts, work must be done to overcome the crystalline bonds between molecules. This work requires an input of energy, but it does not raise the temperature of the substance. Instead, the energy seems mysteriously to disappear. Since energy cannot actually be lost, it is placed in storage in a form known as *latent heat*. Should the water freeze and again become ice, the latent heat will be released as sensible heat. A similar transformation from sensible to latent heat takes place during *evaporation*, when a liquid becomes a gas, because work must be done to overcome the bonds between molecules of the liquid.

When water vapor returns to the liquid state, a process of *condensation*, latent heat is released as sensible heat. Both sensible heat and latent heat are forms of stored energy. (Potential energy is also a form of stored energy.) Sensible heat in rock is important to geologists because the physical properties of rock change as temperature increases. Generally speaking, as rock is heated, it expands and becomes less dense. Heated to a critical temperature–its melting point–rock changes state from a solid to a liquid.

Heat energy is largely responsible for the motions of lithospheric plates and the building of mountain ranges associated with those plate motions. One crucial question we must investigate is the source of the great reservoir of heat in the solid earth. Is it heat left over from some early event in the history of our planet, or is it heat being continually generated in some mysterious way?

Electromagnetic Energy

Sensible heat may be lost directly to the surroundings through conduction, but even in a vacuum objects lose heat. A fundamental law of physics states that all matter at temperatures above absolute zero radiates *electromagnetic energy*. We can think of this radiation as taking the form of waves traveling in straight lines through space. The waves come in a very wide range of lengths, but all travel at the same speed300,000 km/sec-regardless of their length.

Together, the total assemblage of waves of all lengths constitutes the *electromagnetic spectrum*. It includes *visible light*, with its rainbow of colours, and also invisible shorter waves such as *ultraviolet rays*, *X rays*, and *gamma rays*. Besides these, the spectrum includes invisible long waves known as *infrared rays* (sometimes called heat rays), and still longer microwaves and radio waves.

The temperature of the radiating objects determines which part of the spectrum carries the radiated energy. Thus, the sun, whose surface temperature is about three times greater than that of molten steel, radiates most strongly in the visible portion, although all other parts of the spectrum are represented. At normal earth temperatures, however, objects radiate mostly in the infrared part of the spectrum and not at all in the visible part.

The total energy of radiation is also dependent or the temperature of the radiating object. Thus, one square centimeter of the sun's surface emits about 1.5 million times as much energy each second as does one square centimeter of the earth's surface. Electromagnetic energy received from the sun powers a group of important geologic processes constantly at work on the earth's surface. Electromagnetic energy, upon arriving at the earth's surface, is continually converted into mechanical energy and sensible heat, which in turn are consumed in the breakup of exposed rock, in causing its chemical change, and in the transport of the resulting rock particles to new places of rest.

Chemical Energy

Yet another form of energy is *chemical energy*, absorbed or released by matter when chemical reactions take place. These reactions involve the coming together of atoms to form molecules, the recombining of molecules into new compounds, and the reversion to simpler forms of matter.

A familiar example that relates indirectly to certain geologic processes can be felt in the setting of wet cement or plaster. When water is added to these substances, they spontaneously harden into a rocklike mass. At the same time, you will notice that the mixture becomes quite warm to the touch. A chemical reaction is occurring, involving the liberation of heat stored as energy in the reacting chemicals.

In other types of reactions, the material mixture becomes colder as the reaction takes place, showing that sensible heat is being taken up in storage as chemical energy. Geologic processes that involve such changes in mineral matter provide many examples of the absorption, release, and storage of chemical energy.

Electrical Energy

Chemical energy can be converted into another form of *energy—electrical energy*—a process we make use of in the ordinary flashlight cell or auto battery. Within the auto battery, lead plates are immersed in sulfuric acid. A chemical reaction between the acid and the plates produces a flow of electrons through a wire connected to the two battery terminals. Thus, the flow of electrical energy is represented by the flow of electrons through a conducting substance, usually a metal. Later in this chapter, we will describe the electron as a very small particle of matter.

Electrical energy can also be stored temporarily in an *electrical charge*. A familiar example is the static electricity you sometimes generate on your clothing and bodies as you walk across a carpeted floor on a cold, dry day. When you touch another object, the stored electricity is released in a small spark, often accompanied by a rather unpleasant sensation. An electrical charge consists of vast numbers of free electrons collected on the surface of an object. Accumulated charges are of two kinds, positive and negative, designated by plus and minus signs, respectively. It is important to note that objects having charges of the opposite sign (+ –) tend to be drawn together by an attracting force, whereas objects carrying charges of the same sign (+ + or – –) tend to be pushed apart by a repelling force. These concepts will be useful to recall as we investigate the nature of atoms.

Nuclear Energy

Although this list of forms of energy is not complete, we close it with *nuclear energy*, an important subject in geology. To explain the nature of nuclear energy requires a knowledge of the internal structure of the atom and the working of its component particles. Nuclear energy is a subject we will develop quite thoroughly on later pages, because it is generally believed that most of the earth's internal heat originates as nuclear energy. We can only point out here that the internal atomic structure of certain elements present in rock is unstable. These unstable elements spontaneously change into other elements, at the same time releasing energy to surrounding substances. This process of atomic change is called radioactivity, and it involves the steady conversion of matter into energy.

FLOW SYSTEMS OF ENERGY AND MATTER

Understanding the varied processes that form rocks and that shape those rocks into many different configurations requires geologists to think in terms of *flow systems* of both matter and energy. A flow system is simply a series of connected pathways through which either energy or matter moves more or less continuously. An *energy system* traces the flow of energy from a point of entry to a point of emergence. As energy flows through such a system, it may change form many times and may be detained temporarily in storage from place to place.

In this process, the energy flow makes use of matter as the medium of motion and of storage. For example, planet Earth can be regarded as an energy system receiving its energy input in the form of electromagnetic radiation from the sun.

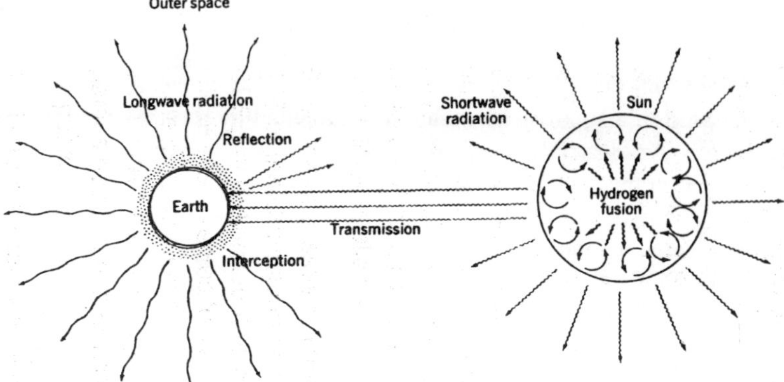

Figure 2.3: A schematic diagram of the sun-earth-space energy system. Energy is produced within the sun by hydrogen fusion and is radiated into space from the sun's surface as shortwave electromagnetic radiation.

As solar energy falls on the earth's surface, some is reflected directly back to space, but the remainder is absorbed by the earth's atmosphere, oceans, and land surfaces. This absorbed energy is now in the form of sensible heat, which, in turn, may be transformed into kinetic energy of moving air, water, and rock particles.

As friction acts on such motions, the kinetic energy is transformed back into sensible heat, and finally back to electromagnetic radiation, which streams outward into space and is lost forever. We refer to this total earth system as an *open system* because energy continually enters and leaves it.

Energy flow systems set in motion and sustain *material flow systems* (flow systems of matter). The matter involved in such systems is not only transported from place to place along certain pathways, but it can also undergo both changes of state and chemical changes. The matter traveling through the system can also be held temporarily in storage at certain points.

A glacier is a simple example of a material flow system. Resting on the floor of a deep mountain canyon, the glacier receives solid nourishment from the atmosphere in the form of snow, which compacts into ice. Gradually the great ice mass flows downvalley to lower levels, where the ice disappears by melting or evaporation (change of state). The glacier is an open flow system because matter both enters and leaves. Some flow systems are said to be *closed systems*, because all, or nearly all, of the matter remains within the system and is simply recirculated (recycled).

The concept of flow systems of energy and matter will appear a number of times in later chapters of this book. As each system is explained, your understanding of natural systems will deepen and your appreciation of our earth as a dynamic planet will heighten.

THE ATOMIC STRUCTURE OF MATTER

In the remainder of this chapter, we move to a different perspective, or vantage point, to take a much closer look at the nature of matter. As Professor Holum explains it, we are "crossing into the realm where individual particles cannot be directly seen and weighed." Even today, with all the sophisticated equipment of the scientific laboratory, we cannot take a photograph of a single individual atom. How small is an atom?

The diameter of a typical atom is on the order of one ten-billionth of a meter (10^{-10} m), which can surely have no meaning to a human being. All present knowledge about the internal structure of an atom and the arrangement of atoms into orderly groupings has had to be obtained by indirect methods using vast numbers of atoms at one time. For example, X rays fired into a mineral crystal are deflected by massed atoms in such a way that the arrangement of those atoms can be calculated from a photographic image of the scattered rays.

Structure of the Atom

The atom consists of a dense but very small central *nucleus* and one or more *electrons* moving at high speed in complex paths enveloping the nucleus. Nearly all the atom's mass is in its nucleus, which is extremely dense. Because the nucleus occupies only about onetrillionth of the volume of the atom, most of the atom is just empty space.

The nucleus of any atom (except hydrogen) consists of two kinds of particles: *protons* and *neutrons*. A proton has a positive electric charge, whereas a neutron is not electrically charged. Otherwise, protons and neutrons are essentially similar, and their masses are almost equal. An electron is a particle with a negative electric charge-opposite, that is, to the charge of a proton.

The mass of an electron is extremely smallonly about 1/2000 that of a proton. The number of electrons orbiting the nucleus of a normal atom is always the same as the number of protons in its nucleus.

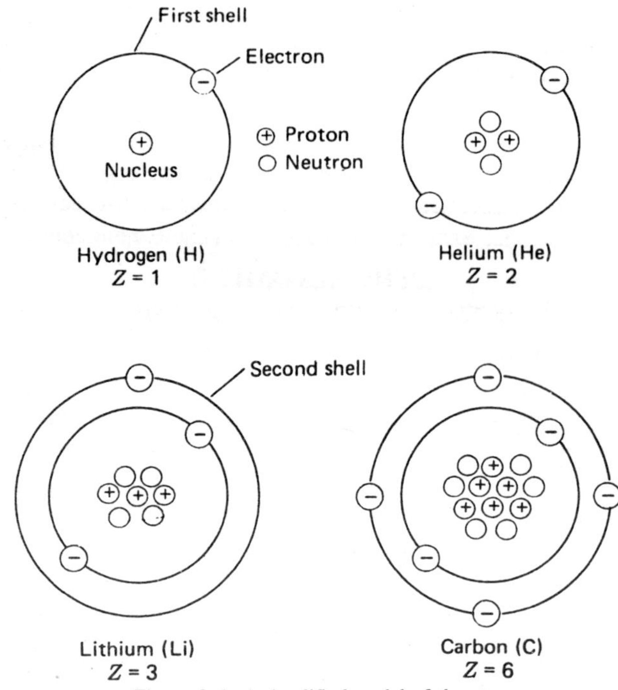

Figure 2.4: A simplified model of the atom, showing the atomic structure of the first four elements.

Therefore, within a stable atom there are as many positive as negative electric charges. The two kinds of charges balance or cancel each other out, and the atom is said to be electrically neutral.

The last paragraph is illustrated in Figure elsewhere in this chapter by schematic diagrams of atoms of four different elements: hydrogen, helium, lithium, and carbon. Each electron has its own path configuration, called an *orbital*, with a certain specified distance, or radius, from the nucleus. We have shown the electrons to be orbiting the nucleus in more or less circular paths (only in one plane) much as the planets orbit our sun, or the moon orbits the earth.

In reality, electrons deviate greatly from such simple paths, so that the actual collection of paths of a single electron describes a space occupying three dimensions. Hydrogen (symbol H) is the simplest of the elements in terms of the numbers of particles; it has a nucleus consisting of only one proton (one positive electric charge) which is orbited by only one electron (one negative charge). To visualise the actual path followed by the single electron we might imagine a small, fastflying insect incessantly circling a central object (the nucleus). Imagine that the insect carries an intense light source the size of a pinpoint, and that in the total darkness of a closed room we photograph its trail of light by leaving a camera shutter open for a long time.

As the pinpoint light source moves around the nucleus, it leaves an image in the form of a line. The travel path varies in distance from the nucleus, being sometimes very close and sometimes far out. Our photograph would show the innumerable flight circuits fused into a disk of light grading in tone from brightest at the center to darkest at the outer edge, where it would merge with darkness. In three dimensions we can picture a spherical ball of light grading inward to its brightest intensity closest to the nucleus. The probability of the electron being at a particular location in the light ball would be proportional to the light intensity.

The imaginary ball of light is called an "electron cloud," but there is only one electron. In Figure elsewhere in this chapter, the orbital of the hydrogen atom is represented by a single circle. Keep in mind that this is only a graphic device to simplify the business of showing that atoms can be bonded together through their electrons. Helium (He) has two protons, two neutrons, and two electrons. Notice in Figure elsewhere in this chpater that both hydrogen and helium have a single orbital, whereas lithium (Li) and carbon (C), have two orbitals, one inside the other. Multiple orbitals of electrons are organised into concentric zones, called *shells*, and these are numbered consecutively outward (from one to seven).

Within each shell, the number of electrons that can be present is limited. The innermost shell (shell one) can hold only two electrons; shell two can hold eight electrons. The concept of electron shells, which we have greatly oversimplified here, will be most useful in understanding how atoms of different elements combine into stable groups.

The atom of each element is uniquely described by its *atomic number*, which is the number of protons in its nucleus. Thus, all the 103 known elements can be arrayed in order of consecutive atomic numbers from one through 103. Atomic numbers of the four elements shown in Figure elsewhere in this chapter, designated by the capital letter Z, are as follows:

hydrogen	(H)	1
helium	(He)	2
lithium	(Li)	3
carbon	(C)	6

The total number of protons and neutrons within an atom is called the *mass number*, symbol A. In three of the four examples shown in Figure elsewhere in this chapter, the mass number is exactly twice the atomic number, because each of these atoms has the same number of neutrons as protons.

Isotopes

Our next step is to take note of the fact that for a given element the number of neutrons may be either fewer or greater than the number of protons. Because neutrons have no electric charge, such differences do not affect the balance of charges between protons and electrons, but only the mass of the atom; its mass number is changed accordingly. Figure elsewhere in this chapter illustrates three forms of the element hydrogen, each with a different nucleus.

While the number of protons (atomic number, Z) remains 1, the number of neutrons is shown to be 0, 1, and 2, from left to right, and the mass numbers are A = 1, A = 2, A = 3. Also shown are three forms of the element carbon. The carbon atom always contains six protons (Z = 6), but its number of neutrons can vary. We have shown three cases: A = 12, A = 13, and A = 14. The varied forms of atoms shown for each element are called *isotopes*. Each isotope of a given element has a different number of neutrons, and hence a different mass number, from other isotopes of the same element.

The number of isotopes that exist for a given element varies greatly. There are, however, only three isotopes of hydrogen. Common hydrogen (A = 1) makes up over 99.98% of all hydrogen atoms occurring in nature, while the two other forms, deuterium and tritium, are very rare. Tritium, which is produced artificially, is extremely unstable and rapidly disintegrates. Carbon has six isotopes, but of these, Carbon12, with its equal numbers of protons and neutrons, makes up 98.89% of all naturally occurring carbon atoms.

Why, then, should we bother to explain what an isotope is? Geologists make use of isotopes to determine the age in years of certain mineral or organic substances. The isotope Carbon-14, although extremely rare, allows us to determine the age in years of many substances ranging from a few centuries old to about 40,000 years old. Isotopes of the elements uranium, thorium, and lead allow us to determine the ages of the oldest rocks on earth—between 3 and 4 billion years old. In other chapter of this book we will explain how isotopes are used for this purpose.

Energy Levels in the Atom

Let us return to the arrangement of shells of electrons within an atom. The numbers one through seven we used refer to energy levels. An electron possesses kinetic energy because of its high speed of travel. Measured outward from one shell to the next, the energy possessed by an electron steps upward to higher values. The increase in energy from one level to the next is referred to as *a quantum* increase; thus, the integer numbers one through seven are called *quantum numbers*. If an electron were to be moved from one shell to the next adjacent shell it would need to gain or lose a certain quantity (or quantum) of energy. In that case, the atom would need to import or export a certain quantity of energy.

Table 2.1 Energy Levels in Atoms

Energy level, or quantum number	*Name of shell*	*Number of electrons*	
		Needed to fill the outermost shell	*Held in all shells including the outermost, when all spaces are full*
1	K	2	
2	L	8	10
3	M	8	18
4	N	8	36
5	0	8	54

While the numbers one through seven designate the energy levels of the electron shells, each shell is named with a capital letter, *K*, *L*, *M*, *N*, *O*, etc. Table elsewhere in this chapter shows the numbers of electrons needed to fill a shell when it is the outermost shell. The final column gives the total number of electrons held in all shells, when the outermost is filled. (The reason the final number is not equal to the sum of the numbers needed to fill the outermost shell is that, once full, some additional electrons can be introduced into orbitals of the earlierformed shells.)

Activity of Elements

Let us now assemble our information about energy levels, shells, and electrons into a two-by-two table, as shown in Figure elsewhere in this chapter. The table is limited to energy levels one through four (shells *K*, *L*, *M*, and *N*), but these include the eight most abundant elements in the earth's crust as well as the important elements making up the atmosphere, the ocean and its dissolved salt, and pure water (a compound of hydrogen and oxygen). A few rather rare elements (gallium,

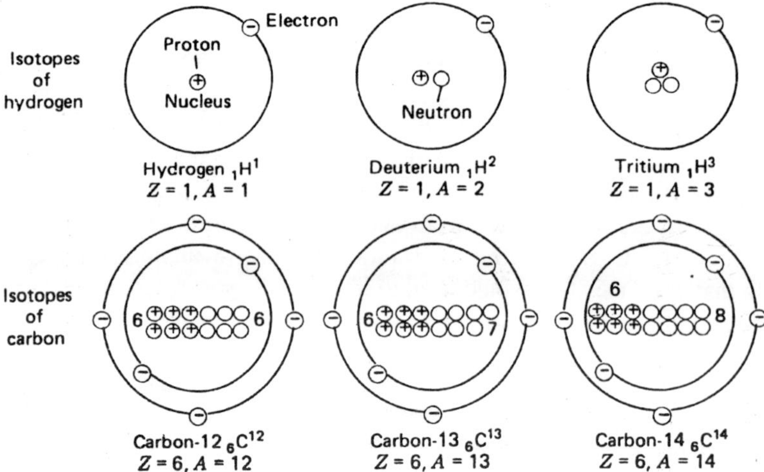

Figure 2.5: The concept of isotopes illustrated by three forms of the hydrogen atom and three forms of the carbon atom.

germanium) are included to fill the blanks in the table. Our goal is to explain how different elements can combine into compounds, and why this event occurs readily with some pairs of elements, but not with others.

First, we note that the last column (extreme right) contains three elements-neon, argon, and krypton—which have a full complement of eight electrons (an *octet*) in the outer shell. When the octet is complete, the atom is extremely stable. The same can be said of helium, with two electrons in the K-shell. All four elements occur as gases; they belong to a group called the *noble gases*. Because none of the four is known to form a compound with any other element, they are described as being chemically *inert*.

Next, we look at the column of elements at the extreme left. Hydrogen, lithium, sodium, and potassium have only one electron in the outer shell. They are highly reactive elements because they readily lose their outer-shell electron to another element in order to achieve a stable condition. In the case of lithium, sodium, and potassium, loss of an electron leaves the next lower shell complete. For example, with potassium the N shell disappears, leaving the M shell with its octet complete. Loss of an electron from the L shell of lithium dispenses with that shell and leaves only the K shell of two electrons, which is stable. The case of hydrogen is somewhat different; loss of its electron leaves only the nucleus of the atom, which is a single proton.

When an atom from the first column (hydrogen, lithium, sodium, and potassium) loses an electron, it is left with a single positive electrical charge. Bear this fact in mind as we go further.

Look now at the elements in the next-to-last column: fluorine, chlorine, bromine. They have seven electrons in the outer shell, one less than a full octet. They, too, are highly reactive elements because they can easily gain a single electron to achieve a full octet. If this gain of one electron occurs, the atom acquires a single negative electric charge. If you have not already guessed, an atom of the first column, upon meeting an atom of the next-to-last column, should immediately pass an electron to it.

IONS

Actually, atoms that have lost or gained an electron of the outer shell are no longer atoms; they are *ions*. (An atom must have a neutral charge; thus, the number of electrons must equal the number of protons.) An ion with positive electrical charge is called *a cation* and is designated by use of a positive sign superscript next to the element symbol, thus:

Sodium atom	**Sodium cation**
Na	**Na^+**

An ion with negative electric charge is called an *anion*, and is designated by a negative sign. In naming the anions, the name of the element is changed to end in "ide." Thus, the anion of chlorine is a chloride ion:

Chlorine atom	**Anion of chlorine (chloride ion)**
Cl	**Cl^-**

Ionic Compounds

This brings us to our first type of compound. As shown in Figure elsewhere in this chapter, the transfer of one electron from an atom of sodium to an atom of chlorine leaves two ions of stable configuration, each with a filled octet in the outer shell. Recall that opposite electrical charges attract one another. Therefore the transfer of an electron, which leaves ions of opposite charge in close contact, causes a strong *chemical bond* between the ions. This particular type of bond is called an *ionic bond*.

The substance produced by electron transfer is called an *ionic compound*. In the example we have used, the compound is sodium chloride, with the formula NaCl. This compound is ordinary table salt; its mineral name is halite. In other chapter of this book we will investigate the arrangement of ions in a crystal of halite. Because the sodium and chloride ions each have a single electrical charge, the compound NaCl will always be formed from equal numbers of atoms of each element.

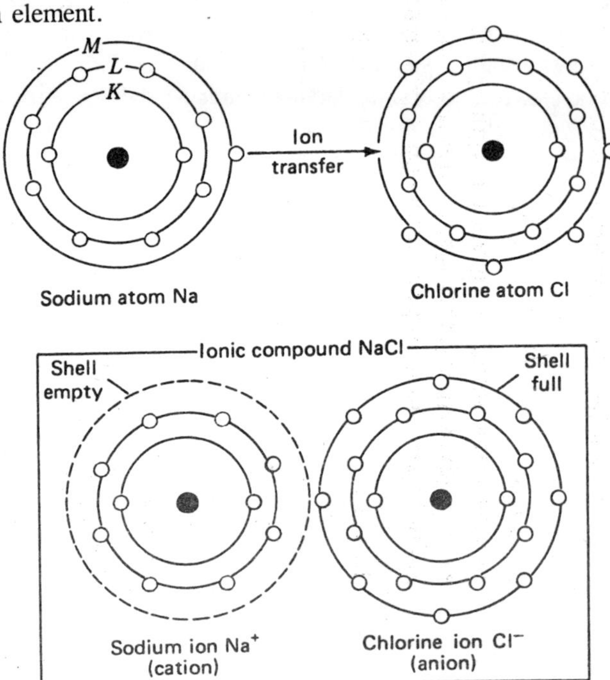

Figrue 2.6: Atoms of sodium and chlorine become ions with the loss and gain, respectively, of one electron from the outer shell. Their combination forms an ionic compound held by an ionic bond.

Returning to our table, consider the elements of the second column, those with two electrons in the outermost shell. These elements can lose two electrons to leave the next lower shell complete. Therefore, they form cations with two positive charges, for example:

Magnesium ion **$Mg\ 2^+$**

Calcium ion **$Ca\ 2^+$**

If magnesium cations are to form ionic bonds with chloride anions, to produce magnesium chloride, the ratio must be two of the latter to one of the former: $MgCl_2$.

Atoms with six electrons in the outer shell (oxygen, sulfur) can also form ions; these would be anions with two negative charges:

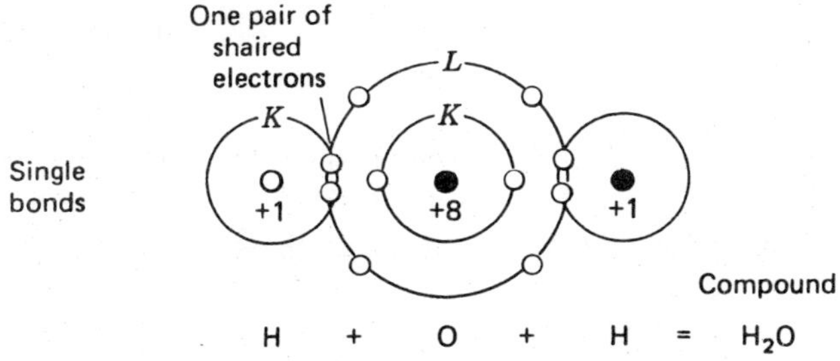

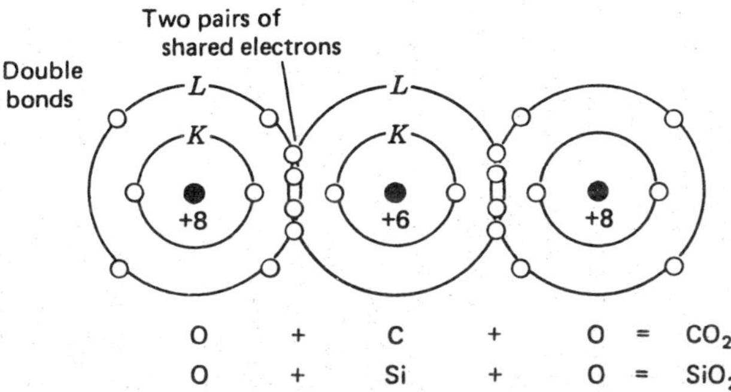

Figure 2.7: Electron sharing, found in the covalent bond, results in the formation of molecules of a compound. Electrons are shared in pairs, completing the octet of electrons in the outer shells of all atoms in the molecule.

Oxide ion	O^{2-}
Sulfide ion	S^{2-}

Cations can also be formed by loss of three electrons in the outer shell. An example is the aluminum ion, Al^{3+}.

Electron Sharing

Atoms with four and five electrons in the outer shell do not readily form ions. Important examples are carbon (C) and silicon (Si). They form compounds by sharing electrons with other atoms, rather than by ion transfer; the process is called *electron sharing*.

Figure elsewhere in this chapter shows how two single bonds join an atom of oxygen (O) with two atoms of hydrogen (H) to form the water molecule, H_2O Each hydrogen atom shares two electrons with the oxygen atom, giving hydrogen a complete pair in its K shell.

The bonds that hold atoms together when electrons are shared are called *covalent bonds* to distinguish them from ionic and other kinds of bonds. The three atoms bonded in this way form a molecule (H_2O) which, as a whole, is electrically neutral. (You should count the electrons in the molecule and make sure that their number equals the sum of the positive charges in the three atomic nuclei.)

Figure elsewhere in this chapter also shows how the common compound carbon dioxide, CO_2, can be formed when each of two oxygen atoms shares four of its electrons in the *L* shell with one carbon atom (C), which needs four more electrons in its *L* shell to complete an octet. We can also use the same diagram for a molecule of the compound silicon dioxide, which is familiar as the common mineral quartz.

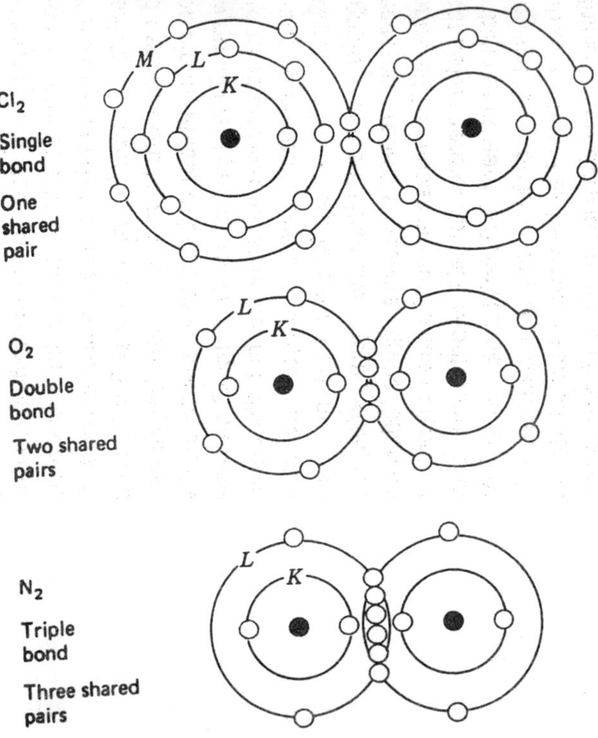

Figure 2.8: Two ions of the same element are joined by a covalent bond to form a molecule. The three molecules shown are normally in the gaseous state.

Several elements listed in Figure elsewhere in this chapter normally exist in the gaseous state at temperatures usually found near the earth's surface. Two of these, oxygen (0) and nitrogen (N), form 99% of the atmosphere. The three elements with seven electrons in the outer shell—fluorine (F), chlorine (Cl), and bromine (Br)—are also normally in the gaseous state. These elements occur as molecules because they are unstable as individual atoms.

When two atoms of one of these elements are paired into a single molecule, the atoms share as many electrons as necessary to complete the octets of both. For example, chlorine needs one missing electron, and thus two chlorine atoms joined in a molecule share an electron, completing both octets, and creating one covalent bond. The chlorine atom is given by the formula Cl_2. Molecules of oxygen and nitrogen have the formulas O_2 and N_2, respectively requiring two and three covalent bonds.

Chemical Reactions

The *chemical formula* for a compound expresses by means of the element symbols the combination of the two or more elements it contains. For example, the formula for the compound iron sulfide is FeS. A knowledge of the number of electrons in the outermost electron shell will allow you to judge the proportions in which the component elements will combine.

Those electrons that can be gained, lost, or shared are called *valence electrons*. Consulting Figure elsewhere in this chapter, we find that iron (Fe) has two valence electrons in its outer shell. This number of electrons is called the *covalence* of the atom. With a covalence of two, the Fe atom can form two covalent bonds. Sulfur (S), with six electrons in its outer shell, can gain two electrons, so its covalence is also two.

In reacting to form a compound, Fe and S will combine in a one-to-one relationship, sharing two covalent bonds. Therefore, *the chemical equation* describing this *chemical reaction* is written

$$Fe + S \rightarrow FeS$$

The arrow indicates that the reaction takes place in the direction from left to right.

A second example is the reaction of aluminum (Al) with sulfur (S) to form the compound aluminum sulfide, $A1_2S_3$. From Figure elsewhere in this chapter, we find that the aluminum atom has three valence electrons, and these can be lost; its covalence is three. We already know that the covalence of sulfur is two. It is obvious that three atoms of sulfur are needed to form a compound with two atoms of aluminum, and a total of six covalent bonds is needed. The equation is written

$$2Al + 3S \rightarrow A12S3$$

Following Dalton's principles, we must always state the numbers of atoms as whole numbers.

POLYATOMIC IONS

The ions we have described so far consist of only a single element. Another important class of ions exists, the *polyatomic ions*, in which atoms of two or three different elements are linked by covalent bonds. Unlike a molecule, which has a neutral charge, the polyatomic ion is charged positively or negatively, as is an ion of a single element.

Figure elsewhere in this chapter shows two polyatomic ions important in geology, the sulfate ion and the carbonate ion. In the sulfate ion, SO_4^{2-}, each of four oxygen atoms has a covalent bond with th one sulfur atom. The entire ion contains 50 electrons (negative charges) and 48 protons (positive charges), so that the ion bears a double negative charge. The carbonate ion, CO_3^{2-}, consists of three oxygen atoms linked by covalent bonds to a single carbon atom. Two of the three bonds are single; the third is a double bond.

Figure elsewhere in this chapter also shows how a polyatomic ion can join with another ion of opposite charge to form an ionic compound. The metallic ion of calcium, Ca^{2+}, has an empty M shell and a full L shell. All outer shells of the polyatomic ion are filled. Because both ions have double electric charges, they will combine in a one-to-one ratio to form calcium sulfate, $CaSO_4$, and calcium carbonate, $CaCO_3$. Calcium sulfate occurs in nature as the mineral anhydrite; calcium carbonate, as the mineral calcite. Both are very common minerals.

Notice that the reactions shown in Figure elsewhere in this chapter use two arrows, pointing in opposite directions. This device means that we are dealing with *a reversible reaction*, and that the compound on the right can readily separate into the component ions. This separation occurs when the mineral is dissolved in water.

POLAR MOLECULES

Although the molecules we have described are electrically neutral, with equal numbers of positive and negative charges inside, those two kinds of charges may tend to concentrate at opposite ends of the molecule. Thus, at one place on the outer surface of the molecule is a positive charge and at another place on the surface, a negative charge. This type of molecule is described as *a polar molecule*. It behaves much like a small bar magnet, in that the negative pole of one molecule will tend to attract the positive pole of another molecule. The charges on the molecule surface are much weaker than the unit charge of an ion and can be referred to as *partial charges*.

The water molecule is strongly polar. One representation of the water molecule shows it as a large spherical object (the oxygen atom) with two small hemispheres attached (the two hydrogen atoms), as shown in Figure elsewhere in this chapter. The hydrogen "*bumps*" are positively charged; the far side of the larger sphere bears a negative

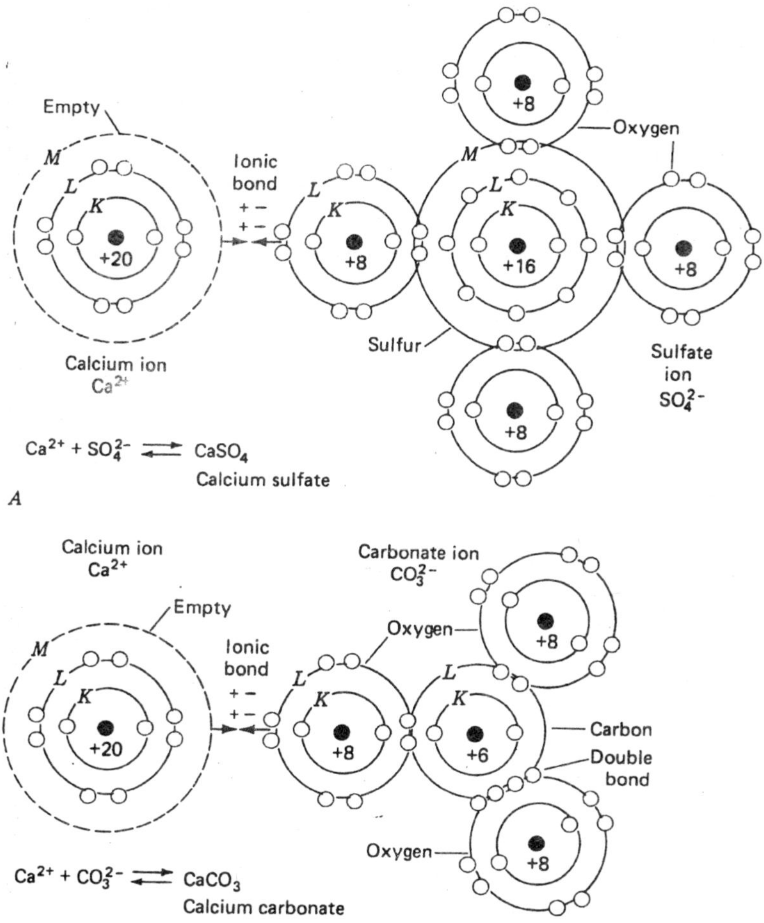

Figure 2.9: Two common polyatomic ions are shown here: the sulfate ion and the carbonate ion. They readily form ionic bonds with the calcium ion to make up ionic compounds abundant in one major class of rocks.

partial charge. As a result, these molecules tend to attract one another, and to attract ions of other elements.

The polar nature of the water molecule explains why water is such a good solvent for many ionic compounds. For example, we know that common table salt, sodium chloride (NaCl), rapidly dissolves in water. What happens is that the water molecules bombard the surfaces of the salt crystals, dislodging individual ions of sodium and chlorine. A free ion is quickly surrounded by several water molecules, which are turned so that their charged surface areas face the ion and attract it with a weak bond.

METALLIC BONDS

Eal'ier in this chapter we referred to the occurrence of certain metals in rock as native elements and we use iron, gold, copper, and silver as examples. Most of the metals have only one, two, or three electrons in the outermost shell, and these are quite easily shared with other atoms. Since their outer electron shells can readily overlap with those of nearby electrons at several places at once, metal atoms are closely packed, with loosely held electrons moving freely throughout the entire mass.

The free-moving electrons are often described as an *electron sea*, and the "*communal*" sharing of electrons characterises the *metallic bond*. These metals have the property of being malleable, meaning that they can be shaped by hammer blows and can be drawn out into thin sheets and wires without fracturing. Although the atoms of the metal maintain an orderly spacing, layers of atoms can easily glide past each other as the metal is hammered or drawn; the bonds remain intact at all times.

We also know that some metals like this are good conductors of electricity, since our transmission lines and household wiring consist of copper or aluminum wire. The conduction of electricity is facilitated by the freely moving electrons in a metal, because electricity is nothing more than a flow of electrons in one direction.

MATTER AND ENERGY IN GEOLOGY

Without an understanding of matter and energy, and their relation to one another, no geologist can hope to carry on useful research in modern physical geology. Without applying basic physics and chemistry, today's geologist could do little more to interpret earth processes than could a geologist who worked around the time of the Civil War.

In fact, some branches of geology were in a state of scientific stagnation by the late 1800s, because physicists and chemists had not yet learned enough about the nature of matter to provide answers to many geological questions. Radioactivity had not been discovered; nuclear energy was yet unknown. Ages of rocks and of the earth itself were being erroneously estimated and varied widely. The source of heat for volcano-building was still largely a mystery. Scientific tools were entirely lacking for probing deep into the earth's crust.

Only in the early 1900s, when physicists began to record and interpret deep earthquake waves, did we begin to form a picture of the states of mineral matter within the earth.

3

MINERALS

The quotation beneath the title of this chapter is a translation from the French of an article by Rene Just Hauy. It was published in the first *year* of the 19th century. Hauy, who had originally trained for the church and entered the priesthood, became interested in minerals when he noticed how large pieces of calcareous spar can be broken into smaller pieces all having a shape similar to that of the original piece.

From his measurements of these and other solids displaying crystal form, he surmised that crystals are composed of small structural units that are of identical shape in all crystals of the same form. Today, mineralogists have demonstrated that Hauy's building blocks are actually groupings of still smaller particles called atoms.

In this chapter, we will discuss the characteristics of atoms, the manner in which atoms are systematically located and repeated in three dimensions within minerals, and the way in which combinations of atoms affect the properties of all earth materials. Before beginning, however, it will be useful to recall that *minerals* are naturally formed solids having threedimensional orderly internal arrangements of atoms and a definite chemical composition or range of compositions.

The phrase "range of composition" is necessary because some atoms are similar in size and properties to one another and, to a limited degree, can replace one another in the internal atomic structure of minerals.

ATOMS

The Architecture of the Atom

If one were to disjoin the constituents of a mineral, one might then be left with certain components that could not be separated further by chemical and physical' methods. These basic ingredients, of which no further separation is possible, are called elements. The fundamental unit of an element is an atom.

There are 105 elements known; 13 of these are manmade and 4 do not occur on earth. Geology students, however, need not be overly concerned about the rather large number of elements, because only eight occur abundantly in the earth's surficial rocks. These eight abundant elements are oxygen, silicon, aluminum, iron, calcium, sodium, potassium, and magnesium. They constitute at least 98 per cent of the earth's crust by weight and over 95 per cent of the elements in the entire earth from core to crust.

Table 3.1: Abundances of Chemical Elements in the Earth's Crust.

Element and symbol	*Percentage by weight*	*Percentage by number of atoms*	*Percentage by volume*
Oxygen (O)	46.6	62.6	93.8
Silicon (Si)	27.7	21.2	0.9
Aluminum (Al)	8.1	6.5	0.5
Iron (Fe)	5.0	1.9	0.4
Calcium (Ca)	3.6	1.9	1.0
Sodium (Na)	2.8	2.6	1.3
Potassium (K)	2.6	1.4	1.8
Magnesium (Mg)	2.1	1.9	0.3
All other elements	1.5		
	100.0	100.0	100.0

The idea that earth materials are composed of the atoms of elements was conceived long before the present atomic age. More than 25 centuries ago, the Hindu philosopher Kanadu proposed that matter was made of tiny, invisible "eternal particles." This same concept was proposed in classical Greek time by Democritus, who taught that atoms were small indivisible particles capable of producing the properties of the substances they formed. Democritus, however, formulated his concept of the atom purely on speculation and intuition. He had no idea about the structure of the atom or that it might contain even smaller constituents.

Knowledge of some of the smaller constituents of atoms and their characteristics began to evolve during the late 1800s and early 1900s, mostly as a result of experiments carried out with radioactive substances. The atom was conceived of as having an extremely small but heavy *nucleus* surrounded by a cloud of rapidly moving particles with negative electrical charge called electrons. Electrons whirl around the nucleus at speeds so great that if permitted to do so, they would encircle the earth in less than 1 second.

They do not revolve in a set path, as do planets around the sun, but swirl around the nucleus rather like a swarm of tiny insects around a light. Thus, it would be incorrect to think of an atom as a solid sphere because in reality atoms consist mostly of empty space. The electrons move so rapidly that they effectively fill that space and thereby give size to the atom.

Located in the nucleus of the atom are closely compacted subatomic particles called *protons*, each of which carries a charge of positive electricity equal to the charge of negative electricity carried by the electron. Protons contain an amount of matter almost identical to that of a hydrogen atom, which has only one proton and one electron of negligible mass.

Associated with the protons in the nucleus are electrically neutral particles having the same mass as protons. These particles are called *neutrons*. Modern atomic physics has provided evidence for the existence of other subatomic particles, but their contribution to the chemical behavior of elements is minimal.

Atomic Number and Atomic Mass

The number of protons within a nucleus determines the type of element to which an atom belongs. For example, if six protons occur in the nucleus of an atom, the element is carbon; if the number is eight, the element is oxygen; 14, silicon, and so on. Thus, all atoms of any one element have the same number of protons. The number of protons in the atom of an element defines that element's *atomic number*. In an electrically neutral atom, there are equal numbers of protons and electrons. The number of protons in naturally occurring elements ranges from 1 in hydrogen to 92 in uranium.

The *atomic mass* of an atom is approximately equal to the sum of the masses of its protons and neutrons. The mass of an electron is so small that it need not be considered in determining the atomic mass. Indeed, it would require 1837 electrons to equal the atomic mass of a single proton. By convention, atomic mass is noted as a superscript preceding the chemical symbol of an element, and the atomic number is placed beneath it as a subscript. Thus 20 Ca is translated as the

Table 3.2: Structure of Some Geologically Important Elements.

Element and symbol	*Atomic number (number of protons innucleus)*	*Number of neutrons in nucleus*	*Atomic mass*	*Electrons in various levels*	*Total of number electrons*
Hydrogen (H)	1	0	1	1	(1)
Helium (He)	2	2	4	2	(2)
Carbon 12 (C)	6	6	12	2-4	(6)
Carbon 14 (C)	6	8	14	2-4	(6)
Oxygen (0)	8	8	16	2-6	(8)
Sodium (Na)	11	12	23	2-8-1	(11)
Magnesium (Mg)	12	13	25	2-8-2	(12)
Aluminum (AI)	13	14	27	2-8-3	(13)
Silicon (Si)	14	14	28	2-8-4	(14)
Chlorine 35 (Cl)	17	18	35	2-8-7	(17)
Chlorine 37 (Cl)	17	20	37	2-8-7	(17)
Potassium (K)	19	20	39	2-8-8-1	(19)
Calcium (Ca)	20	20	40	2-8-8-2	(20)
Iron (Fe)	26	30	56	2-8-14-2	(26)
Barium (Ba)	56	82	138	2-8-18-18-8-2	(56)
Lead 208 (Pb)	82	126	208	2-8-18-32-18-4	(82)
Lead 206 (Pb)	82	124	206	2-8-18-32-18-4	(82)
Radium (Ra)	88	138	226	2-8-18-32-18-8-2	(88)
Uranium 238 (U)	92	146	238	2-8-18-32-18-12-2	(92)

element calcium having an atomic number of 20 and an atomic mass of 40.

Isotopes

Although the atoms of any specific element are essentially identical in the manner in which they behave chemically, they are not always precisely identical in the number of neutrons they contain. Atoms of a particular element that differ in the number of neutrons they contain, and therefore also differ in atomic mass, are called *isotopes*. As an cxample, there are at least three different isotopes of oxygen.

All three have the same atomic number (8) but have mass numbers of 16, 17, and 18, respectively. Of these, oxygen 16 is the most abundant in nature. Isotopes are very useful in geologic studies. The isotopes of oxygen are used to determine the temperature of the oceans hundreds of millions of years ago, and isotopes of uranium, potassium, rubidium, and other elements are used to determine the age of rocks.

Table 3.3: Stable Isotopes of Oxygen.

Isotope	*Number of protons*	*Number of neutrons*	*Atomic mass*
^{16}O	8	8	16
^{17}O	8	9	17
^{18}O	8	10	18

As suggested above, the chemical properties of isotopes of the same element are nearly identical. This is because the chemical behavior of atoms is determined not by protons or neutrons, but rather by the electrons. The number of electrons does not vary among the isotopes of an element.

Most isotopes have a high degree of stability and tend to remain unchanged over long spans of time. Others, however, are called unstable isotopes because they tend to break down into other isotopes. The unstable isotopes are the kind most useful in determining the age of rocks.

Electron Shells

Because of their negative charge, electrons are attracted to the positively charged nucleus of the atom but are prevented from collapsing inward to the nucleus by the centrifugal force associated with their rapid orbital spin. Although electrons may spin along any path, *on the average,* a certain quantity will remain within a particular zone or *shell.* The shells are really areas in which there is a high probability of finding electrons.

It is not possible to know the position of a particular electron at any designated moment. However, scientists have deduced the probable arrangement of the shells from X-ray studies, from atomic spectra, and from the atom's behavior. Such deductions provide a working hypothesis for the number of electron shells and the probable number of electrons in each shell. As the simplest of all the elements, hydrogen has only one proton and one electron. The helium atom has two electrons close to the nucleus-a condition also present in the innermost shell of all elements heavier than helium.

The greatest number of electron shells in even the most complex of atoms is seven. Each of the first four shells may contain no more than the following number of electrons: 2 in the first shell, 8 in the second, 18 in the third, and 32 in the fourth. The shells are always filled outward, so that an inner shell must be full before the next shell can have any electrons. Thus, the first two inner shells must be filled before the third shell can accommodate any electrons. In addition, electrons are distributed in such a way that the outermost shell contains eight or fewer electrons.

Indeed, once an atom has acquired the eight outer-shell electrons, it becomes chemically quite stable. The chemical properties of atoms are thus dependent on their tendency to obtain eight electrons in this all-important outermost (or valence) shell. For example, if the outer shell contains exactly eight electrons (or two in helium, which has only two electrons), the atom will show little tendency to chemically interact. Thus, atoms like neon, argon, and krypton are called *noble* or *inert gases* because their outer shells contain the maximum quota of eight electrons. They are, in a symbolic sense, "*satisfied.*"

For those elements whose atoms have an outermost shell containing fewer than eight electrons, there is a tendency to share, gain, or lose outer-shell electrons so as to assume the stable configuration in which eight electrons occur in the outermost shell. For example, the conceptual two-dimensional diagram of the chlorine atom, shown in Figure elsewhere in this chapter, exhibits only seven electrons in the outer shell.

Chlorine is "inclined" to borrow one electron. Sodium, which has only one electron in its outermost shell, is inclined to give up an electron. Therefore, by combining, both sodium and chlorine achieve a stable configuration. The number of electrons that an atom loses or receives, or the number of pairs it shares, is called its valence. Loss of an electron leaves an atom with a surplus of positive charges. Hence, it is said to have *a positive valence.*

Likewise, a gain of electrons will produce a negative charge on the atom and result in *negative valence.* The electrons gained or lost are called valence electrons and are indicated by small plus or minus symbols, as in O^{-2} for oxygen or Ca^{+2} for calcium.

COMBINING STOMS

As described above, the guiding rule in chemical combination is that when two or more atoms react with one another, they do so in a manner that will promote the attainment of eight electrons in the outer shell, as exemplified by the inert gases. The process is accomplished by accepting and donating electrons or by sharing electrons. Furthermore, atoms tend to achieve a complete outer shell by un dergoing the smallest possible change. Thus, an element having fewer than four electrons in its outer shell will tend to donate those electrons, while an element having more than four will tend to borrow electrons to complete its outmost shell. Once an atom has either borrowed or donated an electron, it is no longer electrically neutral. It is now called an *ion*. More specifically, if the charge on the ion is positive, it is termed a cation, whereas *anions* are negatively charged ions.

EXAMPLES OF CHEMICAL BONDING

Ionic Bonding

The reaction between an atom of sodium (Na^{+1}) and chlorine (Cl^{-1}) can again be used to illustrate one mechanism by which atoms may bond together. In the combination of these two elements, it is easier for sodium to lend the single electron in its outer shell than to borrow seven. Upon lending this electron, the sodium atom becomes positively charged. A chlorine atom, on the other hand, carries 17 electrons: 2 in the first shell, 8 in the second shell, and the remaining 7 in the third, or outermost shell.

Thus chlorine's outer shell is lacking one electron. For chlorine, it is far easier to acquire one electron than to donate seven. In the reaction between the two elements, each sodium atom gives its one outer-shell electron to one chlorine atom, with the result that each has the complete outer shell structure of a noble gas. The acquisition of the extra electron by the chlorine atom converts it into a negative ion. Because unlike charges of electricity attract, the negatively charged chlorine ion is drawn tightly to the positively charged sodium ion to form sodium chloride.

Studies of halite, the mineral form of sodium chloride, reveal that each sodium ion is surrounded by exactly six larger chlorine ions, and similarly, each chlorine atom is in contact with six sodium ions. The cubic crystal form is the result of the manner in which the ions are packed, their electrical properties, and the sizes (atomic radii) of the participating ions.

Covalent Bonding, a Sharing Situation

The completion of the outer shell of atoms by *sharing* pairs of electrons is another way atoms achieve the noble gas configuration of stability. This manner of combination is termed covalent bonding. In covalent bonding, the electrons of each atom orient themselves so that their influence is extended equally between the participating atoms.

The shared electrons become indistinguishable as to their atom of origin and serve both nuclei simultaneously. Examples of covalent bonding are the combining of two hydrogen atoms to form the H_2 molecule, two oxygen atoms to form O_2, or the combination of four hydrogen atoms and one carbon atom to form methane. Diamonds, which consist only of carbon atoms, are also bonded covalently. In a diamond, each carbon atom is surrounded by four other carbon atoms to form a tetrahedron. Within the tetrahedron, every atom shares an electron pair with each of its neighboring atoms.

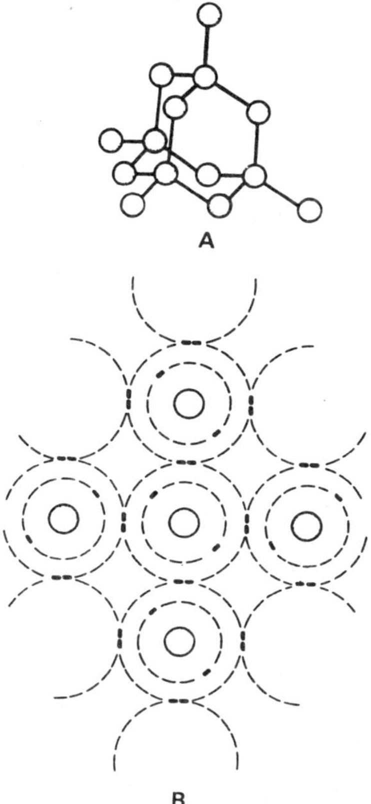

Figure 3.1: (A) The carbon atoms in a diamond are arranged in tetrahedrons consisting of four carbon atoms surrounding a central carbon atom. (B) A twodimensional model of the atoms in a diamond. Note that each carbon atom, which individually has four electrons in its outer shell, shares one outer electron with each of its four neighbors.

Water provides another example of bonding by sharing of electrons. In a molecule of water (H_2O), the single electrons from each of two hydrogen ions are shared by an oxygen ion that needs them to complete its eight-electron outer-shell configuration. The ions, however, do not arrange themselves in a straight line but rather the two tiny positively charged hydrogen ions become fixed at one end of the molecule, whereas two nonbonding or unshared electrons occur at the opposite or oxygen side.

Thus, the molecule is *dipolar* and behaves as if it were positively charged on one end and negatively charged on the other. This characteristic gives water molecules a weak electrical attraction for other ions. If these other ions are at the surface of mineral grains with which water comes into contact, the attraction may pull ions away from the mineral. In fact, the mineral is being dissolved. Such solution work by water molecules is an important aspect of the weathering of rocks at the earth's surface.

Ionic and covalent bonding are the most common kinds of bonds found in minerals. They are not, however, mutually exclusive. Both types of bonding may occur within certain minerals, and in many instances bonding may be intermediate or mixed in character.

Metallic Bonding

Metals can be visualised as closely packed aggregates of positive ions afloat in a "*sea*" of electrons. The valence electrons roam about independently of their atoms of origin. Rather than being associated with only one or two atoms, these electrons are shared by the entire metal ion aggregate. The atoms forming the metal become positive ions by giving their outer-shell electrons to be shared among all other ions. The freedom of those electrons in being able to move about and not be retained by a particular atom accounts for such metallic properties as electrical conductivity, thermal conductivity, softness, and malleability.

ATOMIC AND IONIC SIZES

Another factor that influences which elements will combine to form minerals is the size of the ions and atoms. The ions in minerals tend to be efficiently packed into tight geometric packages that leave little opportunity for any given ion to have so much surplus space that it can wobble about. Combinations of ions of certain sizes can produce only specific kinds of atomic structures.

The number of neighboring anions that are grouped around a central cation is termed the *coordination number*. In the sodium chloride structure, each cation has a coordination number of six. (Every sodium ion is surrounded by six chloride ions.) The size, or ionic radius, of an ion depends on the number of electron shells surrounding its nucleus and on the charge on the nucleus. If an atom loses an electron, the excess positive charge on the nucleus pulls the electron shells inward. Thus, positive ions are much smaller than the neutral atoms from which they were formed. Negative ions are larger than the neutral atoms of the same element. Ions of the same size are sometimes able to substitute for one another in certain minerals and thereby alter their chemical and physical properties.

THE SPACE LATTICE

Crystals are solid bodies bounded by natural plane surfaces. The ions and atoms in crystals are arranged in three-dimensional geometric patterns that we call *space lattices*. The concept of an orderly internal arrangement of the constituents of crystals was accepted by scientists long before the formulation of atomic theory. In the 17th century, Nicolaus Steno observed that similar crystal faces found on different crystals of the same kind of mineral always met at the same angle, regardless of the size and distortions of shape in the crystal. Steno's observation has since been called the *Law of Constancy of Interfacial Angles* and indicates that the intêrnal arrangement of constituent units (later termed ions and molecules) must be orderly and that the components must be present in definite proportions. If this were not the case, the relationships of the crystal faces would not be consistent.

Unfortunately, Nicolaus Steno (and later, Abbe Rene Haiiy) could only infer the existence of internal ordering from exterior form. More direct evidence eluded scientists until 1912. At that time, the German physicist Max von Laue and his associates provided proof of the internal ordering of atoms in a crystal by the use of the mysterious X-rays discovered 17 years earlier by William Roentgen. Von Laue conjectured that X-rays, like light rays, might be wavelike in character, but X-ray wavelengths would be smaller than those in light rays.

Assuming that atoms in a crystal might be systematically arranged, von Laue speculated that the space between atoms would be about equal to the calculated wavelengths of X-rays, and thus, some of the rays would be reflected by striking layers of atoms. A simple apparatus was constructed so that X-rays could be passed through a crystal and onto a photographic plate. Von Laue was rewarded with a photograph showing a uniform geometric pattern of dots (now called "*bon Laue spots*") that reflected a uniform geometric arrangement of atoms.

The interpretation of the von Laue spots was a difficult problem but was solved shortly by W. H. Bragg and his son Lawrence Bragg in England, who showed how the precise wavelength of X-rays could be calculated from the effect that a crystal of known structure has on them.

Then, using X-rays of the known wavelength, the Braggs were able to work out the atomic structure of a mineral. Today, with the aid of high-precision X-ray equipment and computers, accurate crystal determinations have become routine procedures for mineralogists. Steno's belief that the constituent units (ions) that build a crystal are packed into a symmetric three-dimensional array has been amply confirmed.

CRYSTAL FORM

A mineral grows or enlarges by addition of ions to its surfaces from a surrounding solution, melt, or gas. If the growing surfaces do not come into contact with other mineral grains, and provided that the temperature, pressure, and the concentration of ions in the liquid or gas are suitable, then the mineral will acquire a characteristic crystal form. The faces of the crystals will be parallel to uniform sheets of atoms within the crystal.

In nature, the growing surfaces of crystals do come into contact with other mineral grains, and therefore perfect crystals are not abundant. More commonly, growing crystals interfere with one another during growth to form an interlocking system of crystals. Only occasional crystal faces are seen in such crystalline solids, but within individual mineral grains the ions and atoms are nevertheless arranged in a uniform geometric pattern.

When perfect or nearly perfect crystals of minerals do develop, their external form is determined not only by composition but also by such conditions as impurities, temperature, and pressure. The importance of temperature and pressure is clearly evident when one considers the crystal forms of carbon. Under high pressures and temperatures, carbon crystallizes in the octahedral form of diamond. At lower temperature and pressure conditions, soft, platy, six-sided crystals of graphite develop.

On first examining a collection of crystals, one is impressed by what appear to be a bewildering variety of shapes. If these crystal forms are studied further, however, it is evident that most are combinations of simpler geometric forms such as prisms, pyramids, domes, and pinacoids. Crystals having one or a combination of these forms are classified into six categories according to their basic symmetry.

As a clue to the internal arrangement of atoms in a crystal, symmetry is of great importance. We all have an intuitive concept of symmetry. In regard to crystals, it may refer to the repeat ing pattern of similarly shaped faces as the crystal is rotated. Mineralogists often employ a system of imaginary lines passing through the center of the crystal as a reference. These lines are called *crystallographic axes*. For

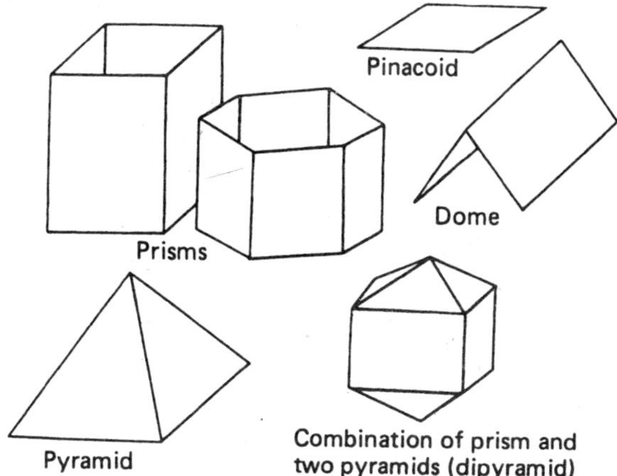

Figure 3.2: Some common crystal forms.

example, in the cube, one can imagine a line drawn from the exact center of each face, through the crystal, to the exact center of the opposite face. One would then have three mutually perpendicular crystallographic axes of equal length. Crystals like cubes and octahedrons that have such characteristics are placed within the isometric crystal system.

In addition to *axes of symmetry*, a crystal also possesses *planes of symmetry*. These are imaginary planes that divide a crystal into halves, each of which is a mirror image of the other. In cubes, for example, there are three mutually perpendicular planes of symmetry.

Geologists are not interested in crystals solely because of their pleasing symmetry. The form of a crystal, either in hand specimen or recognised by optical means, is a valuable criterion for the identification of minerals.

COMMON ROCK-FORMING MINERALS

The rocks that form the earth's crust are composed of aggregates of minerals. Although over 3000 minerals have been discovered and scientifically described, most of these comprise only a small percentage of the crust and are rarely encountered. For our present purposes, it is important to consider only those minerals that compose the bulk of common rocks or that are particularly useful in making interpretations about the earth's history.

Properties Useful in Identification of Minerals

There are a large number of chemical and physical properties used

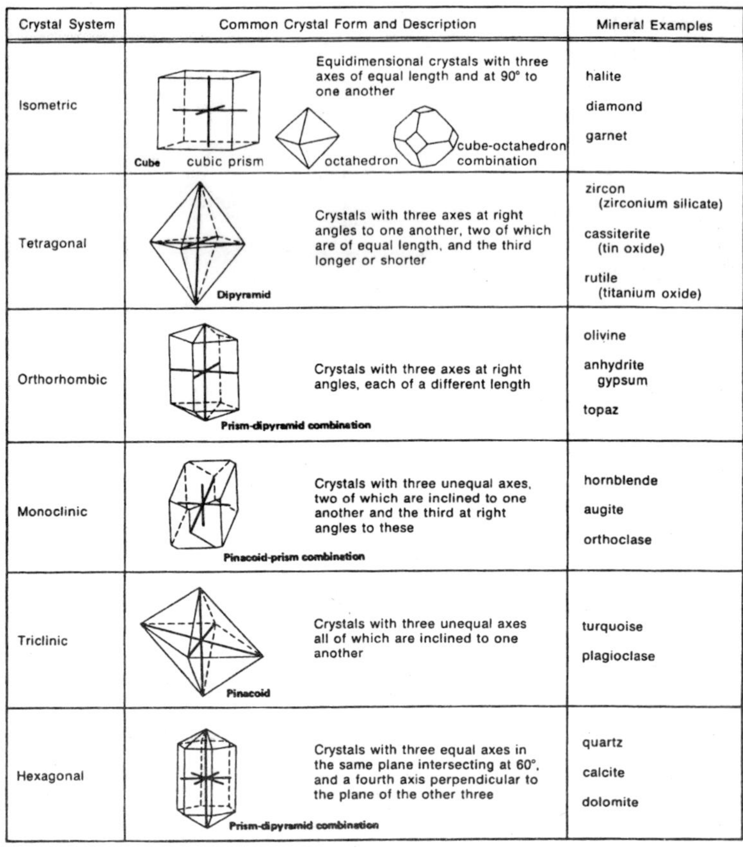

Crystal System	Common Crystal Form and Description	Mineral Examples
Isometric	Equidimensional crystals with three axes of equal length and at 90° to one another Cube cubic prism; octahedron; cube-octahedron combination	halite diamond garnet
Tetragonal	Crystals with three axes at right angles to one another, two of which are of equal length, and the third longer or shorter Dipyramid	zircon (zirconium silicate) cassiterite (tin oxide) rutile (titanium oxide)
Orthorhombic	Crystals with three axes at right angles, each of a different length Prism-dipyramid combination	olivine anhydrite gypsum topaz
Monoclinic	Crystals with three unequal axes, two of which are inclined to one another and the third at right angles to these Pinacoid-prism combination	hornblende augite orthoclase
Triclinic	Crystals with three unequal axes all of which are inclined to one another Pinacoid	turquoise plagioclase
Hexagonal	Crystals with three equal axes in the same plane intersecting at 60°, and a fourth axis perpendicular to the plane of the other three Prism-dipyramid combination	quartz calcite dolomite

Figure 3.3: Crystal systems and crystal forms.

by geologists in the identification of minerals. Some of these properties can only be recognised with the use of microscopes, X-ray equipment, or complex chemical tests. Fortunately, however, most of the common minerals can be identified with a knowledge of only a few easily recognised properties. Color, cleavage, hardness, crystal form, luster, specific gravity, and magnetism are among the easily used clues for the identification of minerals.

Visual Properties

Like most other properties of minerals, *colour* is largely dependent on chemical composition and the arrangement of atoms. Color results from the selective absorption of certain wavelengths of light by the atoms. For example, the transmitted color (or the reflected color for opaque minerals) actually represents white light minus the wavelengths absorbed by particular atoms or ions. Some minerals in which the ions are relatively widely spaced and loosely bonded may tend to absorb nearly all of the light and are dark or black in color. This partially explains why graphite is black and opaque, whereas the more strongly bonded diamond is relatively clear. Iron has ions that are strong absorbers of light. Therefore, minerals rich in iron tend to he darker in color.

Color is a mineral's most conspicuous characteristic. Some minerals possess colors that are constant, such as green in the mineral malachite, blue in lapis lazuli, or red in almandite garnet. Color, however, can be altered by even small amounts of impurities. Quartz and fluorite, for example, may be colorless, white, pink, purple, green, or blue. For this reason, color should always be used in combination with other properties in identifying a mineral.

In order to avoid errors in identification caused by superficial color variations, it is often useful to observe the mineral's streak or color in powdered form. This can be quickly accomplished by rubbing the mineral across the surface of an unglazed porcelain "*streak plate*" and noting the color of the streak of powder that results.

The manner in which a mineral shines in reflected light is its *luster*. Those minerals that reflect light like metals are said to have *metallic luster,* whereas all others have nonmetallic luster. If the luster of a nonmetallic mineral is hard and brilliant like that of a diamond, it is said to have an *adamantine*

luster. A glassy appearance is termed *vitreous luster,* whereas minerals with very low reflectance may simply be called *dull* or, if appropriate, *earthy*.

Hardness

The *hardness* of a mineral refers to its resistance to scratching by another substance of known hardness. To ensure uniformity in measuring hardness, geologists employ a scale formulated by the German mineralogist Frederich Mohs. Mohs arranged ten relatively common minerals in order of increasing hardness. If an unknown mineral can be scratched by a mineral in the scale, it is softer than that mineral; whereas if it cannot be scratched, it is harder. Mohs' scale is only a relative scale of hardness and does not provide exact mathematical relationships. Minerals have relatively small differences in hardness from 1 to 9, but increase 40 × in hardness between 9 (corundum) and 10 (diamond). In the field it is often expedient to make hardness tests with a penny (hardness of 3), fingernail (hardness of 2 to 2.5), or a steel nail (hardness of 5).

A mineral's hardness depends largely on the strength of bonds between the atoms or ions of its space lattice. For example, in diamonds, every carbon atom forms sturdy single bonds with four other carbon atoms to produce a strong three-dimensional lattice.

Table 3.4: Scale of Mineral Hardness.

1	Talc
2	Gypsum
3	Calcite
4	Fluorite
5	Apatite
6	Orthoclase
7	Quartz
8	Topaz
9	Corundum
10	Diamond

In contrast, carbon atoms in graphite crystals are arranged in layers, and the forces between layers are quite weak. Graphite has a hardness of only 1 to 2 on Mohs' scale, whereas diamond has a value of 10. The hardness of some minerals is not uniform throughout. Bond strengths may be greater along some planes in the lattice than along others, causing slight variations in hardness.

Cleavage and Fracture

We have discussed how ions and atoms in a crystal are arranged in definite layers or planes, all of which have a definite angular relationship to one another. If the bond strengths or number of bonds in one set of planes is greater than that of other planes, then it is likely that the structure will separate more readily along the directions of weakest bonding or where the spacing between planes is relatively wider.

The tendency of a mineral to break smoothly along certain directions parallel to those planes of ions across which bonding is weakest is called cleavage. The surface along which the mineral breaks is the cleavage surface. Cleavage is usually described as poor, fair, or good, according to how smooth a cleavage surface is produced.

The number and direction of cleavage planes are important aids to identification. Minerals like mica and graphite, for example, have good cleavage in one direction. If a crystal of halite is broken it will commonly produce three smooth lustrous surfaces at right angles to each other. Thus halite is a mineral having three good cleavage directions. Cleavage planes of minerals may superficially resemble crystal faces. Cleavage planes, however, are formed by breakage along planes of weakness and not by the orderly addition of atoms or ions at the surface of a growing crystal.

A surface of breakage in a mineral other than a cleavage plane is termed a fracture. Fractures tend to be irregular and randomly oriented.

They are described by such terms as *splintery*, *fibrous*, *uneven*, or *conchoidal*. Opal and glass frequently exhibit the shell-like arcs of conchoidal fractures. A *hackly* fracture is a ragged surface with sharp points and edges resembling the broken surfaces of cast iron.

Specific gravity

We are all familiar with the fact that some minerals seem heavier than others and that this can help us to identify them. To assure uniformity in determining whether one mineral is heavier than another, geologists utilise a property called *specific gravity*. Specific gravity is the number of times a mineral is as heavy as an equal volume of water. The iron sulphide mineral pyrite, sometimes called fool's gold, has a specific gravity of 5.0, and is therefore five times as heavy as an equal volume of water. Real gold has a specific gravity of 19.3, and is thus readily distinguished from pyrite on the basis of specific gravity.

In determining the specific gravity of a mineral, we use Archimedes' famous discovery that a body immersed in water is bouyed up by a force equal to the weight of the water displaced. The mineral fragment is first weighed in air and is then weighed again when immersed in water. The difference between the two measurements is equal to the weight of the water displaced. Knowing this, the specific gravity can be derived from the formula:

$$\text{Sepcific Gravity} = \frac{\text{Weight of Mineral in Air}}{\text{Weight in Air} - \text{Weight in Water}}$$

A fragment of pure quartz that is 4.55 g in air will be only 2.83 g when immersed in water. According to the above equation, the specific gravity of quartz is 2.65:

$$\text{S.G.} = \frac{4.55\text{ g}}{4.55\text{ g} - 2.83\text{g}} = \frac{4.55\text{ g}}{1.72\text{ g}} = 2.65$$

The density of a mineral or other body is its mass per unit volume.

$$\text{Density} = \frac{\text{Mass}}{\text{Volume}}$$

Unlike specific gravity, which (like other ratios) is not expressed in units, density must always be expressed in units such as grams per cubic centimeter. Because 1 cm^3 of water at 25°C has a mass of 1 g, the density of a substance as expressed in the metric system has the same value as its specific gravity. For example, quartz with a specific gravity of 2.65 has a density of 2.65 g/cm^3.

As is true for all other physical properties of minerals, density and specific gravity are dependent on the composition and the manner in which atoms and ions are arranged in their space lattice. The more tightly packed the atoms, and the more elements of greater mass, the greater also will be a mineral's specific gravity and density. As an example, both diamond and graphite are composed of carbon atoms. Graphite has a specific gravity of 2.2, whereas the specific gravity of diamond is 3.5 because it has more carbon atoms in a given volume.

Magnetism and Other Properties

In addition to the above characteristics by which a mineral may be recognised, certain minerals have other rather distinctive properties. The salty taste of halite, the malleability of gold, the flexibility of mica, the soapy feeling of talc, and the earthy odor of certain clays when moistened are all useful properties. A few minerals are attracted by a magnet, and, if magnetised, the mineral magnetite acts as a magnet itself. Calcite is easily recognised by the way it effervesces when a drop of dilute hydrochloric acid is applied to its surface.

Silicates

Atomic Structure of Silicates

About 75 per cent by weight of the earth's crust is composed of the two elements oxygen and silicon. For the most part, oxygen and silicon occur in combination with other abundant elements such as aluminum, iron, calcium, sodium, potassium, and magnesium to form an important group of minerals called the silicates. A single family of silicate minerals, the *feldspars*, comprises about one half of the material of the crust, and a single mineral species called *quartz* represents a sizable portion of the remainder.

The fundamental unit in the crystal structure of silicates is called a silica tetrahedron. It consists of a compact tetrahedral arrangement of four oxygen ions around a central silicon ion. This tetrahedral shape is a consequence of the very small size and high charge of the silicon ion and of the relatively large size of the oxygen ions. The silica tetrahedron, however, is not an electrically neutral unit. The combining of four oxygen ions (each with two negative charges) and one silicon ion (with four positive charges) leaves the tetrahedron with four unpaired electrons. Because of the resulting charge, the tetrahedral unit must either bond to one or more additional positive ions (such as magnesium or iron) or covalently share oxygen atoms with neighboring tetrahedra.

Among the abundant silicates, olivine and garnet are minerals that have all of their tetrahedra linked by metallic ions. In the other abundant silicates, the tetrahedra are strongly linked by sharing oxygens. The pattern developed by the connected tetrahedra not only forms the atomic structure of the mineral but also determines many of its properties, such as crystal form, cleavage, and specific gravity. The various silicate structural types are as follows:

Single Tetrahedra Structure

These minerals are constructed of individual tetrahedra which are linked by positive iron and magnesium ions. *Olivine,* a mineral common to the lavas of the Hawaiian Islands, has this structure.

Single- or Double-Chain Structure.

In single chain structure, each tetrahedron shares two corner oxygen atoms of its base with two other tetrahedra; in the double chains, two single chains are combined by further sharing of oxygens. Either kind of chain is bonded to adjacent chains by positive metallic ions. Because the bonds holding adjacent chains together are weaker than the silicon to oxygen bonds within each chain, cleavage develops parallel to the chains. The fibrous nature of amphibole asbestos results from chain structure.

Sheet Structure

In minerals having sheet structure, each tetrahedron shares three corners of its base with three other tetrahedra to form continuous sheets. In these structures, the unshared oxygen of each tetrahedron will stand above the plane of the three shared oxygens. The side with the unshared oxygens is designated the inner side.

In mica, a mineral that exemplifies sheet structure, positive ions bond the inner sides of adjacent sheets together to form a paired set of sheets. The structure reminds one of a sandwich in which the sheets of tetrahedra are like two slices of bread and the positive ions the peanut butter that holds the bread slices together.

Other, often weaker, ions serve to bind one "*sandwich*" to those above or below it. It is this sheet structure that gives mica its single plane of good cleavage.

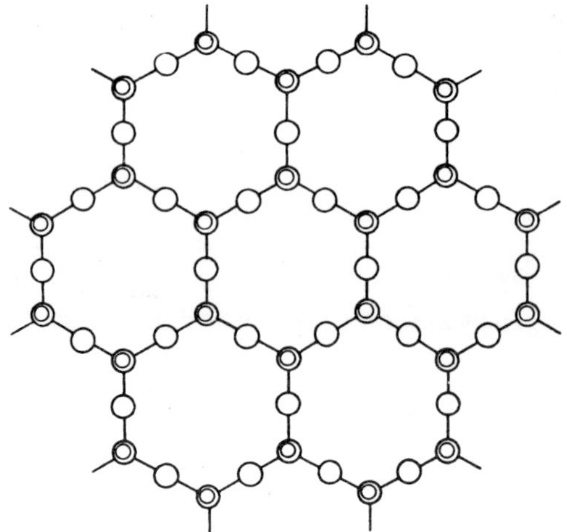

Figure 3.4: Lattice diagram of silicate sheet structure viewed from above. The small spheres represent silicon ions beneath oxygen ions.

Framework Structure

Each tetrahedron shares all four corners with other tetrahedra to form a continuous three-dimensional framework. *Quartz* is a mineral having a framework structure. Because the bonds are so strong in every direction in quartz, the mineral is exceptionally hard and has no tendency to break along preferred directions.

The most abundant rock-forming silicates are included in one or more of the groups described below and in Table elsewhere in this chapter.

Quartz

The mineral *quartz* (SiO_2) is one of the most familiar and important of all the silicate minerals. It is common in many different families of rocks. As mentioned earlier, quartz represents the ultimate in cross-linkage of silica tetrahedra; it therefore will not break along smooth planes. In quartz, the tetrahedra are joined only at the corners and in a relatively open arrangement.

It is thus not a dense mineral, but it is quite hard because of the strong bonding in its framework structure. When quartz crystals are permitted to grow in an open cavity filled with water and dissolved silica, they may develop the hexagonal prisms topped by pyramids that are prised by crystal collectors. More frequently, the crystal faces cannot be discerned because the orderly addition of atoms had been interrupted by contact with other growing crystals.

Such minerals as chert, flint, jasper, and agate are varieties of a form of quartz called chalcedony. *Chalcedony* is composed of extremely small fibrous crystals of quartz. The crystals are so tiny that their study often requires the use of an electron microscope. Spaces between the small crystals are usually occupied by water molecules. Among the varieties of chalcedony, *chert* is exceptionally abundant in many sedimentary rock units.

It is a dense, hard, usually white mineral or rock. *Flint* is the popular name for the dark gray or black variety of chalcedony much used by stone-age humans for making tools. *Jasper* is recognised by its opaque appearance and red or yellow color derived from iron oxide impurities. The term *agate* is used for chalcedony that exhibits bands of differing color or texture. There are many other varieties of quartz minerals than those briefly mentioned here. Properties useful in their identification are provided in Appendix B of this text.

The Feldspars

Feldspars are the most abundant constituents of rocks, composing about 60 per cent of the total weight of the earth's crust. There are two major families of feldspars: the *orthoclase* or *potassium feldspar* group, which are the potassium aluminosilicates, and the *plagioclase* group, which are the aluminosilicates of sodium and calcium. Members of the plagioclase group exhibit a wide range in composition-from a calcium-rich end member called *anorthite* ($CaAl_2Si_2O_8$) to a sodium-rich end member called *albite* ($NaAlSi_3O_8$).

Between these two extremes, plagioclase minerals containing both sodium and calcium occur. The substitution of sodium for calcium, however, is not random but rather is governed by the temperature and composition of the parent material. Thus, by examining the feldspar content of a once molten rock it is possible to infer the physical and chemical conditions under which it originated.

Feldspars are nearly as hard as quartz and range in color from white or pink to bluish-gray. Silica tetrahedra in the feldspars are joined in a strong three-dimensional lattice that is characterised by planes of weaker bonding in two directions at (or nearly at) right angles to each other. Because of this, the feldspars have good *cleavage* (break along smooth planes) in two directions. The resulting rectangular cleavage surfaces and a hardness of 6 are properties useful in the identification of feldspars.

The plagioclase feldspars provide an example of the manner in which ions can be interchanged in a mineral group. A chemical analysis

Table 3.5 : Common Rock-forming Silicate Minerals.

Silicate Mineral	Composition	Physical Properties
Quartz	*Silicon dioxide (silica, SiO_2)*	Hardness of 7 (on scale of 1 to 10); will not cleave (fractures unevenly); specific gravity: 2.65
Potassium feldspar group	*Aluminosilicates of potassium*	Hardness of 6.0-6.5; cleaves well in two directions; pink or white; specific gravity: 2.5-2.6
Plagioclase feld-spar group	*Aluminosilicates of sodium and calcium*	Hardness of 6.0-6.5; cleaves well in two directions; white or gray; may show striations on cleavage planes; specific gravity: 2.6-2.7
Muscovite mica	*Aluminosilicates of potassium with water*	Hardness of 2-3; cleaves- perfectly in one direction, yielding flexible thin plates; colorless; transparent in thin sheets; specific gravity: 2.8-3.0
Biotite mica	*Aluminosilicates of magnesium, iron, potassium, with water*	Hardness of 2.5-3.0; cleaves perfectly in one direction, yielding flexible thin plates; black to dark brown; specific gravity: 2.7-3.2
Pyroxene group	*Silicates of alumin-num, calcium, magnesium, and iron*	Hardness of 5-6; cleaves in two directions at 90°; black to dark green; specific gravity: 3.1-3.5
Amphibole group	*Silicates of alumi-num, calcium, magnesium, and iron*	Hardness of 5-6; cleaves in two directions at 56° and 124°; black to dark green; specific gravity: 3.0-3.3
Olivine	*Silicate of magne-sium and iron*	Hardness of 6.5-7.0; light green; transparent to translucent; specific gravity: 3.2-3.6
Garnet group	*Aluminosilicates of ron, calcium, magnesium, and manganese*	Hardness of 6.5-7.5; uneven fracture, red, brown, or iyellow; specific gravity: 3.5-4.3

of specimens of plagioclase taken from several different rocks would very probably reveal that the proportions of calcium, sodium, aluminum, and silicon (the principal cations in plagioclase) would differ among the specimens. This variability occurs because some ions resemble each other in size and electrical properties and are thus interchangeable in a given crystal. As indicated in Figure elsewhere in this chapter, calcium and sodium ions are large and nearly identical in size. Both aluminum and silicon are small ions and not greatly different in size. Thus, calcium ions might substitute for sodium ions freely if size alone were the only requirement. However, the electrical neutrality of the crystal must also be maintained. The electrical charge of the calcium ion is + 2, whereas that of the sodium ion is + 1. To counteract the surplus positive charge, an aluminum ion (+3) may substitute for a silicon ion (+4) to maintain electrical neutrality. Thus, Ca^{2+} + Al^{3+} can inter change with Na^{2+} and Si^{4+}.

The Mica Group

As noted earlier, mica is a silicate mineral having sheet structure and is easily recognised by its perfect and conspicuous cleavage in one directional plane.

The two chief varieties are the colorless or palecolored *muscovite* mica, which is a hydrous potassium aluminum silicate ($KAI_{2\_}(Al_3Si_3O_0)(OH)_2$) and the dark colored *biotite mica*, which also contains iron and magnesium ($K(Mg,Fe)_3A1Si_3O_{10}(OH)_2$). In muscovite mica, two sheets of tetrahedra are strongly held together along their inner surfaces by positively charged ions of aluminum. These sandwichlike paired sheets are in turn weakly joined to others by positively charged ions of potassium. When muscovite is cleaved into paper-thin layers, the separation occurs primarily along the weaker plane where the potassium ions are located.

In biotite, magnesium and iron ions hold the inner surfaces of the sheets together, but once again potassium ions serve to weakly join each basic set of paired sheets to its neighbor. Identification of large specimens of mica is rarely a problem because of its perfect planar cleavage and the way cleavage flakes snap back into place when they are bent and suddenly released. The micas are common constituents of igneous and metamorphic rocks, in which they can be recognised by their shiny surfaces and the ease with which they can be plucked loose with a pin or pen knife.

Before the manufacture of glass, one of the chief uses of muscovite mica was as window panes. This clear mica was quarried in Muscovy (an early name for Russia), and thus came to be known as "*Muscovy glass*," and eventually muscovite. Today, mica is used in the manufacture of electrical insulators and as a filler in plaster, roofing products, and rubber.

Hornblende

Hornblende is a vitreous black or very dark green mineral. It is the most common member of a larger family of minerals called amphiboles, which have generally similar properties. As can be seen from its chemical formula, $NaCa_2(Mg,Fe,Al)_5(Si,Al)_8O_{22}(OH)_2$, hornblende contains a relatively large number of elements. Because of the presence of iron and magnesium, hornblende (along with biotite, augite, and olivine) is designated *a ferromagnesian* or *mafic mineral.* Crystals of hornblende tend to be long and narrow. Two good cleavages are developed parallel to the long axis and intersect each other at angles of 56° and 124°. The cleavage is a reflection of the location of planes of weaker bonds that exist between the double-chain units of silica tetrahedra in the atomic lattice.

Augite

Just as hornblende is only one member of a family of minerals called amphiboles, augite is an important member of the pyroxene family, in which many other mineral species also occur. Its chemical formula, $Ca(Mg,Fe,Al)(SiAl)_2O_6$, indicates that it too is a ferromagnesian mineral and thus dark colored. An augite crystal is typically rather stumpy in shape, with good cleavages developed along two planes that are nearly at right angles (87° and 93°). Thus, the cross section of a crystal appears nearly square rather than rhombic, as in hornblende). Unlike hornblende, which has a double-chain silicate structure, augite is constructed of single chains, and this accounts for its having differently shaped cleavage fragments.

Olivine

Olivine, another ferromagnesian mineral, has been mentioned earlier as having isolated silicon-oxygen tetrahedra bonded together by iron and/or magnesium ions. Its formula, $(Fe,Mg)_2SiO_4$, indicates that it is a solid solution mineral, containing variable proportions of iron and magnesium. The substitution of these ions for each other is facilitated by their having similar ionic radii and two electrons in their outer electron shell. The ions in olivine are so strongly held by ionic bonding that the mineral has a hardness of 6.5 to 7 on Mohs' scale.

As you might guess from its name, this glassy-looking mineral often has a green color. Frequently, it occurs as masses of small sugary grains or as tiny vitreous crystals in black lavas. It is also an important constituent of stony meteorites. If large unblemished crystals of magnesium rich olivine are found, they may be cut and polished into attractive gemstones called *peridot*.

Garnet

Garnet is the name used for a group of mineral species closely related in composition, crystal form, and physical properties. In general composition, these minerals are silicates of aluminum with iron, calcium, or magnesium. The iron-rich variety, almandite, $Fe_3Al_z(SiO4)_3$, is widely used in jewelry because of its rich dark red color. However, depending on their composition, garnets are found in all colors except blue.

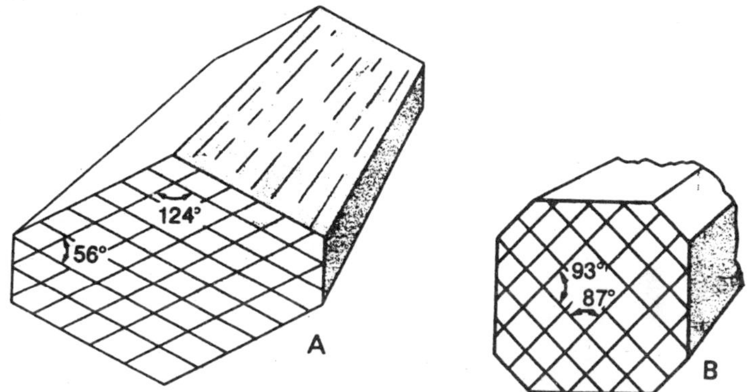

Figure 3.5: (A) Typical shape of a hornblende crystal. Planes of cleavage intersect at 56° and 124°. (B) Typical shape of an augite crystal. Planes of cleavage intersect at 87° and 93°.

For example, if calcium fills the place of iron in the lattice, the garnet is called grossularite and may have a pale green color, although other hues may also develop. Garnets can usually be identified by their well formed equidimensional crystals and their resinous to vitreous luster. Crystals will not cleave but rather break with a conchoidal fracture. Because of the sharp edges produced on fracturing and its hardness (7 to 7.5 on Mohs' scale), garnet is widely mined and sold as an abrasive material.

Chlorite

The chlorites are a group of greenish minerals having a platy form similar to that of micas. In composition, the minerals of the chlorite group are hydrous aluminum silicates of iron and magnesium. Chlorite can usually be identified by its perfect cleavage in one direction, greenish color and streak, and relative softness. It is a common constituent of metamorphic rocks and also occurs as an alteration product of ferromagnesian minerals.

Clay Minerals

The clay minerals are silicates of hydrogen and aluminum with additions of magnesium, iron, and potassium. At one time, clays were assumed to be amorphous mixtures of metal oxides and hydroxides.

Through the use of X-ray equipment, however, it was demonstrated that clays are composed of crystalline units which, like mica, have sheet structure. The individual flakes are extremely small and require the magnification provided by the electron microscope for adequate examination.

Sheets of tetrahedra in clays are weakly bonded, and the potassium, sodium, and magnesium ions may enter and leave exchangeable locations within the lattice structure. Clays are formed through the decomposition of other aluminum silicate minerals, especially the feldspars. Clay minerals are characteristically very soft, have low density, occur in earthy masses, and become rather plastic when wet. These characteristics permit one to recognise clay, but in order to identify the individual species of clay minerals it is necessary to use X-ray diffraction techniques.

THE COMMON NONSILICATE MINERALS

Approximately 8 per cent of the earth's crust is composed of nonsilicate minerals. These include a host of carbonates, sulfides, sulfates, chlorides, and oxides. The nonsilicates that are most abundant at the earth's surface are the carbonates, the sulfates and chlorides known collectively as evaporites, and oxides of iron.

Carbonates

Calcite and dolomite are the two most important members of the carbonate mineral group. Calcite ($CaCO_3$), which is the main constituent of limestone and marble, forms in many ways. It is secreted as skeletal material by certain invertebrate animals, precipitated directly from sea water, or formed as dripstone in caverns. Calcite is easily recognised by its rhombohedron-shaped cleaved fragments and by the fact that an application of hydrochloric acid on its surface will cause effervescence. Clear cleavage rhombs of calcite, called iceland spar, exhibit a property known as double refraction.

For example, if a transparent piece of calcite is held over a dot that has been placed on white paper, two dots will appear when the calcite is held in certain positions. As the mineral is slowly rotated, one dot will appear to rotate around the other. The explanation for this bit of "magic" is that the light entering the specimen is bent or *refracted* and divided into two polarised rays, one of which is bent more than the other.

The image from the more refracted ray appears to move around the image of the less refracted ray as the piece of calcite is rotated. Although double refraction is easily seen in calcite, it is not a unique property of calcite only. All minerals exhibit double refraction, except those in the isometric crystal system. In most of these other minerals, however, optical instruments are required to measure the phenomenon.

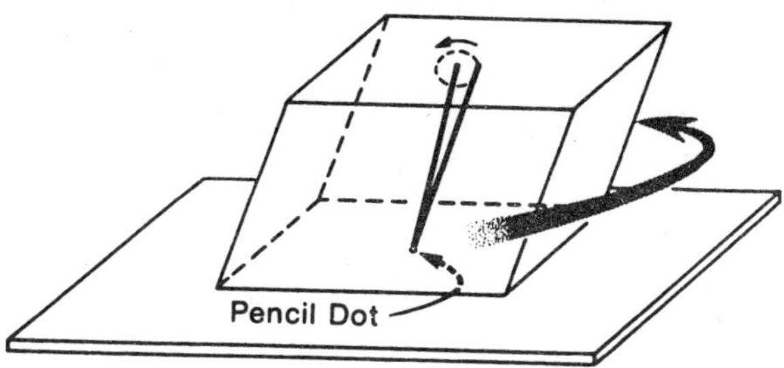

Figure 3.6: If a piece of clear calcite is placed as shown over a pencil dot, two dots will be viewed from a point above the specimen. If one then rotates the calcite as indicated by the arrow, the image of the dot from the more refracted ray will appear to move in a circular path around the other dot.

Dolomite, $CaMg(CO_3)_2$, is a carbonate of calcium and magnesium that occurs as extensive layers of a rock called *dolostone.* In the field, geologists often distinguish dolomite by the fact that it will not effervesce in dilute hydrochloric acid unless it is powdered. Many ancient dolostones are thought to have been formed by recrystallisation of still older limestones. Because calcium and magnesium ions are nearly the same size, it is possible for the magnesium to replace calcium in the atomic structure.

Evaporites

Evaporites are minerals that have precipitated from bodies of water subjected to intense evaporation. They include common rock salt, or halite (NaCl), and gypsum. *Halite is* easily identified from its salty taste and perfect cubic cleavage. *Gypsum* is a soft (2 on Mobs' scale) hydrous calcium sulfate ($CaSO4 \bullet 2H_2O$). The variety of gypsum called satinspar has a fine fibrous structure, whereas *selenite* will split into thin plates. The finely crystalline massive variety known as *alabaster* is widely used in carvings and sculpture because of its uniform texture and softness.

Iron oxide minerals

The chief iron oxide minerals are magnetite (Fe_3O_4), hematite (Fe_2O_3), and goethite ($HFeO_2$). Magnetite is a lustrous black mineral that is found in igneous rocks as well as some metamorphic rocks and clastic sedimentary rocks. It is an important ore of iron. Magnetite has a black streak, but its most distinctive property is related to magnetism. Pieces of magnetite can be picked up by an ordinary steel magnet, and the lodestone variety of magnetite is itself a natural magnet that will attract steel objects.

Hematite is even more important than magnetite as an ore of iron. It occurs as iron ore in thick and extensive beds of sedimentary origin that have been subsequently altered and enriched by solutions. Hematite is very common in a variety of sedimentary rocks, where it often provides a reddish or brownish coloration. The mineral varies in color from steel gray to brownish red but always exhibits a distinctive reddish streak.

Goethite generally forms as a weathering product of other iron-bearing minerals. It is the principal mineral in the amorphous hydrous iron oxide mineral known as limonite ($Fe_7O_3 \bullet 2H_2O$). Goethite has a yellow-brown streak and is generally yellow or brown in color.

MINERALOGIC CLUES TO GEOLOGIC EVENTS

Minerals are studied by geologists not as mere constituents of the crust but as clues to the history of the rocks in which they occur. Some minerals, such as halite, develop exclusively in ocean water and provide the geologist with evidence that a particular sediment was deposited in the sea rather than in a freshwater lake. A thick bed of halite indicates aridity and evaporation so extreme that the brine had become ten times saltier than ordinary sea water.

The magnetic properties of magnetite can, in certain situations, provide clues to the position of continents relative to the earth's magnetic poles. Certain minerals form within a narrow range of conditions and can therefore be used to diagnose the pressures and temperatures involved in the formation of crustal rocks and mountains. Diamonds, for example, form only at high temperatures and extremely high pressures.

Like diamond, graphite is a crystalline variety of carbon. However, it forms at lower temperatures and pressures. Other minerals contain radioactive isotopes that permit us to know the age of the parent rocks. By their size, crystals of feldspar give the geologist information about the rate at which ancient molten mass congealed. Even ordinary grains of quartz may hold clues to their history since being eroded from some ancient granitic terrain. The ancient buried mineral products of weathering may also provide information about past climatic conditions in a region.

MINERALS AS GEMS

Some occurrences of the minerals introduced in this chapter are so attractive, durable, and rare that they are highly valued as gems. Their beauty is dependent on color, luster, and brilliance-properties that are enhanced when the "rough stones" are cut, faceted, or shaped and polished. Gems are sold by weight, and the basic unit of weight is called a gem carat. One gem carat is equal to 1/5 g. (The gem carat is not to be confused with the gold carat, which indicates the proportion of gold in an alloy and is not a unit of weight.)

Such familiar minerals as quartz, corundum, feldspar, olivine, and garnet occur sporadically as materials of gem quality. Gems in the quartz family are particularly well known and include the purple variety known as amethyst, yellow quartz or citrine, rose quartz, opal, the banded microscrystalline quartz called agate, and the mysterious tiger's-eye quartz, which results from the replacement of asbestos by microcrystalline quartz. Jasper is a variety of microcrystalline quartz that is rich in iron oxide.

The two most famous gems in the corundum family are blue sapphire and red ruby. The color differences in these corundum minerals result from small traces of metal oxides. For example, ruby is colored by chromium oxide and sapphire by oxides of iron and titanium. These and other impurities also produce exquisite hues of green, purple, and gold in sapphires. Feldspar gems include the attractive white opalescent moonstone and the green orthoclase called amazonite. Peridot is the gem variety of olivine.

In the garnet family, almandite is prised for its clear wine-red color. There are, however, many other beautifully colored garnets, including pink, orange, and green varieties. Topaz, known to us as the third hardest mineral in Mohs' scale, may occur as a lovely yellow gem, although stones colored pale blue and light green are also prised. Diamond, the most highly treasured of gems, is a nonsilicate composed only of carbon. As mentioned earlier, diamonds are well known for their brilliance and hardness. They are derived from basic plutonic rocks called kimberlites. The principal factors involved in determining the value of a cut and faceted diamond is its weight, color, and degree of perfection. A so-called perfect diamond is one that shows no conspicuous inclusions or flaws when viewed at a magnification of 10 x by an experienced person. The weight of a diamond is usually expressed in a unit called *a point,* which is 1/100 of a carat. Thus, a 25 point diamond would weigh 0.25 carat.

4

SOILS

To geologists and soil scientists, soil is a surface layer of variable thickness formed in the open atmosphere by the physical, chemical and biochemical alteration of pre-existing rock materials, to which humic substances are added. In many instances, soils are underlain by unlithified, partly weathered rock known as regolith, but this is not considered as being true soil because the breakdown is predominantly inorganic in character.

The further back one goes in geological time the rarer true soils become. In the beginning of the Earth's history, and until 3 billion years ago, soils were lacking in biochemical activity and were more of the nature of partly surfaceweathered regolith (protosoils). In the succeeding 2 billion years soils remained at this primitive level of development despite the advent of algae, bacteria and a much more oxygenated atmosphere. Palaeozoic times saw a major proliferation of animal and plant life, and the beginnings of true herbaceous soils.

Forest, shrub and grassland soils were initiated widely for the first time in the late Mesozoic period, and modern soils in late Pleistocene times, about 150000 years ago. Some of the most important changes in soils have occurred in the last few hundred years due to the activities of man. Protosoils and soils are the temporary resting place of inorganic and organic materials en route to the sea.

While they are in residence the original source components undergo differing degrees of change from a little, as in the case of quartz, to much, as in ferro-magnesian and clay minerals, and this change is ultimately reflected in the constitution of any derivative sediment. Red muds accumulating off the mouth of the Amazon and brown goethitic oolite shoals adjacent to the Chari delta feeding into Lake Chad seem to relate directly to rubified (reddened) and brunified (browned) soils, some lateritic, being eroded in the hinterlands.

However, it is unusual for such a direct linkage to be established and in most ancient successions it can only be surmised that terrigenous detritus has passed through a soil-producing phase. Soils take several thousands of years to develop without human interference and in more climatically and physically stable parts of the world may even have taken hundreds of thousands of years. The lateritic soils which cover extensive regions of peninsular India, Indonesia, Africa, Brazil and Australia appear to have been mainly initiated in Tertiary times, though there is evidence for Mesozoic beginnings in southern Australia.

The soil processes at work over such lengths of time are rarely continuous, as long-term climatic changes can effectively terminate or interrupt their progress. Most of the Tertiary laterites of northern Australia and India are now relict due to climatic fluctuations in Quaternary times. They are considered now as palaeosols on whose surface slightly different modern soils are forming. Palaeosols are soils formed under conditions which no longer obtain in that place and can be recognised comparatively easily in ancient successions.

Soils are readily eroded given suitable wet climatic and physiographic conditions, so that their long-term preservation-potential in those circumstances is not very high. Some soils are transported to other sites and regain either similar soil qualities or different qualities if the climatic or vegetational regime has significantly altered. Superimposed soil profiles, pre- and postWisconsin (Quaternary) and exhibiting differing soil characteristics, are found in the central United States.

Other soils are given a greater coherence due to internal diagenetic changes which delay, if not negate, erosive processes. Duricrusts are of this type. Very few soils, partly lithified or not, can withstand the effects of stripping by ice sheets, intense wind activity (in arid regions) or the sea during major marine transgressions. On the other hand, there is ample evidence from many fluviomarine successions of Carboniferous, Mesozoic and Tertiary age that gentle inundation and penetration by the sea across afforested marshy areas do not inevitably destroy the whole of the original soil profile. Rapid covering by sub-aerial lava and ash-flows and ash-falls has also often ensured the preservation of parts of soil profiles, as in the interbedded lithomarge, bauxitic and lateritic deposits of Carboniferous and Tertiary volcanic sequences of Northern Ireland and Scotland.

SOIL-FORMING PROCESSES

Climate always exerts an important control upon the relative effectiveness of the various weathering processes, upon the biological factors in soil formation, and especially upon the distribution of soluble substances in the soil. It is found that in regions of maturely formed soils, such as eastern Europe or the central United States, the influence of the parent rock is minimised, and uniform soil conditions tend to be developed over wide areas within definite climatic belts.

The influence of topography is felt in several ways. In regions of appreciable relief, erosion of the upper layers of partially developed soils prevents the soil from reaching a mature condition, so that there is always a considerable proportion of incompletely weathered material, and the influence of the parent rock is strongly evident. In peneplaned regions, on the other hand, mature and uniform soils develop over wide areas. The process of weathering, which provides the mineral basis of soils, is the first and perhaps the most important stage in soil formation.

It is a function of interdependent mechanical and chemical processes which produce fragmented and chemically modified constituents from solid rock. Water in pores and fissures is the most important agency by which most of these changes are brought about. The water when frozen or warmed causes contraction and expansion of the rock leading to exfoliation. It can carry dissolved within it high concentrations of salts, which can create internal stresses when the salts either precipitate out or aggressively react with the existing rock minerals.

Salt weathering is common in modern coastal arid desert areas. Underless extreme circumstances, the waters are still chemically reactive, and solution, hydrolysis, carbonation and oxidation occur either singly or in combination to bring about rock disintegration. Thus calcite gradually dissolves, mica and other minerals become hydrated, ferrous and sulphide ions are oxidised and silicate minerals, such as feldspars, are hydrolised.

The most important property of soil depends upon the presence of a complicated system of colloidal substances, both organic and inorganic, by virtue of which the soil is enabled to retain considerable resources of moisture and exchangeable bases within itself. Organic matter, derived mainly from plants but also partly from animals, is added at the surface, and mineral substances are supplied from below by subsurface weathering; the resulting mixture of organic and residual matter is subjected to certain further changes which lead to the construction of soils as opposed to undifferentiated mineral residues.

As soils develop, mineral and humic particles aggregate into crumblike units known as peds. This aggregation is due to the activities of burrowing organisms, root growth, physical compaction and fissuring. The shape, size and distribution of peds, which are well developed in calcareous clay soils but less so in acid sandy soils, produce soil horizons roughly paralleling the ground surface. Well-formed soil profiles consist of several soil horizons recognisable by differences such as colour, texture, structure, root activity and humic content.

In soils with normal drainage, water passes downwards during rainy periods, and upwards in times of drought, when evaporation from the surface exceeds the rainfall. In humid climates, such as that of the British Isles, the movement is predominantly downwards, so that soluble salts, colloids and small mineral particles tend to be removed from the surface layer, known as the A horizon, and transported into the underlying layers of the soil. This process is known as eluviation.

The soluble salts are completely washed away into the groundwater, but certain of the colloids, especially those consisting of minute particles, are largely redeposited within the soil at some lower level, which is known as an illuvial horizon. Eluviation of this kind leads to the formation of heavy clay subsoils with sandy or loamy topsoils. The operation of this process depends upon the dispersion of soil-colloids in the eluviated layer, and is hindered in soils where surface conditions are not favourable to dispersion.

Thus the presence of exchangeable calcium, by keeping the claycolloids in a flocculated condition, prevents them being dispersed into a mobile form, and eluviation can only become effective after the calcium has been leached away from the sphere of activity.

Eluviation of soluble substances in a moist climate leads to the removal of these materials to the groundwater level, and the upper layers of the soil become leached. Calcium carbonate and salts of sodium and potassium tend especially to be removed in this way. This process is called podzolization.

In arid and semi-arid climates, especially if the level of ground water is not far below the surface, strong evaporation causes an upward migration of soil moisture, and the process of eluvation works upwards instead of downwards. As a result of this, we find in some regions that the more soluble salts, instead of being leached away, become concentrated at the surface, giving saline or alkaline soils. There is also a tendency for calcium carbonate, calcium sulphate, alkali silicates, and iron compounds to migrate towards the surface, where they form hard-cementing crusts or duricrusts of limestone, gypsum, silica or ferric oxide up to several metres thick. These varieties are commonly referred to as caliche or calcrete, gypcrete, silcrete and ferricrete. In less extreme situations, the salts may be precipitated as grain coatings, veins, concretions and thin layers within the upper part of the soils.

For example, pedocals are soils enriched in this fashion with carbonates. In climates with seasons of rain and drought, downward and upward eluviation alternate. Since some constituents are irreversibly precipitated, while others migrate up and down with alternating movements of the soil water, rather complicated and highly individualised soils may be produced.

When the water table is high and the lower soil layers saturated, so that a reducing environment exists, the process known as gleying occurs. This creates a mottled brown, olive-grey and greyish-blue colouration, the soil becoming admixed with varying amounts of amorphous organic matter, ferrous iron complexes and salts.

SOIL CLASSIFICATION

A detailed discussion of the many kinds of soils and soil classification is beyond the scope of this book. Briefly, the bases for such include field profile characteristics, soil-formation processes and soil capability for food production. For large-scale maps the classification produced by the United Nations Food and Agriculture Organisation in 1974 is in international use. This broadly divides soils into 26 major types, such as podzols, ferralsols, chernozems, rendzinas, gleysols and fluvisols, then subdivides these further on subordinate qualities such as humic and calcic content.

On larger domestic-scale soil maps additional classifications are used in which soil series are a fundamental mapping unit. A soil series is a group of soils identifiable in the field on the basis of comparable grain size, profile characteristics and lithologically similar parent materials. It is usually named after a type locality where it is well developed. For example, the Windsor soil series in Britain is a variety of clay-rich soil in which the drainage is impeded at moderate depths by a relatively impermeable substratum, in this instance the Eocene London Clay.

On a different climate zonal basis, soils have been grouped into pedalfers, which are leached, acid varieties found in moist regions, and pedocals, which are lime-accumulating varieties associated with drier climates. The former includes podzols, lateritic and bauxitic soils and the latter chernozems (black earths), kastanozems (chestnut and brown soils) and caliche.

MODERN AND ANCIENT SOIL TYPES

Podzolic Soils

These are yellow, brown and pale-grey acid soils in which alkalies and alkaline earths have been removed by intense leaching. Their total thickness is usually less than a metre. Together with closely related types, they cover enormous areas in northern Europe, and also spread over a considerable area in Canada and the northeastern United States. The podzols thus occupy a cold to temperate humid climatic zone lying to the south of the tundra.

In the development of a typical podzol, the surface layers are leached of all soluble substances such as calcium carbonate and the silicates are hydrolysed to form colloidal-clay minerals, ferric hydroxide and various soluble salts. Intense leaching removes all but the original quartz, so that a bleached sandy layer is formed, which near the surface is mixed with black peaty material. A considerable quantity of humus is usually carried downwards into the subsoil with the other products of leaching. Clay, ferric oxide and humus tend to become concentrated in the subsoil and can form hardpans. This layer may become so impervious to water that the course of soil development is altered.

We thus have three horizons in the profile of a typical podzol:

A. Dark peaty surface layer, underlain by bleached quartz sand (eluviated horizon).

B. Subsoil enriched in humus, clay and ferric oxide (illuvial horizon).

C. Disintegrated parent rock.

Laterites and Bauxites

Laterites and bauxites, which are commonly interbedded in ferralsol profiles up to 35 m in thickness, are the extreme end product of the weathering cycle and are formed under oxidising conditions in areas of low or moderate topographic relief, but with a minimum of erosion. The climate is persistently warm or even subtropical with temperatures above 25°C. Rainfall exceeds evaporation almost continuously and there is a progressive leaching of silica and alkali ions plus a destruction of humus by microflora.

The presence of Al-rich parent rocks is preferable because it is the selective retention of aluminium which eventually creates the laterites and bauxites. The important aluminium hydroxides concentrated in these rocks are gibbsite, diaspore and boehmite and in the highest quality bauxites these minerals predominate. In general, gibbsite typifies Tertiary and Recent deposits, and diaspore Palaeozoic deposits. In laterites and lateritic and bauxitic clays there is a variable dilution by clay minerals, especially kaolinite, and iron oxides.

The red and brown iron oxides in laterites sometimes form more than *45%* of the deposit and make the rock worth working as an iron ore. In south-east Asia the iron oxides frequently occur as pisoliths, spherules and nodular concretions embedded in a matrix mainly formed of kaolinite. The exposed duricrust surfaces present a slaggy or cavernous appearance and a brecciated structure is produced by the collapse and recementation of the superficial crusts.

Beneath the top soil (ferralsol) and duricrust layers there is normally a gradation into a mottled zone, then a pallid zone before the decomposing parent rock is reached. The parent rocks vary considerably and include basic lavas and volcaniclastics as in India, or sandstones, shales, slates, gneisses and granites, as in Australia and Africa. The pallid zone is mottled yellowish-red, has a blocky structure and consists of a high proportion of secondary clay minerals, despite which the parent rock texture is often preserved. This rock product is known as lithomarge.

Laterite and bauxite soils are capable of being eroded and their constituents transported into a range of environments including the sea. Under these circumstances they can show bedding and be composed of fragments from boulder size downwards.

Terra rossa is a humus-deficient, decalcified red bauxitic soil, sometimes relict, found resting on chalks, limestones and dolomites in many karstic temperate regions. It commonly occupies poljes and sinkholes (dolinas) into which the materials have been washed. Rendzinas, in contrast to terra rossa soils, are rich in humus at the surface.

They are grey to black, thin calcimorphic soils derived from the degradation of marly and shaley limestones, including caliche, in temperate humid climates. The parent limestone fragments maintain high levels of calcium carbonate in the soil profile, effectively preventing podzolization.

Soils of Subhumid and Semi-arid Regions

In temperate to cool regions, where the annual precipitation is between 250 and 600mm, the rainfall is roughly balanced by evaporation. From the surface, effective leaching cannot take place, and there is a tendency for the mobile products of weathering, including some of the soluble salts, to remain within the soil. Calcium carbonate and gypsum are the commonest of the substances which accumulate in this way, but various other carbonates, sulphates and other salts are also found. During the dry season water is drawn through the pore spaces to the surface of the soil, where it evaporates and deposits its dissolved salts.

In the case of calcium bicarbonate, calcium sulphate and colloids or soluble salts containing iron or silica, concretionary structures tend to be formed in the upper soil and these substances thus accumulate in a fixed condition, without being washed down again during the moist season. In general, the clay fractions of semi-arid soils are basesaturated, usually with calcium, and consequently tend to have a mildly alkaline reaction.

The chernozems of Russia owe their black to grey colour to the presence of dark organic matter, which usually amounts to 8% or 10% of the soil. The more soluble salts are leached out of typical black earths, but calcium carbonate and gypsum are incompletely removed, and tend to form concretionary masses at some level below the surface. Soils of this kind are characteristically developed on the extensive plains of the Russian steppes, where the parent material is principally loess, but they may also be formed from other rocks such as clays, chalk or granite.

Similar black soils are found under comparable climatic conditions in Germany and the Great Plains of the United States of America. The black cotton soil or regur of India is developed on the Deccan basalts under warmer and wetter conditions, though with a pronounced seasonal dry period. They are not subject to the long and severe winters of the American and European plains.

This kind of soil contains calcareous concretions known as kunkar, but differs from genuine chernozems in containing rather less organic matter. Similar soils are found also in Kenya and other parts of Africa. The chestnut-brown earths are so named from their dark brown colour, which appears to be due to 3-5% organic matter and not to ferric iron. They are found in parts of Russia and North America where the climate is rather drier than in the areas covered by chernozems.

Soils of this kind contain less organic matter than the black soils of similar latitudes, and calcium carbonate concretions accumulate nearer to the surface. Under drier, though not arid, conditions with seasonal rainfall amounting to about 400 mm per year, vegetation is sparse and the organic input into soils is low. This gives them a pale greyish-brown to grey colour. Limeaccumulating horizons in these soils are close to the surface or even become emergent where the soil has been stripped away. Given an adequate length of time, measurable in hundreds of years for thin deposits, though probably thousands for thicker, the horizons develop a concretionary, crusty structure, which can be several metres in thickness, known as caliche or calcrete.

The whole soil profiles in modern terminology are calcic yermosols or xerosols. Caliches in some eastern Mediterranean countries, New Mexico, Argentina and Chile are well developed as cappings resting on a wide range of unlithified and solid rocks including gravels, marls, limestones, sandstones and basic igneous rocks. Internally, the structure of these duricrusts is complex due to the repeated dissolution and precipitation of salts, mainly calcium carbonate, so that cross-cutting veins, drusy cavity fills and corrosion of original soil and rock constituents are usual.

With the passage of time, the upwards movements of alkaline capillary water and downwards movements of rain water bring about a greater degree of lithification and a firmly cemented, crudely laminar structure is created. Caliche layers and nodules are found in some ancient red-bed successions, such as the Old Red Sandstone, Lower Carboniferous and Triassic of north-west Europe and in various semi-arid Pleistocene deposits in the western USA. In Britain they are sometimes called cornstones. In most of these old deposits the upper horizons of the original soil profile have been partially destroyed, leaving behind the caliche horizon.

Occasionally, the whole profile is destroyed by flash-flooding and undercutting, and the more resistant caliche is redistributed as a bouldery conglomerate. Such deposits have been referred to as conglomeratic cornstones.

Soils of Arid Regions

Regions where the annual rainfall is less than 250 mm are typified by light grey, brown and red alkaline yermasol-type (desert) and solonchak-type (salt) soils. The soils are very thin and very immature (raw) with little soil profile development. The paler colour of these soils is due to their poverty in organic matter, which often amounts to less than 1%.

Red colours are usually due to haematite coatings on grains and these are probably deposited from capillary waters carrying iron from hydrolised minerals in the soil and regolith. This process is comparable with that which caused the reddening in ancient red-bed successions. Carbonates and sulphates accumulate a short distance below the surface of some of these soils, the sulphates usually below the carbonates, and sodium and potassium salts are usually present.

Where the water table is not far below ground level soluble salts are brought through capillary pores to the surface, where they become concentrated as the water evaporates. In this way efflorescences of soluble salts and nodular masses or encrustations of calcite and gypsum may be formed either at the surface or within the superficial layers.

In saline soils the preponderant salt is sodium chloride, which forms a white crystalline efflorescence at the surface during drought. These soils are most characteristically developed in arid or semi-arid regions such as Turkestan, southern Russia, the Dead Sea basin, and parts of western North America. Sodium sulphate is usually present in saline soils and in some areas is the preponderant salt, as in the 'white alkali' soils of North America.

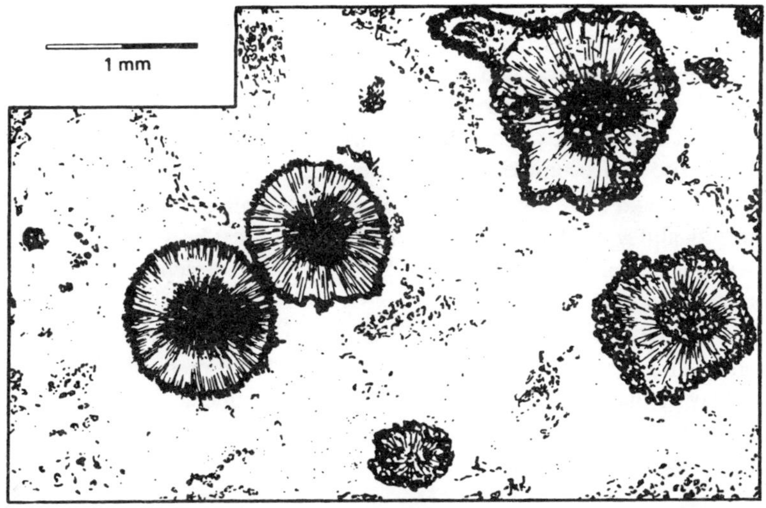

Figure 4.1: Fireclay, Carboniferous, S cotland. An aggregate (clear) of aluminium oxides and hydroxides, within which early diagenetic sphaerosiderite has formed. Polarised light.

Alkaline soils contain sodium carbonate, and owing to the completely dispersed condition of their clay particles, they develop deep shrinkage cracks on drying; this imparts a characteristic prismatic structure to the soil. The black alkali soils of the arid regions of western North America are of this kind, and similar soils are found in Hungary and the Ukraine. In the Kenya Rift Valley the reaction between clay minerals in the soils and sodium carbonaterich ground water often produces zeolites. Analcite forms up to 40% by volume of some soils.

Seatearths

Seatearths are palaeosols which commonly contain fossilised roots in position of growth and often underlie coal seams. They vary in lithology, some being brown quartz arenites, others pale to dark grey mudrocks, and this reflects differences in their manner of origin. The quartz arenitic seatearths probably formed on sheet sand bodies extending into swampy terrains and afe of a gley podzolic type. Acid leaching and eluviation occurred in the upper layers of the sand, whereas the lower parts were partially waterlogged. In Britain these hard, brittle rocks are called ganisters.

In contrast, the mudrock seatearths, known by such names as underclays and fireclays, are humic gleysols formed in waterlogged acid marshy and boggy environments. They vary considerably in plasticity, some being very plastic and others hard and brittle. The very plastic varieties, often carrying diagenetic siderite and limonite nodules, are full of slickensides due to compactive stresses, and consist of large kandite clay mineral and aluminium hydroxide complexes. The non-plastic flint clays are smooth and white, fracture conchoidally and are without slickensides. They are formed of smaller complexes in which halloysite is a prominent clay constituent.

It should be realised that a fireclay, technically, should have refractory qualities, such as the ability to withstand great heat (up to 1600°C) without fusion or disintegration. Many so-called fireclays do not have these properties.

The frequent occurrence of sideritic and sphaerosideritic nodules and aggregates implies a genetic connection with soil-producing processes. It may be that leaching of iron in the soil by organic acids was the main mechanism of iron concentration at the local water table where stagnant reducing conditions prevailed.

5

WEATHERING THE SOIL

Over 3 billion years ago, rain and snow fell upon the earth much as today. Solid rocks were reduced to grains of sand by the action of water, ice, and chemically active solutions. We know this from clues left in ancient rocks whose great antiquity has been established by radioactive dating. The processes by which rocks are broken into smaller particles and chemically decomposed have continued unabated down through the eras of geologic time.

We term these most enduring of geologic processes *weathering*. Weathering is of enormous importance to all of us. It causes solid rock to crumble and thereby facilitates erosion. Weathering provides much of the sediment for the making of sedimentary rocks. Without weathering there would be no soil for the growth of plants that are vital to ourselves and all other forms of life.

Weathering is a phenomenon that combines two categories of processes. On the one hand are those processes that result in the fragmentation of once solid rock without chemical change, as when water freezes and expands in crevices so as to cause the rock to break apart, or when rocks are wedged apart by the growth of roots from trees and shrubs. These processes of physical disintegration are designated *mechanical weathering*.

In contrast, *chemical weathering* involves the chemical decomposition of rocks as their minerals react with water and air. Mechanical and chemical weathering work together in the destruction of rocks at the earth's surface, although one may predominate over the other. In areas of extreme cold or aridity, for example, mechanical weathering may predominate, but in warm wet regions, chemical decomposition is the predominant kind of weathering.

Although mechanical breakup and chemical decay of a rock are partners in weathering, it is advantageous to discuss these two components of weathering separately, because disintegration in-volves largely mechanical processes, whereas decomposition entails the chemical interaction of water, gases, and minerals.

MECHANICAL WEATHERING

The chief agents of *mechanical weathering*, or *disintegration*, are frost action, temperature changes, unloading, crystal growth, and the wedging action of plant roots. Many different factors influence the efficiency of these agents in causing disintegration. The composition and texture of the rocks being weathered and the presence of joints, fractures, and voids clearly affect the rate at which solid rock can be reduced to rubble.

Mechanical weathering is also affected by climate, topography, and the length of time over which weathering agents have been operating. In general, mechanical weathering predominates over chemical weathering in regions characterised by extreme cold or in warm arid regions. By definition, mechanical weathering involves the physical disintegration of solid masses of rock into loose fragments. There is little chemical change in the rock itself. Thus, a chemical analysis of the disintegrated rock would be similar to that of the parent rock.

Frost Action

Water expands by about 9 per cent when it freezes. If the water freezes in a confined space, pressure caused by the expansion may cause rock masses to be pushed apart and ruptured. Such frost action has an important effect in mechanical weathering in that the freezing water is capable of exerting thousands of pounds of pressure per square inch. Many of us have become aware of the effects of frost action during severe winters when concrete roads and sidewalks are cracked by alternate freezing and thawing.

The way freezing water disrupts solid rock is similar to the manner in which it breaks up streets and sidewalks. After water has filled a crack in the rock, the water at the lip of the crack may freeze. Once this has occurred, the water deeper in the crack is sealed off, and as it begins to freeze and expand, it pushes against the walls of the fracture, and thereby widens the space between those walls. In a

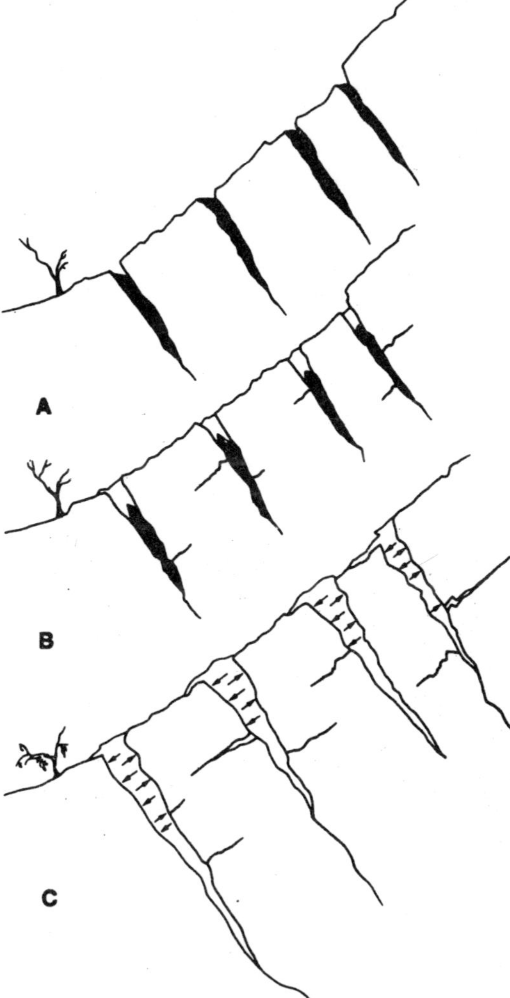

Figure 5.1: Three steps inteh process of widening fractures and joints by the action of freezing water.

subsequent thaw, fragments of rock may slip down into the crack and act as wedges to hold it open.

These processes are most prevalent in areas where water is abundant and where temperatures drop to below freezing at night and then warm during the day. Mountainous regions in temperate zones have this kind of daily temperature change, and in such rugged regions frost action is an important weathering process.

Insolation

Rocks, like many other solids, expand when they are warmed and contract when they are cooled. The heat is provided by the sun, and so geologists refer to the weathering that may result as insolation weathering. It would seem logical that repeated daytime heating and nighttime cooling of rocks (with concurrent expansion and contraction) would weaken the boundaries between grains and eventually cause them to separate from the rock mass.

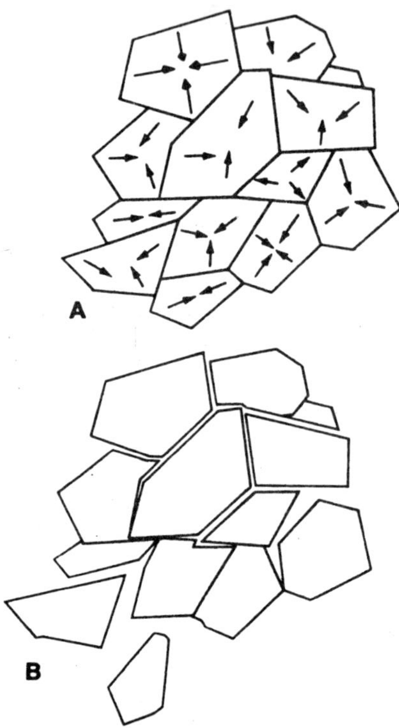

Figure 5.2: A theory to explain fragmentation of rock by alternate heating and cooling. Expansion during daytime heating and contraction during rightime cooling.

In laboratory experiments conducted by David T. Griggs at Harvard University, however, repeated heating and cooling did not cause fragmentation of the rock. Unfortunately, the experiments were unable to assess the amount of strain induced by millions of expansions and contractions over thousands of centuries.

Most geologists believe that insolation is only a minor contributor to mechanical weathering. Its effect is usually obscured by other weathering processes. Perhaps in desert regions where there are large-scale fluctuations in temperature, insolation may have some effect in weathering rocks.

Unloading

Quarrymen and miners are well aware that when the heavy weight of rock is removed from a mine with consequent loss of support to the surrounding rock, the outside pressure may sometimes cause a veritable explosion of rock fragments into the excavations. In nature, erosion may similarly remove large volumes of surface rock, thus lightening the load on deeper rock and allowing it to expand. Because the mass of rock is still confined on all sides, it can only respond to the release of pressure by expanding upwards.

As it expands it will rupture and form joints that are roughly parallel to the surface topography. The process is called *unloading*, and the joint systems are referred to as sheeting. Sheeting is often seen in granite and other massive crystalline rocks that have been laid bare by glaciation or running water. It can also be readily observed along the upper walls of quarries, where it may even facilitate the removal of the stone.

Sheeting provides planar passageways along which solution and frost action may proceed vigorously. Half Dome at Yosemite National Park and Stone Mountain, Georgia, are examples of mountains whose shape has been controlled, at least partially, by sheeting.

Saline Crystal Growth

The growth of crystals, especially crystals of sodium chloride, calcium sulfate, or magnesium sulfate have been found to cause scaling in exposed rock surfaces and to dislodge grains and crystals from their parent rock. The process is particularly effective in porous rocks in which crystals exert large expansive stresses as they grow. Disintegration of building stone because of the growth of crystals in pore spaces has been a vexing problem for architects involved in the preservation of historically important buildings.

The Temples at Luxor, Egypt, for example, have been seriously damaged by alternate solution and crystallisation of salt. During dry spells, magnesium (and calcium) sulfate have been known to crystallise along zones of weakness in the dolomitic building stones of London's Houses of Parliament, thereby causing spalling and crumbling at a rate that has caused considerable anxiety among restoration specialists. The magnesium sulfate is formed as a result of a chemical reaction between coal smoke and the magnesium in the dolomite.

Root Wedging

When one observes the growth of a plant from a seed, it is apparent that even a frail seedling is capable of exerting sufficient force to push aside relatively firm soil. As they grow into rock fractures and expand, the roots of large plants such as trees can exert correspondingly greater forces and are capable of widening cracks and accelerating the rate of disintegration by significant amounts.

When the plant dies and the roots decay, they leave openings in which freezing water may accumulate and further widen the void space. Plants also react with rocks chemically, as will be described in the following section.

CHEMICAL WEATHERING

Chemical weathering refers to the decomposition of rocks at or near the earth's surface and under relatively low temperatures. During chemical weathering, water and chemically active water solutions as well as oxygen and carbon dioxide from the atmosphere attack the minerals comprising rocks.

The susceptible parent minerals are frequently altered to softer secondary minerals, as when feldspars are converted to clay minerals and lose certain ions to solution. Chemical weathering is an exceedingly important process, for it leads to the formation of soils and in some parts of the world has resulted in the concentration of iron, aluminum, uranium, gold, and tin.

Relation of Disintegration to Decomposition

The processes of chemical weathering and mechanical weathering are not truly independent of one another. Because of climatic factors, especially the amount and frequency of rainfall, one or the other process may be dominant in a particular location. For the most part, however, decomposition and disintegration are interacting processes. For example, disintegration greatly promotes decomposition by reducing large intact rock masses to accumulations of smaller fragments.

This fragmentation provides more total surface area for chemical attack than was originally present in the larger mass of rock. One can see what happens readily by imagining a block of rock measuring 2 meters on each edge. Such a block would have a total of 24 square meters of surface area.

If the block is split once along each of three perpendicular planes, each resulting block would have 6 square meters of surface area or a total of 48 square meters for all eight pieces. Chemically reactive

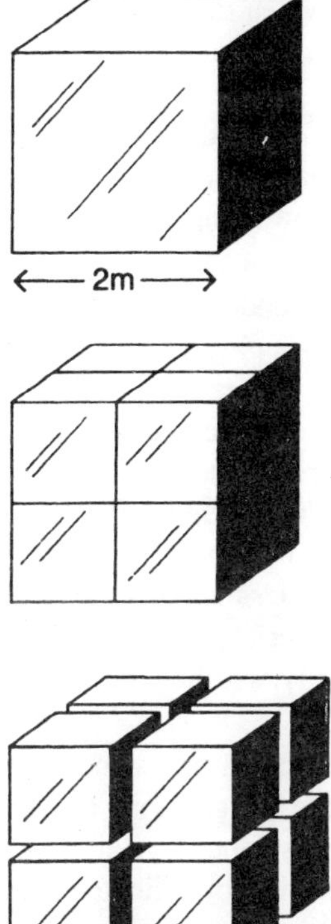

Figure 5.3: By splitting a cube of rock as shown here, the surface area exposed to chemical agents is doubled. It is for this reason that finer-grained materials decompose more rapidly than coarser-grained materials of identical composition.

solutions are only able to decompose surfaces they can reach, and so the rate of chemical attack is enhanced by increasing surface area. One can demonstrate this fact by applying a few drops of dilute hydrochloric acid first to an unfragmented piece of calcite and then to a similar piece of calcite that has been pulverised. Because of the greater amount of total surface area, the pulverised rock will display the more vigorous effervescence.

Processes of Chemical Weathering

People sometimes are surprised to learn that many rocks exposed at the earth's surface are really not in chemical equilibrium with their environment but rather are unstable and are slowly but continuously reacting with atmospheric components to dissolve or change to new substances that are more nearly stable at the earth's surface.

Geologically, weathering proceeds in accordance with a generalisation called the "*rule of stability*," which states that a mineral approaches stability most closely in an environment similar to that in which it formed. Intrusive igneous rocks, of course, are formed under conditions of high temperature, high pressure, and a deficiency of free oxygen and fresh water. When such rocks are laid bare on the continents, they find themselves in an environment of low temperature, low pressure, and abundant oxygen and water.

Chemical change is inevitable, and the minerals most susceptible to change are those that formed under physical and chemical conditions most removed from conditions at the earth's surface. For example, among the common rock-forming silicate minerals, olivine forms at high temperatures and pressures early in the crystallisation of a magma. Consequently, it rapidly weathers in the environments that exist at the earth's surface. Quartz forms much later under less extreme conditions of temperature and pressure and is less susceptible to weathering. In reference to the Bowen Reaction Series, it is apparent that minerals nearer the top of the series generally weather more rapidly than those near the base.

In other chapter of this book, we noted how silicon-oxygen tetrahedra are joined together by sharing some oxygen ions and that the number of metallic ions needed to neutralise the crystal is reduced by this sharing. This increased oxygen sharing by silicon results in a greater number of strong covalentlike bonds between silicon and oxygen, and this in turn greatly increases the ability of the mineral to resist weathering. For example, the ratio of oxygen to silicon in olivine is 4, in pryoxene 3, in hornblende 2.7, in hiotite 2.5, and in quartz 2. The diminishing ratios correlate nicely with greater resistance to weathering. All of the oxygen atoms are shared by silicon atoms in quartz, and this partially accounts for its great resistance to weathering.

The chemical weathering of rocks involves reactions that are interrelated, that may occur simultaneously, and that utilise water, oxygen, carbon dioxide, and organic acids. These processes include hydrolysis, oxidation, carbonation, solution, hydration, and chemical changes induced by the growth of plants.

Hydrolysis

Hydrolysis is a chemical reaction between a mineral and water. It involves a reaction between the H^+ or OH^- ions in the water and relatively active metallic ions such as sodium, calcium, potassium, and magnesium. Hydrolysis is particularly important in causing the decomposition of silicate minerals. Although it may occur in the presence

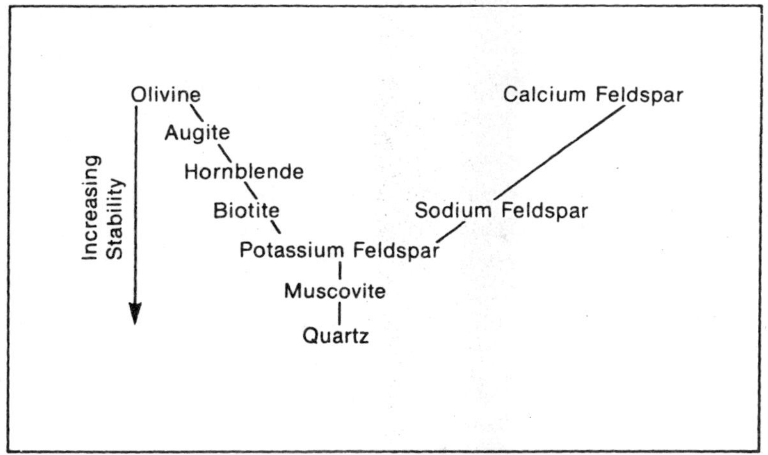

Figure 5.4 : From a study of the weathering of various kinds of igneous rocks, the mineralogist S.S. Goldich in 1938 determined that the weathering sequence for the common rock-forming minerals coincides with Bowen's order of crystallisation of minerals from a silicate melt. Thus quartz is the most resistant to decomposition of the common silicate minerals.

of pure water, in nature hydrolysis nearly always involves carbon dioxide. To illustrate, small quantities of carbon dioxide from the atmosphere or soil are dissolved in water to form carbonic acid.

$$\underset{\text{water}}{H_2O} + \underset{\text{carbon dioxide}}{CO_2} \rightleftarrows \underset{\text{carbonic acid}}{H_2CO_3}$$

The carbonic acid may then ionize to form hydrogen ions and bicarbonate ions.

$$\underset{\text{carbonic acid}}{H_2CO_3} \rightleftarrows \underset{\text{hydrogen ion}}{H^+} + \underset{\text{bicarbonate ion}}{(HCO_3)^-}$$

Next, the hydrogen ions penetrate into the crystal lattices of silicate minerals such as potassium feldspar. They have little difficulty in gaining admission because of their small size, and once within the lattice, they disrupt the charge balance as they displace cations such as potassium.

$$\underset{\text{water}}{H_2O} + \underset{\text{hydrogen ions}}{2H^+} + \underset{\text{potassium feldspar}}{2(K\,Al\,Si_3O_8)} \rightarrow$$

$$\underset{\text{potassium ions}}{2K^+} + \underset{\text{kaolinite clay}}{Al_2Si_2O_5(OH)_4} + \underset{\text{silica}}{4(SiO_2)}$$

The potassium ions released from the feldspar may be carried away in solution, utilised by plants, or become incorporated into clay minerals. A small part of the silica is removed in solution, although the greater part remains in the clay-rich weathered residue.

Carbonation

As implied by the term, carbonation involves the chemical addition of carbon dioxide to earth materials. Carbon dioxide in the atmosphere (and in the air trapped within soils) is readily absorbed by water to form carbonic acid. Although relatively weak, carbonic acid nevertheless has a pervasive cumulative effect in the chemical weathering of a variety of different kinds of rocks.

It is involved in the dissolution of common silicate minerals and is particularly effective in dissolving limestones and dolostones. The reaction for limestone is indicated below.

$$\underset{\text{water}}{H_2O} + \underset{\text{carbon dioxide}}{CO_2} \rightarrow \underset{\text{carbonic acid}}{H_2CO_3}$$

$$\underset{\text{carbonic acid}}{H_2CO_3} + \underset{\substack{\text{calcite}\\ \text{(in limestone)}}}{CaCO_3} \rightarrow \underset{\substack{\text{soluble calcium}\\ \text{bicarbonate}}}{Ca(HCO_3)_2}$$

In order for carbonation to occur, water must be readily available. For this reason carbonation is most vigorous in moist climates.

In the weathering of silicate minerals, carbonation and hydrolysis work together as the earth's most important processes for achieving decomposition of rocks. The hydrolysis component provides clay minerals and takes silica into solution, while simultaneously carbonation removes metallic elements as ions in solution.

Oxidation

Oxidation, the addition of oxygen to a compound, is one of the main kinds of changes produced in rocks by chemical weathering. Oxygen has a strong affinity for iron, which may be present in such silicate minerals as hornblende, and olivine, as well as sulfides such as pyrite (FeS_2). The oxidation of the iron (essentially what we call rusting) takes place chiefly in the presence of atmospheric moisture and results in the range of red and brown colorations we see in soils and weathered rocks. In the oxidation process, oxygen gas dissolved in water reacts with iron to form hematite (Fe_2O_3) or limonite ($Fe_2O_3 \cdot H_2O$). The process is illustrated by the following formula:

$$\underset{\text{iron}}{4Fe} + \underset{\text{oxygen}}{3O_2} + \underset{\text{water}}{nH_2O} \rightarrow \underset{\substack{\text{"limonite" (iron}\\ \text{hydroxide or "rust")}}}{2(Fe_2O_3) \cdot nH_2O}$$

('n means a variable amount)

The oxidation of a common nonsilicate such as pyrite ("*fool's gold*") involves combining oxygen with both iron and sulfur as follows:

$$\underset{\text{pyrite}}{4FeS_2} + \underset{\text{water}}{nH_2O} + \underset{\text{oxygen}}{15O_2} \rightarrow \underset{\text{"limonite"}}{2Fe_2O_3 \cdot nH_2O} + \underset{\text{sulfuric acid}}{8H_2SO_4}$$

Here, the relatively insoluble iron compounds may remain as a coating on the rock, whereas the sulfuric acid is leached away and becomes available for chemical reactions with other minerals.

Hydration

Hydration is a process whereby water is absorbed by a mineral and incorporated into the weathering product. For example, the mineral anhydrite ($CaSO_4$) may take in water to become alabaster gypsum ($CaSO_4 \cdot nH_2O$), or hematite (Fe_2O_3) may be converted to limonite ($Fe_2O_3 \cdot nH_2O$). Hydration is an important process in the development of clay and accounts for the presence of water within many clay minerals.

Another aspect of hydration is that the hydrated mineral, because of the water it has taken up, is larger than the parent mineral. The increase in volume causes growing hydrated crystals to exert pressure on the walls of the spaces they occupy, and such pressure may contribute to rock disintegration.

Solution

We have seen how the dissolving power of water for certain rocks and minerals is increased when carbon dioxide is present. Even without the addition of carbon dioxide, however, water has the ability to dissolve rocks and minerals. The dissolving away of thick beds of salt and gypsum is an example of simple solution weathering. A quartz sandstone also may weather by solution, although the process is exceedingly slow because of the low solubility of quartz.

The ability of water to dissolve substances is related to the configuration and electrical properties of the water molecule itself. A water molecule consists of a large strongly negative oxygen ion and two hydrogen ions. The hydrogen ions have contributed their electrons to the molecule and thus exist as two small positively charged protons. Both of the protons are located on the same side of the molecule, so that it has an electrically positive side and an electrically negative side. It is termed a dipolar molecule. The electrical charges at either end of the water molecule not only attract their opposites in other water molecules but also attract compounds in rocks and minerals that likewise have separate centers of positive and negative electrical charge. These compounds are taken into solution as weathering proceeds.

Biochemical Weathering

Rooted plants also play a role in the weathering of materials at the surface of the earth. During photosynthesis, plants utilise the energy of the sun's radiation to combine water and carbon dioxide and produce tissue necessary for growth and maintenance. Part of the hydrogen from moisture taken in by the plant is released at the surface of rootlets where it is exchanged for ions of calcium, magnesium, and sodium from adjacent particles of clay and other minerals.

These exchanged ions are required by the plant for nutrition. The hydrogen ions transferred to the clay particles render the clay slightly acidic, and this triggers weathering reactions with neighboring feldspars and other silicates. Eventually, some of these minerals also decompose to clay, thus perpetuating the weathering process.

RATES OF WEATHERING

The Role of Climate

How rapidly a particular exposure of rock will weather is dependent on several different interacting factors in the environment. Climate is important because it controls temperature and rainfall. Warmer temperatures, of course, promote chemical reaction. Even an increase of as little as 10°C can double reaction rates. Similarly, high rainfall accelerates rates of weathering by increasing the possibilities for solution and reaction and by helping to wash away surface debris and expose fresh surfaces for chemical attack.

Both higher temperatures and an abundance of moisture promote the growth of plants, and luxuriant plant growth assists in weathering by extracting ions from minerals and adding chemically reactive compounds. The importance of temperature and rainfall is at least partly evident in the thickness of loose material or regolith that develops over bedrock under various climatic conditions. In temperate regions having level topography, the layer of regolith is usually only about a meter deep, whereas in similar topography of the tropics, the loose material may form a blanket tens of meters thick. Arctic areas are only thinly covered by regolith.

Parent Rock

Depending on their composition, texture, and such features as fractures and joints, rocks exposed at the earth's surface may exhibit

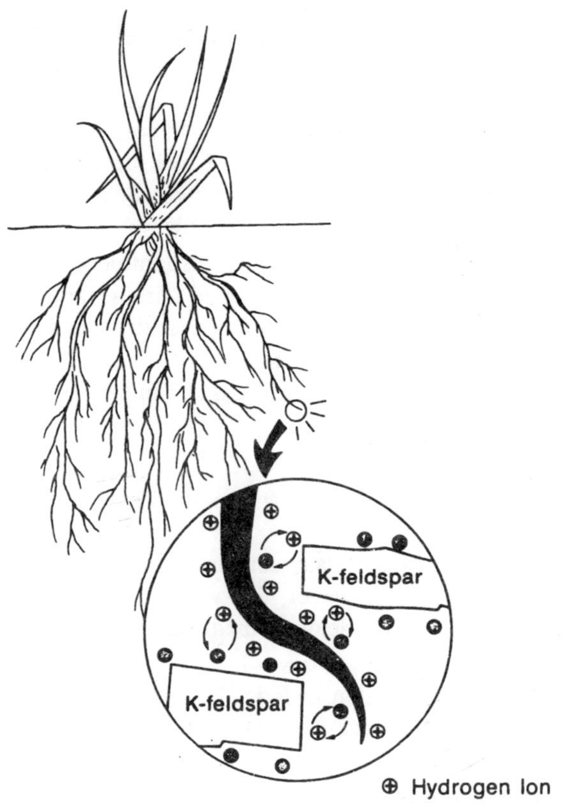

Figure 5.5: Plant rootlets provide a continuing supply of hydrogen ions, which are exchanged for cations from adjacent mineral particles.

a wide range of resistance to weathering. Limestones and dolostones, for example, are highly susceptible to solution. In regions having high rainfall, these rocks weather more rapidly than in arid regions. This observation suggests that it is difficult to list rocks in the order of their resistance to weathering because a rock might be more durable than another under one set of conditions and the same rock less durable under different conditions.

As a general rule, igneous and metamorphic rocks are more resistant to weathering than sedimentary rocks. A silica-cemented quartz sandstone, however, is among the most enduring of rock types. Statues and other sculptured edifices that have actual dates carved into the stone permit one to assess rates of weathering. In cities, because of the higher levels of carbon and sulfur dioxides, chemical weathering rates are generally higher than in nonindustrialised areas. Figure elsewhere in this chapter provides an example of damage to statues in cities. The bust of Beethoven was erected in a St. Louis park in 1884. In spite of many attempts to treat the Italian Carrara marble from which the bust was carved, it continued to decompose. The composer's gruff expression has been so altered that St. Louisans refer to the statue as "*poor Beethoven.*"

The texture of the parent rock will also influence its susceptibility to weathering. In general, coarsergrained porous varieties of rocks of the same composition tend to weather more rapidly than dense finegrained varieties. Rocks that have a large number of fractures provide a greater amount of surface area for chemical attack and therefore are also more readily weathered.

Acid Rain

Rates of chemical weathering are, to he sure, affected by the acidity of the solutions that attack rocks and minerals. As described above, even unpolluted rain-water contains carbon dioxide and is slightly acidic. In recent years, however, this weak natural carbonic acid has been augmented by strong acids formed where rain has fallen through air polluted with oxides of nitrogen and sulfur.

This acidic precipitation includes both nitric and sulfuric acid. It is called *acid rain.* Acid rain commonly exceeds the natural acidity of rainwater by more than 200 times, and in a few instances by over 1000 times. If not neutralised by alkaline rocks and soils, the acid accumulates in lakes and streams where it is often lethal to fish and other forms of aquatic life. Acid rain also damages the foliage of plants and reduces the quality of soil by leaching away nutrients.

The acid solutions attack limestone, marble, and concrete building materials and damage valuable works of art. Presently, millions of dollars are being expended in acid rain research in the hope that a solution to the problem can be found. Aside from local chemical treatment of particular areas, however, a long-term solution must involve reduction in atmospheric pollutants that form acid rain.

SPHEROIDAL WEATHERING

Spheroidal weathering is a term used to describe the spalling away of concentric surficial shells of the rounded surface of a boulder or rock mass. Such "*onion-skin*" weathering is believed to result primarily from the mechanical effects of chemical weathering. When feldspars decompose, the clay product has a greater volume than the parent feldspar. The increase in volume disrupts the interlocking texture of mineral grains in the rock and causes breakage and separation of the layer of partially weathered rock near the surface. Corners of roughly rectangular blocks broken loose along joints are decomposed along three surfaces simultaneously and are progressively rounded by spalling until the rock mass assumes the spheroidal form.

Spheroidal weathering operates most effectively on relatively small rock masses such as those of boulder size. The kind of spheroidal weathering involving a larger-scale breaking off of concentric plates from bare rock surfaces is referred to as *exfoliation*. Rocks that have experienced exfoliation may resemble those having sheet structure, but while sheet structure is caused by release of pressure when overlying rock masses are removed by erosion, all forms of spheroidal weathering result from physical and chemical weathering.

THE DECAY OF GRANITE

The mineral alterations involved in chemical weathering are nicely illustrated by the decay of a granitic parent rock. As described in other chapter of this book, granite is a common igneous rock composed of about two parts orthoclase, one part quartz, one part plagioclase, and small amounts of ferromagnesian minerals. Where masses of granite have been exposed for a long period of time, one may find places where the weathered products of the granite have not been washed away but have accumulated as a clayey granular residue.

On close examination, this material is found to consist of quartz grains, partly decayed and clay-coated feldspars, rust-colored particles of partially decayed ferromagnesian minerals, and clay. The quartz grains in the weathered material are relatively unaltered because of the great resistance quartz has to chemical attack. As other minerals in the granite decompose around them, quartz grains are freed from the rock matrix, later to be transported and deposited as sand, and perhaps ultimately to become components of sandstone. Granite provides a good example of *differential weathering*, by which its various components weather at different rates.

We have noted that feldspars are *aluminosilicates* of *potassium*, *sodium*, and *calcium*. In the weathering process, potassium, sodium, and calcium are largely dissolved and carried away in solution. Subsequently they may be combined with other elements and incorporated into sedimentary rocks. At least some of the potassium is retained in newly forming clay minerals. The remaining aluminum and silicon in the feldspars become the chief ingredients of clay, and this accounts for the coating of clay frequently found around feldspar grains and the clay residue found adjacent to bodies of weathering granite. Later, this same clay may find its way into the making of sedimentary rocks like shale and claystone.

During the decomposition of the ferromagnesian minerals present in the granite source rock, potassium, sodium, and calcium are dissolved in the same way as in the feldspars. Again, aluminum and silicon go into the construction of clay minerals.

The ions that remain are iron and magnesium. The iron combines readily with oxygen to form iron oxide minerals such as hematite (Fe_2O_3) and hydrous iron oxide minerals like goethite, FeO(OH). These iron minerals color sediment and rocks in tints of yellow, orange, and brown. Finally, the magnesium that is derived from parent ferromagnesian minerals may find its way into limey sediments or become a component of certain clay minerals.

SOIL

In terms of importance to human populations, the most significant aspect of weathering is in the development of soil. Soil can be defined as weathered material that will support the growth of rooted plants. A heap of quartz sand is not soil according to this definition, for it will not support vegetation. Nor is the so-called "*lunar soil*" of the moon a true soil. It is merely disintegrated mineral matter and should more properly be termed regolith. Regolith refers to the debris of rock and mineral matter that rests on solid bedrock. It includes soil, alluvium, and weathered fragments of the bedrock.

Soil is not a mere accumulation of particles derived from mechanical and chemical weathering. It is a vital natural material that would not develop were it not for the activities of a myriad of *bacteria*, *fungi*, *worms*, and *insects*. Actually, soil can be considered an intricate mixture of mineral solids, water, gases, dissolved substances, remains of dead organisms, and multitudes of thriving organisms.

Why is a certain amount of organic material necessary in order for weathered material to support the growth of rooted plants? One reason is that the organic material supplies nutrients required for vigorous plant growth. Plants need nitrogen in order to synthesize protein, but they are unable to take free nitrogen (N_2) from the air. Certain soil bacteria, however, are able to use this free nitrogen and convert it into usable fixed nitrogen compounds (nitrates).

Also, the decaying vegetation and animal waste contain ammonia (NH_3), the nitrogen of which can be utilised by most plants. The material that colors the upper parts of many soils gray or black is called *humus*. Humus is organic material that is so thoroughly decayed that one cannot discern the nature of the parent organisms. This essential organic mixture increases the porosity and water-holding capacity of the soil, provides a buffer against rapid changes in acidity, and assists in the retention of chemicals needed as plant nutrients.

Carbon dioxide liberated by humus will, as we have noted, combine with water to form a weak acid capable of dissolving calcium carbonate. Once the calcium is separated from the carbonate part of a compound, chemicals in humus known as chelating agents, bond to the calcium and prevent it from again forming $CaCO_3$. Most plants are unable to extract the calcium they need directly from $CaCO_3$, but they have evolved mechanisms for releasing the chelated calcium that exists in humus.

Factors Governing Soil Development

Five factors are involved in soil formation. These are climate, parent material, topography, time, and biologic activity. Of these factors, climate is of greatest importance. Indeed, soil types and climate are so closely related that maps showing global distribution of climates resemble maps showing the occurrence of different soil types.

Temperature and rainfall, of course, influence not only rates of weathering but the nature of vegetation. Given enough time, soil-forming processes operating within a given climate will prevail over differences in parent material and provide a basically climate-controlled soil type. The reverse is also true. Identical rocks in temperate high-rainfall regions and in semitropical arid regions will produce quite different soils.

The influence of parent material on the formation of soils is of secondary importance but can be observed in soils that have recently developed from unaltered bedrock. Such soils still retain textural and mineral components that are derived from the parent rock but which may become obscured after the soil has evolved to a further stage. In some cases, various influences of parent material will persist in the developing soil.

Soils developed above quartz sandstones may tend to be more acidic than soils formed on limestones, and these differences will be reflected by the natural vegetation living on each soil type. Soils forming on rocks that are deficient in certain elements such as magnesium or boron will ordinarily also lack these constituents. Not all soils are developed from underlying consolidated bedrock. Many fine agricultural soils are formed on loosely consolidated sediment laid down by streams, winds, glaciers, or the waters of lakes.

Topography also influences the ultimate nature of soils. On steep slopes, for example, erosion is usually more vigorous and the soil layer generally thinner. Also, water falling on slopes is partially lost to runoff, and therefore there is less water available for percolation into deeper layers of soil. Even the direction of slope may have a bearing on soil development.

In the northern hemisphere, slopes facing toward the south tend to be warmer and dryer and therefore support quite different kinds of plants than do the northern slopes.

The factor of time in soil formation is often not fully appreciated by those who clear away soil for construction or mining and expect nature to quickly restore this important resource. The development of soil is an exceedingly slow process. Depending on climatic conditions, several hundred to several thousands of years are required to produce a fertile soil. Agricultural soils formed during this lengthy natural process may be lost to erosion in only a few years, particularly in areas where soil conservation measures are not followed.

The problem is not trivial, for an estimated 3 billion metric tons of good agricultural soil is lost each year from agricultural fields in the United States alon.

Soil Horizons

Soils are not uniform in texture and composition from bedrock to the surface but rather consist of a number of fairly distinct layers or soil *horizons*, which differ in composition and physical characteristics. The stack of soil horizons is called a *soil profile*. A simple profile, such as might develop in a humid region on granitic rock, has three distinct divisions. Soil scientists have named these layers the A, B, and C horizons.

The A horizon, also known as "*top soil*," which lies immediately beneath the surface, is characterised by a high content of organic matter. It is also a zone in which soluble compounds are dissolved and, along with fine clay particles, carried downward to be deposited in the underlying B horizon. Because of these losses, the A horizon is referred to as leached or eluviated, whereas the B layer is sometimes called the washed-in or illuviated zone.

Iron and clay minerals tend to accumulate in the B layer. In humid regions one can often recognise the B horizon by its more brownish color and clayey texture. The C horizon of the soil profile consists of partially altered parent material. In it one finds evidence of chemical weathering, but soil development has not progressed to the level at which the original characteristics of the bedrock are unrecognisable. Beneath the C horizon lies unaltered parent material.

Soils developed on limestones in humid regions often do not exhibit a clear division between the B and C horizons. In such soils, the A horizon consists of dark humus-rich material that changes downward into relatively light colored clay and then stained and partly dissolved limestone. The clays in these soils are residues of clay impurities left when the limestone was dissolved.

How does the sequence of soil horizons develop in a humid region underlain by granitic rock? Imagine that glaciation has recently stripped away all loose sediment over a granite outcrop. Initially, disintegration and decomposition of the granite will produce a sandy layer composed mostly of grains of quartz and feldspar along with clay derived from the weathering of silicate minerals.

As the layer of granular material thickens, plants establish themselves, and organic material from a variety of sources becomes incorporated into the surface layer. Meanwhile, with each rainfall, soluble ions and clay particles are flushed downward from the surface laver into a lower level. This lower level soon takes on the clayey character of the B horizon. While these events are taking place, solutions reaching still deeper continue the chemical attack on the parent material, thus perpetuating the C horizon.

Major Kinds of Soils

Soil scientists employ a rather complex nomenclature for the great variety of soil types present at one place or another around the globe. Fortunately, most of the names in these lengthy classifications can be avoided if one desires only a general understanding of a few major kinds of soils. For example, the soils that predominate in the eastern United States and southeastern Canada are called pedalfers.

The name was fabricated to emphasize the fact that in such soils aluminum *(al)* and iron *(fer)* have been leached from the A horizon and deposited in the underlying B horizon. Pedalfers are clayey soils that develop in regions having an abundance of rainfall. As this water percolates through the humus, it becomes acid, and this accounts for its ability to leach aluminum and iron from the topsoil.

It is also effective in dissolving carbonates, which are then carried away by ground water. The acidity of some pedalfers may diminish their fertility, and so farmers often spread finely ground limestone (agricultural lime) on their fields to combat the acidity.

As a consequence of the fact that the top soil or A horizon in pedalfers is leached, it is usually a lighter color than the underlying subsoil where iron has accumulated. Pedalfers include several lesser categories of soils, which differ in their profiles. Most important among these are *podzols,* which typically have an ashy gray A horizon.

In the more arid regions of the world, less humus develops in soils, and there is less opportunity for solution and leaching. In the absence of water, chemical weathering is slower, and less clay is produced. Usually, there is insufficent ground water to flush soluble materials such as calcite out of the B horizon, so it accumulates there as soil moisture is lost by evaporation. Because of the persistence of calcite in these soils, they are called *pedocals* (*pedon*, *soil*, and *cal* for *calcium carbonate*).

Frequently in dry areas, there is not only insufficient water to cause downward leaching, but there is an upward movement of water because of the high rate of evaporation at the surface. Mineral matter dissolved in the diminishing soil water is precipitated at the surface as a hardpan or caliche layer.

In the hot and humid regions of the tropics, the characteristic soils are laterites. The term "*laterite*" is derived from the Latin word *latere* or brick, and originally referred to the use of this material to make bricks in India and Cambodia. Laterites have a relatively thin organic layer covering a reddish leached layer, which is often underlain by a still darker red layer. In *laterites*, *oxidation* and *hydrolysis* have been so intense that feldspars and *ferromagnesian* minerals are completely decayed. Not only is calcium carbonate removed, but also silica. Only the most insoluble compounds, mainly aluminum and iron oxides, accumulate in these soils.

Although laterites may support a lush growth of natural tropical vegetation, they are not good agricultural soils. When forested areas with lateritic soils are cleared and plowed, the thin organic cover tends to be rapidly oxidised in the prevailing warm climates. A thick accumulation of organic matter such as that found in rich black soils of more temperate regions cannot develop. After a few years of tilling and planting, the organic component of the soil is so depleted that fields must be abandoned.

In some laterites, the concentration of either iron or aluminum may reach levels that permit the deposits to be profitably mined. The iron-rich laterites originate from the weathering of parent materials that also contained iron, although not concentrated into ore bodies. Extensive lateritic ores of iron occur in Cuba, Columbia, Venezuela, and the

Philippines. If there is little iron in the parent material and an abundance of aluminum, then laterites rich in hydrated aluminum oxides may form. Such materials are called bauxites. Ancient bauxite deposits are mined in Guyana, Ghana, northern Queensland, and Arkansas. Concentration of bauxite takes place in tropical areas of low relief where temperatures exceed 25°C most of the time and where there is an abundance of water for leaching and chemical reactions. At the present time, bauxite is the only ore from which it is economically practical to extract aluminum.

Soil Erosion

It takes nature thousands of years to produce a fertile soil. That same soil can be lost to erosion in only a few decades. Clearly, as we attempt to feed our expanding world population, care must be taken to prevent unnecessary erosional losses of this vital resource.

Soil erosion results primarily from uncontrolled runoff of surface water. In areas that have never been tilled, soil is protected by a cover of plants, and there is an approximate balance between the slow loss of soil by erosion and the development of new soil from underlying parent material. Whenever the protective shield of vegetation is broken, however, erosion will follow. There are, of course, natural events that can destroy the *vegetative cover*. These include droughts, plagues of insects, epidemics of plant disease, and fires caused by lightning.

Agriculture, however, has been a far more potent factor in causing erosional loss of soils. Some of the erosion resulting from farming is unavoidable. We must have food, and so we must break into the natural vegetative cover. There are, however, farming methods that reduce the loss of valuable topsoil to erosion, and these methods are practiced widely.

For example, farmers now plow and plant their crops in rows that follow contours so that rainwater is retained rather than allowed to run off. Efforts are made to reduce the amount of time between preparation of the soil for planting and planting itself so that the bare soil will not remain exposed to erosion for long periods.

Strip-cropping is also practiced. This technique involves planting alternate bands of erosion-resistant crops such as clover and alfalfa with open-spaced crops like corn. Soil losses in the erosion-susceptible corn strip are trapped in the adjacent strip of denser crops. Where fields are temporarily not utilised, soilholding "cover" crops are planted, not only to retard erosion but to improve the soil by adding nitrogen. Because ordinary grass is an effective retardant for erosion, overgrazing by cattle and sheep must be prevented.

One of the most troublesome erosional problems faced by farmers is the control of gullies. *Gullies* start in either natural depressions or plow furrows and generally extend themselves toward higher ground by a process called headward erosion. The gully head receives rainwash from a wide area. This water converges so as to flow rapidly into the steep depression at the upper end of the gully, eroding *vigorously* as it enters the channel.

It is this rapid erosion that causes the gully to cut ever farther in the headward or upslope direction. To halt this destructive process, steps can be taken that prevent water from entering the gully. Normally this is done by excavating a shallow diversion ditch around the gully head to trap its potential water supply and "*starve*" the gully. Because gullies grow by headward erosion, disposing of old refrigerators and *automobiles* in these ditches has no effect at all in retarding their growth.

6

SEDIMENTARY STRUCTURES

The complexities of transport and depositional processes, about which much is known and much remains unknown, are subjects for more advanced texts. Nonetheless, physical factors contributing to the origin of certain common types of sedimentary structure are introduced here to give a flavour of the relationships between the two. Structures formed entirely by sedimentary processes are difficult to classify satisfactorily and most attempts have resulted in mixed morphologic-genetic groupings.

Dunes have a characteristic shape (or bed form) created by movement of loose particles. They can be regarded as representative of many structures formed by external (*exogenetic*) forces. In contrast, structures such as load casts, generated within sediments after deposition, but prior to *lithification*, can be regarded as *endogenetic*. The contemporary activities of organisms produce biogenic structures.

BEDDING

Most sediments are laid down in layers, known as bedding or stratification, which have a degree of lateral *continuity* and *compositional* unity. Exceptions include some reef-like bodies and thick boulder clays. Bedding planes separate individual beds and are either sharp or diffuse depending on whether they represent a pause in deposition, with or without scour, a change in the nature of the sediment input, or subsequent variations in the diagenesis of the sediment. In some limestone successions, what appears to be bedding is in reality caused by a parallelism of stylolitic contacts; this is not true bedding. Beds vary considerably in thickness and many classifications, not always consistent, have been devised to characterise the variability. Table elsewhere in this chapter is one example.

Laminations (*laminae*) are thin layers, less than 1 cm, though several such can constitute a thicker bedded lithological unit. Banded *ironformations*, *lacustrine oil-shales* and major evaporite bodies commonly consist of thick beds defined by sharp bedding planes, but which are also internally well laminated. Laminations also typify deep-water hemipelagic and pelagic (*open-sea derived*) deposits and frequently resemble varves or couplets produced by some form of cyclical control.

Table 6.1: Bedding Scale.

Scale	Thickness
Very thickly bedded	> lm
Thickly bedded	30-100cm
Medium bedded	10-30 cm
Thinly bedded	1-10cm
Thickly laminated	0·3-1 cm
Thinly laminated	< 0·3cm

Changes in the thickness of beds is usual when they are traced laterally for any great distance so that most, in practice, are lenticular, though in a restricted outcrop they may appear planar or flat. Wedging-in and out of beds, on the other hand, can be observed even in restricted exposures and is especially noteworthy in rocks of fluvial and deltaic origins.

Bedding is emphasized by sedimentary structural differences between adjacent layers. Cross-bedding and graded bedding are cases in point. In contrast, certain beds, not necessarily very thick, are structureless internally and are said to be massively bedded. Limestone beds are more often massive than sandstones.

Ripples, Dunes and Cross-Bedding

Considerable progress has been made over the last decades in understanding how sub-aqueous ripples and dunes form by studying unidirectional flow in alluvial channels and flumes, whose beds are composed of non-cohesive materials such as quartz sands and silts. Within these channels the flow conditions, such as current velocity, resistance to flow, shear stress at the bed, viscosity and the effect of gravity can be approximately determined and assessed in terms of flow

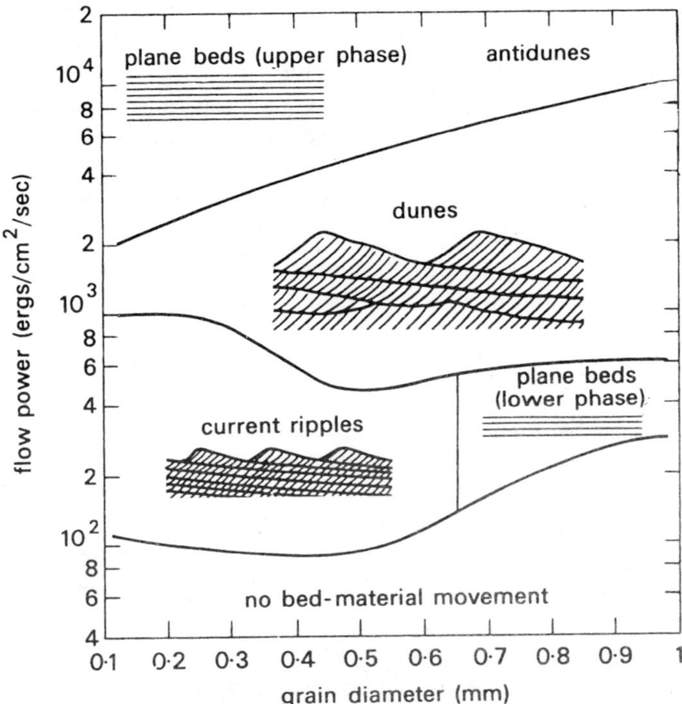

Figure 6.1: Bed forms, flow power and grain size. The diagram is based on sand movement in flumes under unidirectional flow conditions.

regime. Flow regimes can be defined as a range of flow conditions having similar sediment transport and resistance characteristics that produce similar bed forms.

Under these conditions loose sand is moulded into *plane beds*, *ripples*, *megaripples*, *dunes* and *sandwaves*. These structures only broadly reflect progressively increasing current velocities and it should not be inferred that the bed forms are a simple evolutionary sequence. For instance, sands coarser than about 0.6 mm, initially redistributed as lower phase, evenly bedded or laminated plane beds, are moulded into megaripples and dunes at higher velocities without necessarily passing through a period of current ripple formation.

In different flow media, such as air, and with grains of different density, such as *calcareous ooids* or *gypsum flakes*, the field boundaries are likely to be different from those for quartz. The sediment particles appear to move intermittently, rolling up the gentle upstream stoss-slope of the structures and sliding or *avalanching* down the steeper downstream *lee-slope*. The highest flow conditions are typified by plane beds and antidunes. In general the particles appear to move much more continuously than in the lower regimes.

All the structures reproducible in flume experiments are recognised in a range of environments other than alluvial channels. Ripple marks, for example, are well-known features of sandy deserts, littoral and sublittoral zones, and have been photographed at depths of several kilometres in oceanic basins.

The extrapolation of the information on Figure elsewhere in this chapter to these other identical bed forms, but which often contain multidirectional elements, has, therefore, to be done cautiously. Plane beds formed under relatively high current velocity conditions, including beach swash, have no surface irregularities larger in amplitude than a few grain diameters. On the upper surface of flat laminae in many ancient fine- to medium-grade *fluvial* and *littoral sheet* sandstones, the high velocities are expressed by very faint parallel ridges and hollows, known as parting or primary current lineations.

The long axes of the constituent grains roughly parallel the elongation of the structures and the largest grains are invariably concentrated in the ridges. Also at the highest velocities some ripple and dune-like structures take on regressive migratory qualities, actually tending to move in tidal channels and in streams in spate, upstream against the current flow; they are then called antidunes. Their presence below the water surface is commonly indicated by upcurrent-moving surface waves.

Ripple marks are environmentally unrestricted and are formed by wind activity as much as water movement. In the Bahamas and Persian Gulf windblown rippled deposits on the coast are often composed of clastic lime particles. Wind-generated ripples formed entirely of gypsum grains are known from inland playa situations. Ripples are commonly small scale, asymmetrical or symmetrical, with successive crests less than 30cm apart and with amplitudes less than 30 cm. Wind ripples are usually flatter and less symmetrical than *current ripples*; their crests are straight or gently curved.

The trend of the crests of ripples depends on the trend of the water or air currents in any particular environment. In shallow water areas the crests of small-scale ripples generally subparallel the nearby strandline. This is particularly well seen in sandy and silty areas of deposition, such as bays and intertidal flats. Away from the influence of strandlines, ripple crest trends are more random.

The characteristic internal structure of most ripples is cross-lamination developed during down current movements. Sand is eroded from the gentle up current side and mostly deposited towards the top of the steep lee-slope. At intervals this sand avalanches down the lee-slope, in doing so producing the laminated, forward-growing structure.

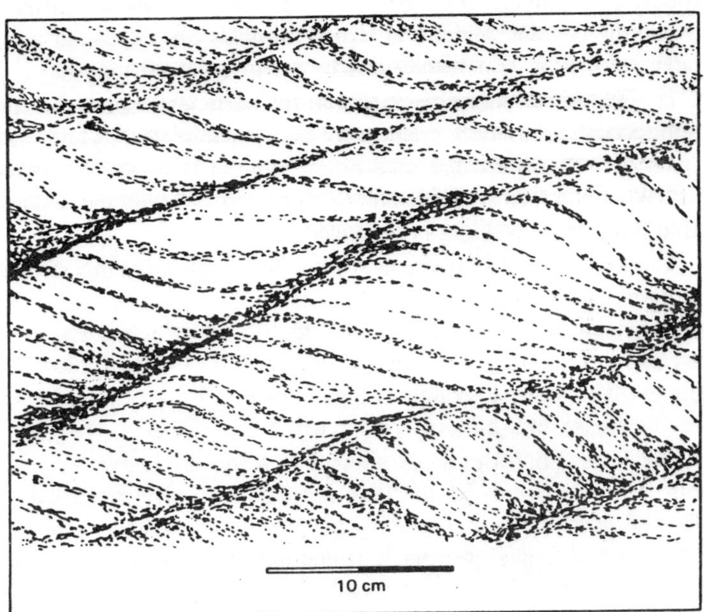

Figure 6.2: Ripple-drift cross-lamination. Successive ripples have moved from left to right, mildly truncating their immediate predecessor.

In some situations the ripples seem to climb on to the gentler stoss-slope of the ripple immediately downcurrent. The angle of climb is commonly between 5 and 20 degrees from the horizontal. The term describing this arrangement is ripple-drift cross-lamination and beds composed of piled-up sets of this type, each set a few centimetres thick, are common in the sediments of fluviodeltaic successions and some *basinal turbidite* successions.

They are also described from glacial outwash deposits. The surfaces of modern sandy and muddy intertidal and adjacent subtidal areas are commonly rippled in patterns and symmetries reflecting the ebb and flow of tidal currents, and the local dominance of one over the other. During quieter periods of deposition, which can extend over several weeks or months, fine mud is deposited in greater volume. Settling lag and scour lag play some role in this accumulation. Settling lag refers to the time lag between the moment at which a current of decreasing velocity can no longer hold a particle in suspension and the moment at which the particle reaches the bottom. The smaller the particle the greater is the settling lag. During quiet phases, in the absence of much turbulence, the settling lag time interval for the deposition of mud particles is relatively low. Scour lag is the *measurable* difference between the current velocity at which particles are deposited and the greater velocity at which they can be set in renewed motion.

The additional energy required to lift particles against the forces of adhesion and gravity is greatest for clay grade materials and, again, in quiet periods means that mud on the bottom is likely to remain in place. When the formation of sand ripples alternates with the deposition of fine mud flaser bedding is produced.

A mild degree of subsequent compaction emphasizes this structure. In certain major sandy river channels and at the mouths of some sandy estuaries ripples reach an amplitude (height from crest to trough) varying between 30 cm and 2m and are best referred to as megaripples. Their crests in plan view are straight, linguoid or lunate in form and may be traceable for several hundreds of metres.

Modern examples in the Brahmaputra River are known to be very mobile and movements up to 200m per day have been recorded. The surface of *megaripples* is commonly mantled by *small-scale ripples*. The size of structures having a basic ripple-like *morphology* can be even greater than that of megaripples and at this point different terms are used, namely *dunes* and *sandwaves*.

Water-laid dunes have an amplitude varying between 2 m and 8 m and wavelengths which often reach and can exceed 500 m. They have a high mobility especially during periods of enhanced current velocity. Wind-laid dunes, in contrast, can be immense, reaching heights of .30m commonly and even as much as 250m in Saudi Arabia, though the individual cross-bedded units or sets constituting each dune is rarely much more than 10m.

The windward surfaces of such dunes are frequently covered by very mobile small ripples. Sandwaves are well known on many present-day and ancient *epicontinental shelves*. Amplitudes commonly reach 15 m and wavelengths may be as little as 200m or as much as 1000m. The crests tend to be straight and sometimes can be traced laterally for several hundreds of metres. Megaripples may be superimposed on these larger structures.

Large-scale sandwaves formed of calcareous ooids and peloids, but now wholly *dolomitised*, have been recognised in Upper Permian succe-

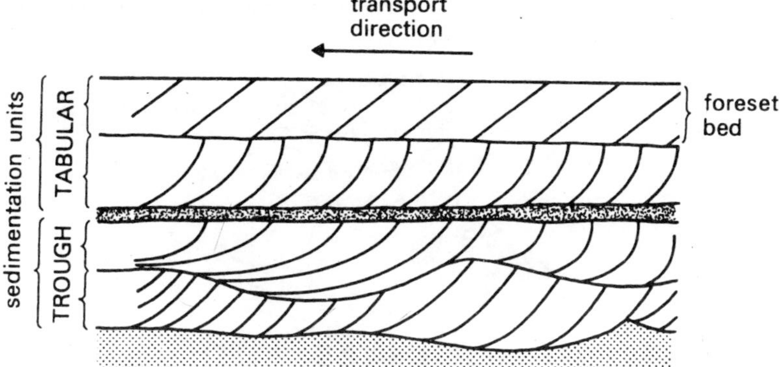

Figure 6.3: Simple cross-bedding terminology.

ssions in eastern England, where they appear to have been localised by *topographic highs*, *possibly buildups*, on the sea floor.

The movement of particles in migratory dunes and sandwaves is similar to that of ripples, but creates a larger-scale type of internal structure referred to as cross-bedding. Cross-bedded units can be five or more metres thick. The thickest are most often found in wind-blown deposits, whereas the units in shallow water deposits (*gravels*, *sands*, *calcarenites*) are generally less than 3 m.

Several varieties of cross-bedding are recognised such as tabular with relatively planar surfaces bounding the unit (or set) and trough with curved surfaces beneath the unit. Sometimes the bounding surfaces in either type are non-erosional but, more often, are erosional in origin. All varieties of cross-bedding tend to grade into each other and can be seen closely intermingled in some thick sandstones which comprise several units or sets.

Some types of cross-bedding can be useful as an indicator of current and sediment transportation directions. In practice, as many foreset directions as possible are measured within a given bed or group of closely associated beds and, after making corrections for tectonic dip, the data are plotted on a suitable base map.

It is generally believed by most workers that the distribution of cross-lamination and cross-bedding directions within sandstone and siltstone beds can indicate the source direction and regional slope during the period of deposition. This assumption is probably valid in most aeolian, fluvial and deltaic sequences where the dominant foreset direction can be taken as pointing away from the source area, but it has to be realised that in marine situations the transporting currents often

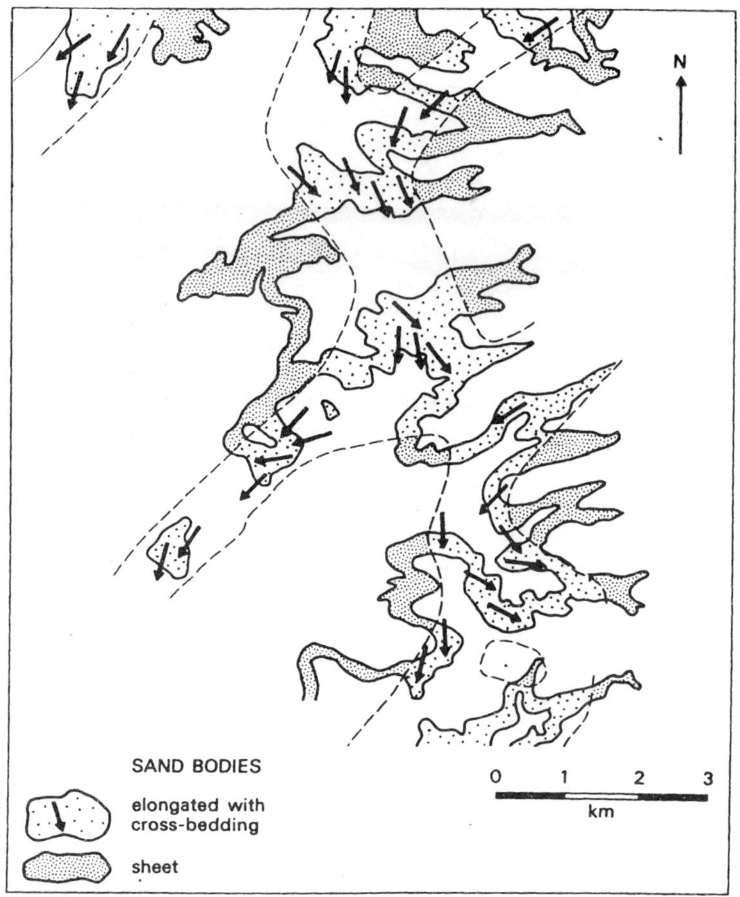

Figure 6.4: Fluvial channel cross-bedding in a sandstone sheet. The southwards direction of sediment transport and the location of the channel can be inferred from the cross-bedding and external form of the sandstone bodies.

flow longitudinally parallel to nearby coastlines or basinal slopes. In some fluvial and strongly tidally affected sublittoral situations sand bars and banks can develop *herring-bone* cross-bedding, in which the sets have mutually opposed dip directions.

The cause of this bipolarity is flow reversal of the water either by flooding or by tidal ebb and flow. Contourites are thinly bedded, sometimes cross-laminated sands, silts and clays laid down from deep sea currents moving roughly parallel to basin flanks. Towards the base of the continental rise and on adjacent abyssal plains the sediment is sometimes moulded into sandwaves with amplitudes of several tens of metres.

TURBIDITES, GRADED BEDS AND SOLE STRUCTURES

Fining-upwards clastic graded beds, referred to as turbidites, are frequently associated with thick basinal flysch-type deposition but are also recognised in thinner basinal, shelf and lacustrine successions. The deeper parts of the Black Sea are characterised by some 400 m of late Pleistocene to Holocene terrigenous turbidites.

The thickness and lithology of individual beds is variable, up to a metre or more in some *cyclical greywacke* and *pelagic limestone* basinal sequences, to a few centimetres or less in shelf and lacustrine sandstone, siltstone and limestone cyclical sequences. Turbidity or density currents are unsteady gravity flows of suspended sediment, whose movement depends on an excess of specific weight over the surroundings.

They usually have very large *Reynolds* Numbers, indicating high turbulence, and almost certainly are in a supercritical or rapid flow state. The causes of the initiation of turbidity currents are many. In shelf situations, storm wave activity plays an important role in suspending bottom sediment and redistributing it, often by turbidity currents.

For example, graded sandstone sheets up to 20cm thick in the Silurian of the Welsh Borderland are believed to have been deposited by such currents ebbing towards the western edge of the shelf. Some deposits, known as tempestites, are found within the Middle Lias shelf succession of north-east England and indicate two very marked periods of storm or hurricane activity.

In the Gulf of Mexico, storm-surge currents during recent hurricanes have moved sand and silt as much as 30km further off shore. Sediment instability on bottom slopes as low as a few degrees may also be a function of high rates of deposition leading to *underconsolidation* of the material. This means that the material is laid down so quickly that *interstitial* water is trapped in quantity and excessive pore pressure is developed.

Hence the sediment has a lower shear resistance than when normally consolidated and is likely to slump or slide by 'spontaneous liquifaction' when abnormally loaded. Earth tremors may be an important contributory cause of failure. As the material slumps, water may be gradually taken up, so reducing viscosity and increasing velocity. At this stage, when the water content is still not too high, the moving sediment may be in a 'fluxo-turbidite' state with little turbulent mixing of the particles. Deposits of this type can be seen in the marine Pakhna Formation (*Miocene*) of *southern Cyprus*. They consist of thickly bedded, graded

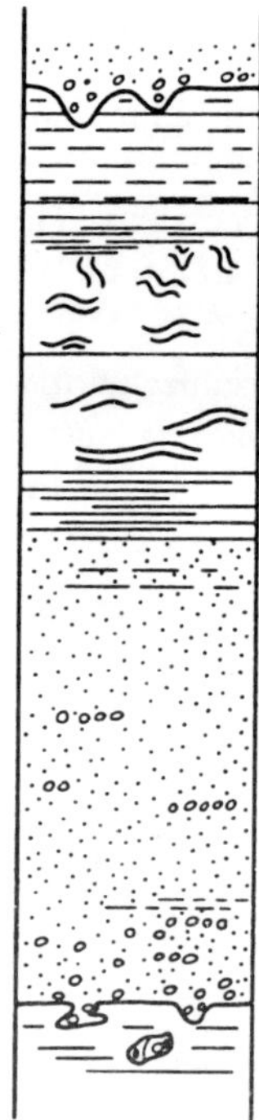

Figure 6.5: Ideal Bouma turbidite sequence for clastic sediments, reflecting very turbulent rapid flow initially diminishing to quiet conditions as the flow ceases.

sandstone sheets carrying '*rafted*' limestone slabs, up to 1 m in size, eroded from older strata exposed at the time only a few kilometres away. If acceleration continues there is further admixing of sediment and water, especially at the head of the flow, and the density decreases and turbulence increases; a turbidity current has now formed.

The coarser material is concentrated towards the head and bottom of the current, hence the specific weight of these parts is higher. This results in the coarser material in the lower head region of the flow moving relatively faster than the material in the upper rear parts. Velocities at the flow head are likely to be of the order of one to several metres per second.

The variations in velocity and sediment dispersal within the turbulent head of flows are considerable. The slowing of turbidity currents brings about a *diminution* in their thickness, density and turbulence, so that deposition with vertical grading ensues. Lamination and current ripples form as velocities fall and grain size decreases. Ultimately a proportion of the finest particles from the upper and tail parts of the current are deposited. The pauses between successive marine turbidity currents are irregular and measurable in tens of thousands, rather than hundreds, of years.

In the intervening periods the '*normal*' style of sedimentation persists appropriate to the situation. Near to the edge of continental shelves and near to the surface of the Atlantic and Polar Oceans there is a slow access of terrigenous sediment on to the bottom via the nepheloid layer. This layer is a relatively permanent mobile suspension of clastic detritus (< 12 um), about 1 km thick and extending over vast areas, in which the particles are held in suspension by *turbulence* induced by storm waves, frictional interactions between surface and deep sea currents, and occasional turbidity currents.

Linear (sole) structures occur at the base of many turbidite and tempestite units and most are caused by the scouring action of currents from which the material is being deposited. The coarser fraction of the load is carried along at the base of currents as a traction carpet, and this undoubtedly enhances the scouring capacity of the current.

The scour marks in the underlying cohesive muds take on many forms. Some are gutter casts, which are relatively deeply incised, narrow and linear channel fills. Others are *bulbous*, *triangular* or *corkscrew* shaped with steep walls at the upcurrent end and are collectively called flute marks. They grow where small irregularities have already been initiated on the mud floor, which promote strong turbulence, eddying and increasing erosion during the period of turbid flow.

Some scour marks are probably eroded by pebbles, shells or wood fragments suspended within the flow (*tool marks*). Current directions can be deduced with reasonable accuracy from flute marks present within ancient successions. The technique has been succesfully demonstrated for the Martinsburg Formation (*Ordovician*) of the Central Appalachians by measuring the trend of marks at the base of greywacke beds.

It is concluded that turbidity currents frequently flowed down the slopes leading away from a southeastern land mass, then swung through a wide arc so as to travel longitudinally along elongated basins running parallel to the land mass. It is now generally recognised from such studies that longitudinal movement of sediment along the axes of deep troughs is a commonplace event during their rapid infilling with turbidites.

A considerable proportion of the 6000 m of Cretaceous-Oligocene flysch greywackes of the Polish Carpathians was transported longitudinally. Studies of Quaternary fills of deep sea troughs bounding the western flanks of South and Central America, and offshore basins within the continental borderland of California, also confirm this sediment dispersal pattern.

DEFORMATIONAL STRUCTURES

Post-depositional deformation of sandstones, siltstones and limestones is commonplace irrespective of the environment in which the beds were formed. Many of the beds affected by deformation give every indication of a *penecontemporaneous* origin for the structures prior to renewed deposition of overlying undisturbed beds. Other beds have been affected by downwards bulging into underlying beds. Whatever the mode of origin, the deformation appears to have occurred while the rocks were in a plastic, water-saturated state.

Slumping is widespread on various scales in shallow water and deep water sediments and is characteristic of the flanks of basins and troughs. It can produce highly contorted slump sheets, balled-up masses and pebbly clay deposits which can be traced for some tens of kilometres away from the presumed source. The Eocene pebble beds of Ancon in south-west Ecuador are almost certainly of this origin, as are the Precambrian tilloids of Congo.

Some of the most spectacular thick slumped sequences occur as sedimentary melanges, in which a chaotic assemblage of blocks up to 1 km across become emplaced in a clay matrix by sliding and slipping down the margins of troughs. The blocks, though disoriented and occasionally overturned, often retain their original depositional structure without much evidence of internal deformation.

In some melanges, however, the blocks are often markedly sheared internally with many slickensided and broken surfaces; the clay matrix is very scaly. These features imply considerable tectonism affecting the deposits during or soon after emplacement. It is now recognised that a spectrum of melange-type sediments is associated with deposition near subduction zones during plate convergence.

Many were well developed in Tethyan collision zones during Mesozoic-early Tertiary times and can be traced intermittently from the eastern Mediterranean, through Oman and the Himalayas, into Indonesia. In some of these and other melanges a crude stratigraphical ordering of the clay-enwrapped blocks (olistoliths) is apparent, suggestive of the systematic disintegration of continental and oceanic crust during convergence. The term olistostrome can then be used for the slumped

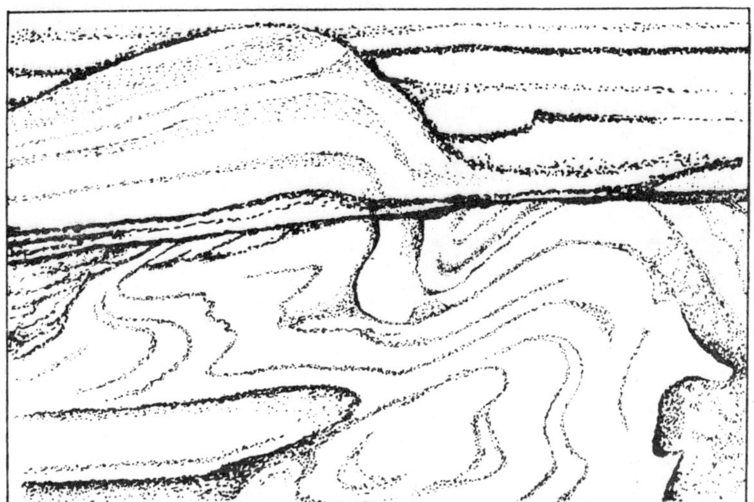

Figure 6.6: Slumped pelagic chalks, Tertiary, Cyprus. Penecontemporaneous upright, overturned, recumbent and sheath folds cut by near horizontal shear faults (detachment surfaces).

sequence. On a smaller scale, individual beds within normally bedded successions do show slumping with internal contortions consisting of open to tight upright, inclined and overturned folds, and refolds, in which the fold axes roughly parallel the strike of the palaeoslope. Slumps of this type occur within Tertiary pelagic and turbiditic chalks, marlstones and sandstones of Cyprus and the Pyrenees. This type of slumping is probably initiated by some short-lived localised event, such as an earth tremor.

Convolute lamination sometimes resembles slump bedding in internal structure. However, the convolutions are confined to a layer within an individual bed and the layer is rarely more than 30cm thick. In the upper parts of Bouma turbidite sequences the layer can sometimes be traced with little or no change in thickness for several hundreds of metres. The complexity of the convolutions increases from the bottom to the top of the layer but even in the highly folded upper parts micro-faulting is absent.

The structure is possibly produced by the surface drag of eddying currents of waning strength flowing over loosely consolidated sand and silty clay laminae. An origin by load deformation, simultaneous with deposition and perhaps localised by deposition on an irregular surface, is also feasible.

Load structures (casts) are disoriented, bulbous projections, extending

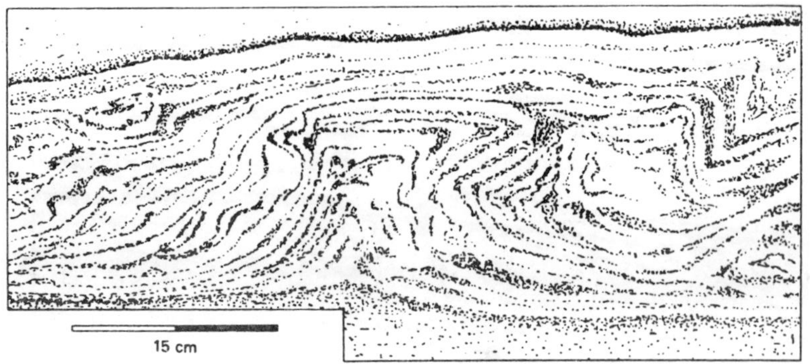

Figure 6.7: Convolute lamination in Cenozoic sandstones of the Carpathians. The degree of penecontemporaneous convolution increases upwards and the top of the affected layer is invariably truncated by a later undisturbed bed.

from a few millimetres to more than a metre downwards from the base of a sandstone or clastic limestone bed into an underlying softer rock, commonly a mudrock or marl. They are caused by the differential loading of soft, fine-grained sediment when it is overlain by a layer of coarser sediment of different density, and can be initiated on the site of pre-existing sole structures.

The downwards movement appears to require some minor vibration of the sediments, such as by an earth tremor or by microseismic shocks during abnormal wave and current activity, in a range of marine and non-marine environments. The vibration causes the soft mud to partly liquify, modifies pore pressures by dewatering, and the coarser, denser layer sinks into the mud, sometimes even disintegrating into discrete pseudo-nodules.

Within load structures the original bedding or lamination is preserved, roughly paralleling the outer morphology. The softened mud is squeezed between the coarser projections, sometimes into sharply pointed, obliquely oriented, narrow wedges a few centimetres long, known as flame structures.

Under similar dewatering circumstances, sand is mobilised and injected into immediately overlying sediment as dykes and veins, occasionally erupting at the surface as small-scale '*volcanoes*', up to a metre across. Such features are recognised in Carboniferous beds of Ireland and Scotland.

Dish structures are laminae disrupted by vertical movements of sand-choked water, sufficiently forceful to bend the broken laminae into a concave-upwards shape. Thin pipes or columns of injected sand separate the vertically stacked dishes.

BIOGENIC STRUCTURES

The post-depositional disturbance of sediment by living organisms is called bioturbation. The disturbance may be caused by organisms moving over the surface or by their penetration into the top metre or so of the sediment, and is usually contemporary with deposition. The structures formed vary in scale and type from the surface trails (epichnia) of large terrestrial vertebrates to burrows (endichnia) of marine organisms, such as *Chondrites,* which are a few millimetres in diameter.

From the sedimentological viewpoint the importance of organisms which grow *in situ* or crawl, forage, burrow, rest and dwell on and in sediment is twofold. First, they physically and chemically affect the original deposit, and secondly they can give some slight indication of bottom conditions at the site of accumulation.

It is not always appreciated how intensive is the physical modification of soft sediment by organisms. It has been suggested that some 80-90% of the comminuted shell sand fringing the Bermuda islands has passed through the gut of sea urchins and holothurians. These, and many other organisms, such as parrot fish and gastropods, which graze on reefs or bottom sediment, excrete vast quantities of fecal pellets, changing the physical character of the original sedimentary particles considerably.

On the bottom of reef lagoons the modern burrowing shrimp *Callianassa* has the capacity to produce extensive, 5-60cm thick layers and mounds of well-sorted lime sands and silts. During its activities the shrimp ejects the finer particles on to the surface, leaving behind the coarser (> 2mm) materials in the burrows, thus effectively sorting the bottom sediments. On some modem European intertidal flats the degree of disturbance may be such as to destroy completely primary lamination and linear orientation of particles within the sediment. *Corophium volutata,* a small but energetic arthropod often present in large numbers, is particularly adept at this.

A considerable amount of churning in sands may be caused by the activities of the lugworm *Arenicola marina* when present in large numbers. Clearly the degree of biological disturbance is dependent on the density of colonisation (or persistence of ecological suitability) of a site. If the density is low, possibly because of high sedimentation rates, then the amount of disturbance is correspondingly less.

In ancient sediments a mottled or grumose (clotted) texture, due to variations in grain size between the mottles and matrix, may be the only evidence that bioturbation has occurred. In pelagic chalks, which are frequently intensively churned, though not always clearly to the

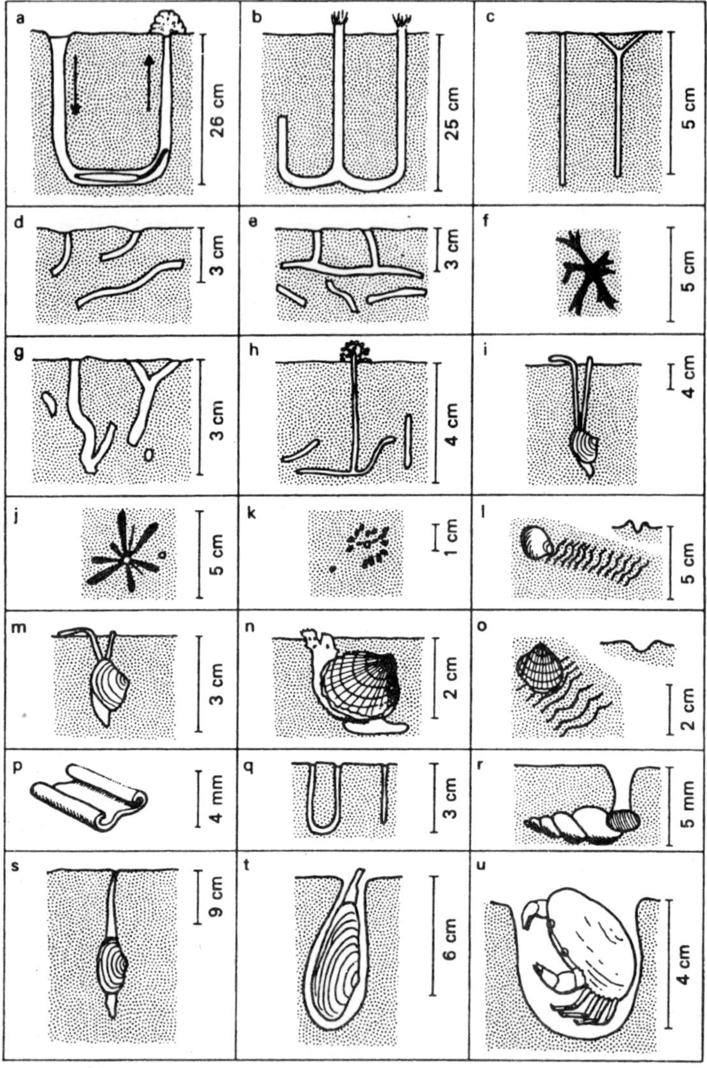

Figure 6.8: Organic structures on modern temperate zone intertidal flats. Examples from the silty and sandy muds of The Wash, east England. (a) Arenicola marina, burrow; (b) Lanice conchilega, burrow; (c) Pygospio sp, burrow; (d)Nepthys sp, burrow; (e) Scoloplos armiger, burrow; (f) Nereis diversicolours, surface trails; (g) Nereis diversicolours, burrow; (h) Heteromastus sp, burrow; (i) Scrobicularia plana, burrow; (j) Scrobicularia plana, surface trails; (k) Macoma balthica, surface trails; (1) Littorina sp, surface trails; (m) Macoma balthica, burrow; (n) Cerastoderma (Cardium) edule, burrow; (o) C. edule, surface trail; (p) Mytilus edulis, faecal pellet; (q) Corophium sp, burrow; (r) Hydrobia ulvae, burrow; (s) Mya arenaria, burrow; (t) Pholas candida, burrow; (u) crab, burrow.

naked eye, the loci for diagenetic chert nodules are often mottles caused by burrowing organisms.

The textural qualities of a sediment may be considerably modified by burrowing, with marked changes in porosity and permeability. On submarine slopes changes induced by organisms and affecting pore pressures and shear strength may be such as to make the sediment more susceptible to slumping and sliding. Conversely, burrowing activities may promote cementation and effectively stabilise the sediment. Chalk hardgrounds are a case in point.

Although the organism causing a specific type of bioturbation may be preserved in place, it is more usual to find no trace at all. In general, this creates little problem in modern deposits as there is always the possibility either that the sedimentary effects and their causative organism are known or that further biological research will identify one with the other. This contrasts

with ancient bioturbation structures where the chances of successfully identifying the causes, in the absence of organic fragments, are slight. One difficulty with fossil examples is the knowledge, gained from present environments, that the morphology of certain structures can be duplicated by a variety of unrelated organisms.

There is also the possibility of successive occupation of certain burrows and tubes by a number of unrelated organisms, each modifying the structure. Nonetheless, by analogy, the originators of certain varieties of burrow have been reasonably designated. For example, *Rhizocorallium* and *Thalassinoides* are believed to be crustacean burrows and *Teichichnus* a burrow formed by an annelid.

An interesting aspect of some burrows is the way in which they are distributed among particular lithological associations. In general, the greatest variety and highest density are to be found in deposits of marine shelf origin. It is there that crawling traces (repichnia) of animals, such as trilobites *(Cruziana, Walcottia),* simple narrow feeding tubes (fodinichnia) of *Chondrites* and *Scolicia,* and larger feeding and dwelling tubes (domichnia) of *Planolites, Skolithos, Corophioides, Teichichnus, Rhizocorallium* and *Thalassinoides* are found in abundance.

Diplocraterion is a Ushaped tube and seems to be an escape structure (fugichnia) formed soon after phases of rapid sedimentation. Internally, the tube is marked by concave upwards laminae or spreiten indicating the temporary resting place of the original animal as it adjusted its position to the new sediment surfaces. *Skolithos* is a straight tube penetrating vertically by as much as a metre into the sediment and

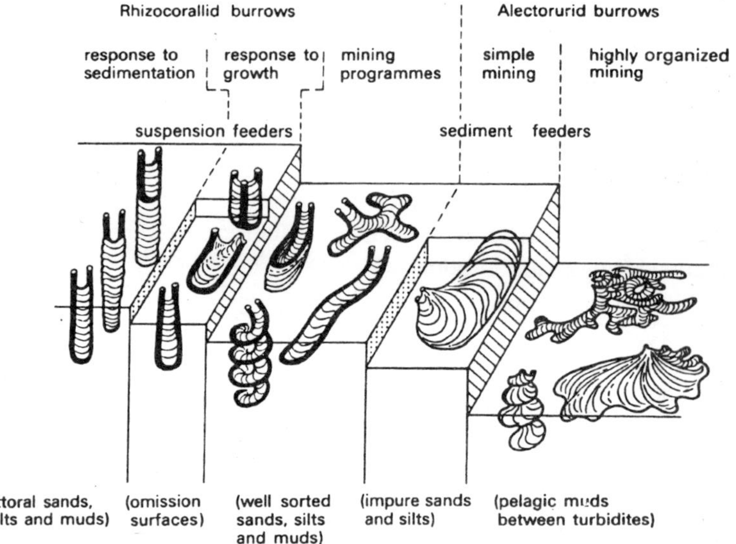

Figure 6.9: Bathymetric zonation of fossil burrows. Illustrates the general rule that suspension feeders prevail in shallow, highly agitated waters and elaborate sediment feeders in deeper, quieter waters.

seems to have some affinity with *Monocraterion*, which is also a simple straight vertical burrow, but with a funnel top up to 4cm wide. The last two are prominent in some sheet sandstones, such as Cambrian '*Pipe Rock*', a basal *quartz arenite*, in north-west Scotland.

Away from marine shelf environments trace fossils are less well preserved or less abundant for various reasons. In terrestrial situations the chances of preservation are diminished by climatic, weathering and erosional factors. Nevertheless, under certain circumstances, traces are preserved. *Beaconites* is a conical escape burrow (of some crab-like creature), measuring about 7cm wide and tapering downwards to a depth of 10-15 cm, found in Devonian fluvial sheet-flood sandstones. The roots of land plants also create their own particular brand of structure in ancient soil profiles.

In deeper water oceanic situations, burrowing and mining become more complicated, and in the deepest pelagic environments, where oxygen and light-penetration conditions are low and nutrients less freely available, the trace fossils are fewer in quantity and type and are often structurally complex. *Palaeodictyon,* a horizontal net-like feeding trace found in turbidite successions, is probably the least complex.

Trace fossil (ichnofossil; *Greek ichnos,* footstep) assemblages in marine sediments to some extent reflect environmental factors, including

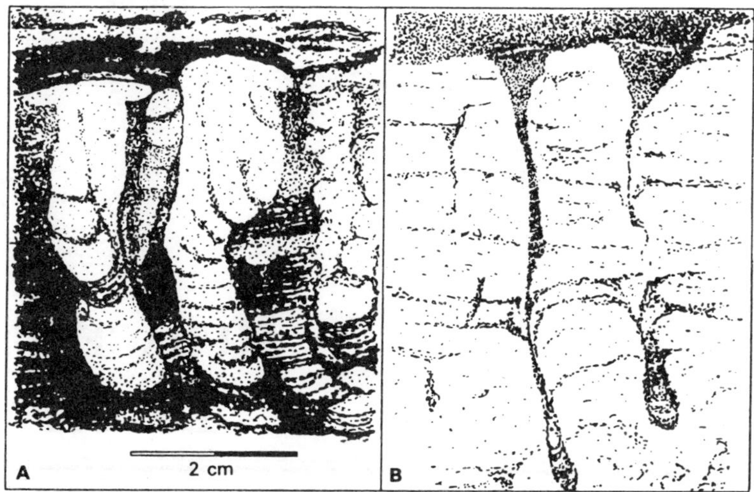

Figure 6.10: A, Diplocraterion. The burrows are U-shaped and internally laminated. The internal structure (spreite) is caused by the upwards movement of the organism in response to sedimentation.

relative depth of accumulation, making them useful in facies interpretation providing the evidence is used in conjunction with all the available *sedimentological* data. Hence, in ancient successions it is possible to recognise a *Cruziana* facies (*burrows*, *tracks*, *resting traces*) taken to be broadly indicative of shallow waters, a *Zoophycus* facies (mining structures, including spiral forms) for medium depths, and a *Nereites* facies (worm trails, complex mining forms) representing deep waters. Even so, the true depth of water existing for any particular facies can only be surmised.

7

Rudaceous Deposits

Rudaceous deposits are the coarsest members of a continuous spectrum of clastic or fragmental deposits composed of detrital material which has been transported *mechanically*. Figure elsewhere in this chapter defines the position of rudaceous deposits (*gravels*, *conglomerates*, *breccias*) in that spectrum in terms of grain size. When the fragments are approximately of uniform size the deposits are said to be well sorted and when very variable in size, poorly sorted.

In the poorly sorted state, suitable qualifiers are used in naming the sediment, such as sandy gravel, pebbly sand or silty clay. The coarser clastic sediments form a somewhat heterogeneous group. They present little of the mineralogical and mechanical uniformity imparted to many of the finer-grained clastic rocks by long-continued action of transportation, selective chemical weathering and mechanical sorting.

In general, the pebbles or analogous components of rudaceous deposits consist of rock fragments, removed from the parent mass by mechanical agencies; occasionally differential chemical weathering leaves residual masses of resistant material which go to form rudaceous deposits.

CLASSIFICATION

Unlithified sedimentary rudaceous rocks, excluding the products of volcanism, are called gravels irrespective of the size and shape of their *conspicuously* large fragments, which are known as pebbles, phenoclasts or clasts. *Lithified rocks* with a *dominance* of rounded pebbles are termed conglomerates; when the fragments are *predominantly* angular

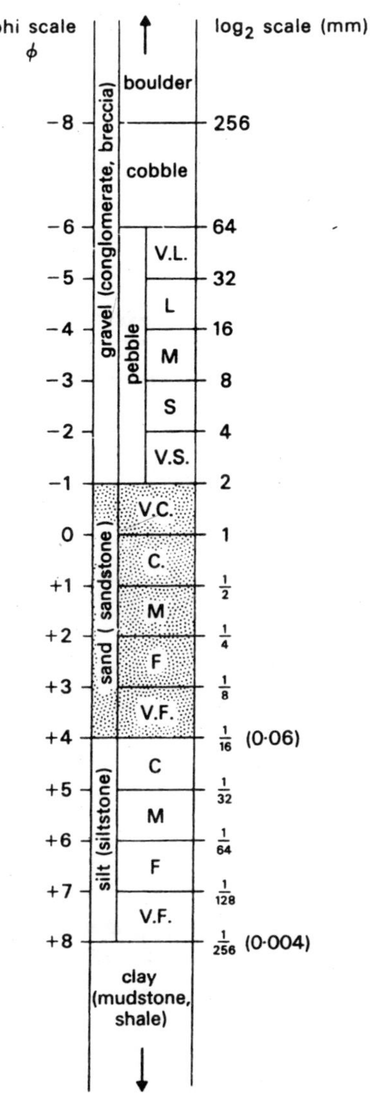

Figure 7.1: Size classification and nomenclature of fragmental deposits. Within each main grade the fragments vary from very fine (V.F.) to coarse (C) or very coarse (V.C.), or in the pebble gravels from very small (V.S.) to very large (V.L.).

they are called sedimentary breccias. Variable degrees of roundness result in hybrid rock names such as breccio-conglomerates.

In theory, gravels and their lithified equivalents are distinguished from other fragmental rocks on the basis of the overall grain size being greater than 2 mm. In common practice, the names are used if the proportion of pebbles exceeds 25%.

Frequently the matrix in rudaceous rocks is either sand and silt, as in *fluvial deposits*, or silty clay, as in glacial and slump-slide deposits. Secondary infiltration of matrix into pores and cavities is usual. Precise description of the rock then becomes difficult, though there have been many attempts to establish suitable schemes of classification. One such, based on *texture*, is shown in Figure elsewhere in this chapter. On the other hand, if acute refinement is unnecessary, much simpler descriptive

Table 7.1 : Conversion Table, Millimetres into Units.

ϕ	*mm*	ϕ	*mm*
-3.00	8.00	+2.25	0.21
-2.75	6.73	2.50	0.18
-2.50	5.66	2.75	0.15
-2.00	4.00	3.00	0.125
-1.75	3.36	3.25	0.105
-1.50	2.82	3.50	0.088
-1.25	2.38	3.75	0.074
-1.00	2.00	4.00	0.0625
-0.75	1.68	4.25	0.0526
-0.50	1.41	4.50	0.0442
-0.25	1.19	4.75	0.0372
0.00	1.00	5.00	0.0313
+0.25	0.84	5.25	0.0263
0.50	0.71	5.50	0.0221
0.75	0.53	5.75	0.0186
1.00	0.50	6.00	0.0156
1.25	0.42	6.25	0.0131
1.50	0.35	6.50	0.0110
1.75	0.29	6.75	0.0092
2.00	0.25	7.00	0.0078

schemes can be used as follows:

I. Texture a. Orthoconglomeratic: framework of gravel and sand bound together by chemical cement; matrix < 15%; pebbles not supported by matrix; commonly bimodal

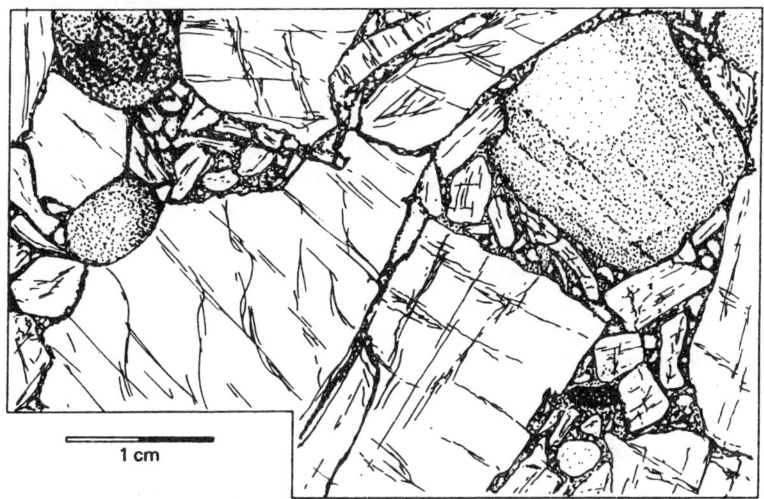

Figure 7.2: Sedimentary breccia, Old Red Sandstone, Scotland. The angular fragments are predominantly sheered vein quartz and haematitised sandstone set in a fine-grained silica clastic matrix infiltrated by haematite.

	b. Paraconglomeratic:	matrix >15% pebbles matrix supported; commonly polymodal
II Composition	a. Polymictic:	pebbles of several rock types, one of which may predominate
	b. Oligomictic:	pebbles of few rock types
III Source	a. Extraformational:	pebbles originate from extra basinal sources
	b. Intraformational:	pebbles originate from within the basin

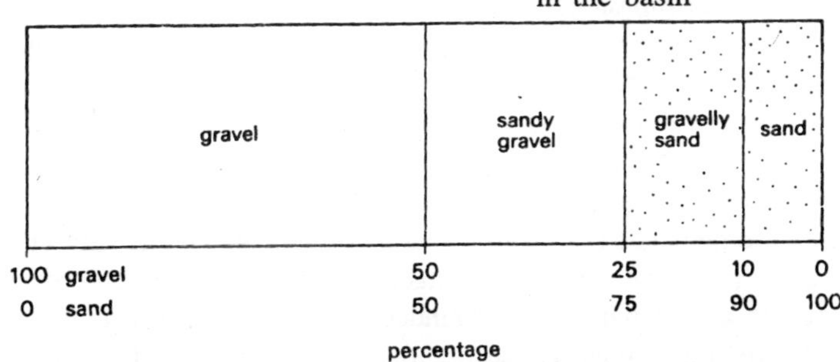

Figure 7.3: Simple classification of gravel-sand mixtures. The lithified equivalents are conglomerate, sandy conglomerate, pebbly sandstone and sandstone.

COMPOSITION OF PEBBLES

In *rudaceous* deposits which have been newly derived from older rocks by mechanical disintegration there is often very little alteration in the proportion of the various lithological types involved, but if the debris undergoes water transportation or chemical weathering for any appreciable length of time, the fragments of the softer and less stable rocks are reduced in size and begin to be eliminated.

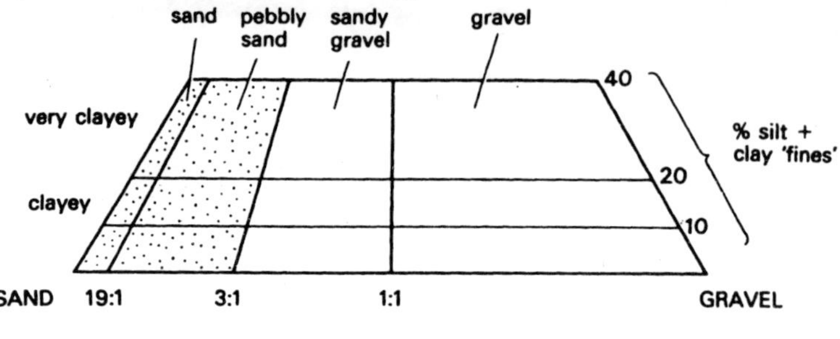

Figure 7.4 : Textural classification of gravel-bearing sediments.

Thus the newly formed debris at the foot of a chalk seacliff can consist principally of angular pieces of chalk, with a small proportion of flint; when this material is picked up by the waves, and washed to and fro in the beach, the chalk is slowly ground away, leaving an oligomictic flint gravel. A similar kind of flint gravel is, produced when chalk has undergone chemical weathering, with solution of the calcium carbonate and concentration of the relatively *insoluble flint*.

This deposit shows no sorting or mechanical abrasion. In common with other sedimentary rocks, conglomerates are liable to be broken down during weathering, as a result of which the pebbles of less stable rocks, such as *limestones*, *calcareous sandstones* and *basic igneous rocks*, in part share the fate of the cement and readily disintegrate. Highly *siliceous pebbles*, however, are not eliminated in this way since they are almost indestructible by ordinary chemical weathering, and at the same time are extremely hard and tough.

For this reason, once pebbles of these materials have been formed, they tend to be handed on from one conglomerate to another throughout geological time, with very little change, except possibly a slight reduction in size, or a somewhat more rounded outline. *Conglomerates* which have derived their pebbles at second or third hand from older rudaceous deposits without access of fresh material tend, therefore, to contain a high proportion of such resistant materials.

The principal rocks which are found to behave in this way are *vein-quartz*, *quartzite*, *flint*, *chert*, *jasper*, *rhyolite* and *quartz-aggregates* derived from acid igneous rocks and gneisses. It is *seldom* that a gravel is composed entirely of fragments of few kinds of rock (*oligomictic*); much more frequently it is polymictic, or composed of many kinds. The nearest approach to an oligomictic composition in this country is found in some of the flint gravels of the south and south-east of England, where a considerable search is sometimes necessary before a pebble of any other kind can be found.

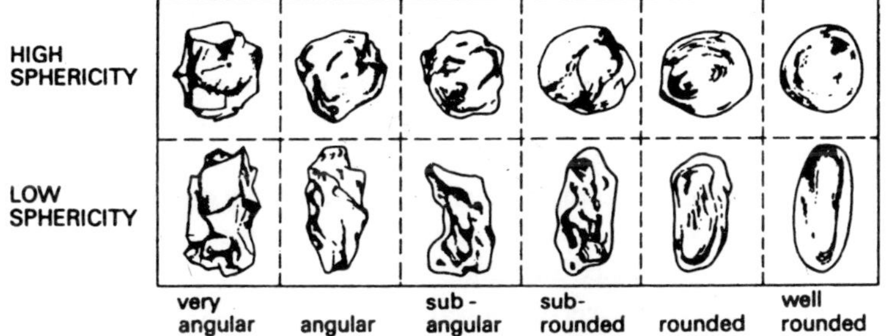

Figure 7.5: Roundness and sphericity chart for boulders, pebbles and grains in fragmental deposits.

Gravels of glacial origin, and gravels containing reworked glacial material, generally show the greatest variety of pebbles, since they often include material transported for very long distances and derived from varied sources; for example the large number of *Scottish* and *Scandinavian* pebbles in the glacial gravels of eastern England.

SHAPE AND ROUNDNESS OF PEBBLES

The shapes assumed by pebbles naturally vary according to their manner of origin. Many gravels and conglomerates contain second-hand pebbles derived from older rudaceous rocks. It is only by considering pebbles which have been newly derived from *non-rudaceous rocks*, and have been shaped entirely within one environment, that we can obtain any true idea of the shapes characteristic of abrasion under given conditions.

Technically it is necessary to distinguish between the shape (or form) and roundness of fragments. Shapes may be described as tabular (*oblate*), bladed (*roller*) or spherical (*equant*), or may be described in terms of degree of sphericity. Roundness concerns the sharpness of the corners and edges and is considered to be independent of shape. There are five classes recognisable:

angular	little or no evidence of wear; sharp corners and edges
subangular	worn with corners and edges beginning to be rounded off
subrounded	considerable wear; corners and edges rounded to smooth curves
rounded	all corners smoothed off to broad curves
well-rounded	no flat areas; entire surface consists of broad curves

The original shape of a rock fragment has a very strong influence upon its shape during a long part of its history, and controls the form of the pebbles even when rounding of the edges is far advanced. Thus the *fissility* of a *laminated sandstone*, or of a slate or schist, is important principally in its influence upon the shape of the original fragments, rather than in providing planes of easy fracture during abrasion. Sharp edges are worn down comparatively quickly, but flat surfaces are worn much more slowly, so that *discoidal pebbles* tend to retain this form even after long-continued abrasion.

The surface of a water-worn pebble is usually smooth and free from pitting or striation. Pebbles which have been shaped during long-continued glacial transportation tend to show faceted surfaces, separated by slightly rounded rather than sharply angular edges. The flattened surfaces are frequently scratched, and the edges not uncommonly show signs of bruising.

Glacial striations, however, do not develop well on all rocks. They are specially conspicuous on boulders of compact, fine-grained rocks, such as micritic limestone, which are moderately hard, yet still soft enough to be readily scratched. In *fluvioglacial gravels* the pebbles are generally much like those of ordinary rivers, but some faceted or striated blocks are usually present, the degree of rounding depending upon the amount of water transport which the material has undergone. Glacially shaped pebbles show surfaces which, apart from the characteristic bruises and striations, are smooth, and may even show a slight polish. They never show *pitting* or *etching*.

Wind action produces two entirely different pebble shapes. Small rock fragments which are not too heavy to be brought into motion by the wind become rounded, much as they would by water transportation, and often show a remarkable approach to a spherical form. Pebbles which are too large to be transported tend to become faceted by the constant abrasion of winddrifted sand. In this way are produced the well-known dreikanter, with two wind-faceted surfaces meeting in a sharp edge, the third face (base) of the pebble often being formed by part of the original surface of the rock fragment. The edges are sharp, and the abraded faces are not usually flat, but somewhat curved. Wind action produces smooth surfaces on *homogeneous rocks*, but if the material is irregular in hardness, the surface is pitted and etched. In all except the softest rocks the surface acquires a high polish.

SEDENTARY RUDACEOUS DEPOSITS

Where consolidated rocks are exposed at the surface of the earth to rigorous climatic conditions, mechanical weathering leads to the *accumulation* of loose blocks of broken rock, which are sometimes of large size. If the *disintegration* of the parent rock is carried out principally by frost action, the resulting debris usually consists of irregularly angular fragments, as, for instance, in the great masses of shattered rock which form the summits of many of the higher mountains throughout the world. Where exfoliation, resulting either from temperature *fluctuations* or from *hydration*, is the dominant disintegrating process, the resulting deposits consist of more or less rounded blocks, in extreme cases showing a superficial resemblance to large water-worn boulders.

This type of deposit is especially characteristic of the disintegration of well jointed igneous rocks in arid or semi-arid countries. The granite kopjes of Mashonaland, Zimbabwe, and the dolerite kopjes of the Karroo in South Africa, for instance, are covered with tumultuously heaped masses of huge boulders and partially rounded. blocks of exfoliated rock. *Coarse-grained* residual deposits are also formed by weathering from heterogeneous rocks, some parts of which are more resistant to weathering or to erosion that the rest. *Boulder clay* frequently behaves in this way, the soft clay being readily removed from the harder glacial erratics, which then accumulate as an irregular mass of boulders.

Many of the older conglomerates possess a somewhat soft matrix or cement, so that the component pebbles are easily set free. This occurs, for instance, in the *Old Red Sandstone* conglomerates of Scotland, which are often very coarse in texture. Even where the cement consists of a resistant material such as silica, prolonged weathering often results in the release of the original pebbles, as, for example, at the outcrops of the gold-bearing conglomerates (*Precambrian*) of the Witwatersrand. Concretions of harder material in soft sedimentary rocks are often left behind as a boulder deposit.

TRANSPORTED RUDACEOUS DEPOSITS

Transported deposits of coarse material are very abundant; their origin may be attributed to gravity alone, or to the action of waves, running water or ice.

Mountain Scree and Solifluction Deposits

The screes or talus *breccias* of mountain regions consist of angular blocks of rock of all sizes and shapes that have been loosened by *weathering* and have fallen or slipped by their own weight. In cold regions the scree material owes its origin to the expansion of water on freezing in the joints of rocks. This process is responsible for the formation of screes in the Arctic zones and in the mountains and hills of temperate latitudes, and, of course, in the highest parts of high mountains in all parts of the world.

In Britain the screes of Wast Water, in the western part of the Lake District, are very well known. They rise to a height of about 450m above the lake in one continuous slope at an angle of about 30°. Well known, too, are the light-coloured screes on the slopes of the Dolomites in the Austrian and Italian Tirol, which from a distance look like patches of snow. In screes which are still in course of active formation, the surface layer, especially at the upper part, is almost at the maximum angle of rest for the material involved.

A slight disturbance will cause general *downhill* movement of the scree in the immediate neighbourhood. The deeper parts of wellestablished screes, however, are usually subject to chemical weathering, which allows the material to settle, with a certain amount of *interlocking* or cementation of the blocks, so that a considerable degree of stability is achieved. Many of the inland scree slopes in northern Europe date from *Pleistocene glacial* or early Holocene times when frost action was more vigorous. *Talus breccia* also accumulates at the base of seacliffs, where it almost immediately becomes affected by storm wave activity which removes the soft fragments and slowly rounds the residuals.

Hillside waste, which would remain at rest if the force of gravity alone were operative, may continue to move by the processes known as creep and solifluction. If a rock fragment on a slope is moved by its own expansion or increased pore pressure in the materials on which it is resting, it may be exposed to the pull of gravity. It will then creep slightly downhill. Such movements, of course, are normally infinitesimal, but since they are always directed downhill, they result in a gradual but general flow of the surface layers.

Similar, but more powerful, movements (*solifluction*) take place as a result of alternate freezing and thawing in Arctic and sub-Acrtic regions. In certain parts of the south of England there are extensive deposits of a *peculiar* character, which are known as head. They consist of local talus or scree-like material, mixed with an earthy matrix, and sometimes showing a structure resembling bedding. These deposits are believed to owe their peculiar characters to movement by solifluction during the Pleistocene glacial period when the ground was permanantly frozen to a considerable depth, as in the tundras of Siberia and North America.

In regions where great extremes of temperature occur in the course of a day, the chief agent in scree formation is the sudden expansion and contraction of the rock, the unequal stresses thus set up causing fractures. Screes are therefore well developed on steep hills and mountain ranges under desert conditions, and excellent examples can be seen in the deserts of Egypt, the Sinai Peninsula, and in many other hot and dry regions of the world. Under these conditions scree formation is much aided by a peculiar type of spheroidal weathering, for which the term desquamation is employed. A notable feature of desert screes is, very commonly, the perfect freshness of the material, since chemical weathering is in abeyance, except under special circumstances.

Torrential Deposits and River Gravels

In the steep upper courses of rivers draining a maturely dissected country, coarse torrential gravels are transported and deposited. As the gradient of the stream becomes gentler, there is a corresponding decrease in the diameter of the particles which are transported under normal conditions of flow, and the gravels deposited are of finer texture. In the lower reaches, where the river is flowing across a well-developed flood plain, the sediments deposited on the valley-flat are usually silts and muds; gravel, if present at all, being confined to the stream bed.

The distribution of gravel deposits along the course of a river is, however, strongly modified by local conditions peculiar to each drainage basin, such as

its topographical peculiarities and the seasonal or other fluctuation in discharge. Rivers which discharge from mountainous country on to flat plains usually deposit the bulk of their coarser debris in a more or less well-defined piedmont belt, consisting of the more or less confluent alluvial fans laid down by each stream where its velocity is checked at the foot of its steep mountain course.

These accumulations come under the general heading of torrential deposits, and they are found for the most part in upland regions; sometimes, however, when a river has a rapid fall throughout its whole length, they may even extend down to sea-level. In Britain modern and ancient examples are well developed in Wales, in the Lake District, and in Scotland. Torrential gravels are also abundant in the valleys and on the flanks of the Alps.

They are of enormous thickness, and have a very wide extension over the low ground north of the *Alps* and *Carpathians*. From a very early period to the present time their formation has been more or less continuous. Of these the freshwater Molasse of Upper Tertiary age consists of sandstones and conglomerates formed by stream action; the Miocene Nagelfluh is a thick series of conglomerates, partly re-sorted by water.

The overlying Pleistocene Deckenschotter includes glacial deposits and the gravels of river terraces. The Triassic conglomerates of the eastern United States are interpreted as having accumulated in the form of piedmont gravels, brought down by torrential streams from the Appalachian Mountains. The pebbles are more or less rounded, unweathered, and, in spite of an enormous range in size, show little or no sorting.

The material of the pebbles varies considerably along the outcrop, but always consists of easily recognisable older local rocks. In different areas the main constituents are *limestone*, *quartz*, *granite*, *gneiss*, *schist* and *basalt*, principally derived from Cambrian and Precambrian formations.

The introduction of gravel into the middle and lower reaches of relatively fast moving rivers can create gravel-rich and braided floodplains. This happened during Pleistocene times in the Thames Basin of southern England. Most of the gravel is composed of flint pebbles derived from Tertiary and Chalk sources, but some pebbles are of more exotic, far-travelled origin.

During the Pleistocene oscillations in sea-level, solifluxion processes and fluvial reworking variably remobilised the gravels, so that they now rest on a series of terraces and slopes elevated above the present floodplain.

Marine Gravels, Conglomerates and Screes

Usually these deposits are the result of the denudation of the coast, and the form and degree of rounding of the pebbles must obviously depend on many causes, the most important of which are the state of the fragment when separated from the parent rock and the amount of abrasion it has since undergone. Here, as elsewhere, boulders may be obtained ready made from older gravels and conglomerates.

Gravels which are now accumulating in marine environments, whether *offshore* on *shelves* or in the *littoral zone*, invariably have had a complex evolution. The offshore gravels may well be reworked relict deposits laid down by *fluvioglacial* or *fluvial* processes during periods of Pleistocene or older low sea-levels. Two well-known littoral deposits in southern England, the Chesil Beach and Dungeness, reflect several episodes of reworking of gravels originally laid down on the terraces of a network of rivers occupying the site of the English Channel and its flanks during Pleistocene low sea-levels.

The great majority of the pebbles are of flint and they are, for the most part, extremely well rounded. The source of the flint was the Cretaceous Chalk outcrop extending from south-east England to France, and well exposed during low stands of sea-level. But the original angular hard flints could only have become so well rounded and concentrated as a consequence of several phases of marine transgression during sea-level rises, of which the *Flandrian* (*Holocene*) Transgression, which commenced about 10,000 years ago, is the latest.

The older marine conglomerates are normally found as comparatively thin deposits, often associated with an erosional break or unconformity in the succession. For this reason they are spoken of as basal or transgressive conglomerates with respect to the succeeding conformable strata. Marine basal conglomerates are abundantly represented in the geological column, especially in shallow water shelf successions. Excellent examples are afforded by the coarser beds of the Precambrian Ingletonian Series of northern England.

The pebbles include quartzite, granulite, schists, gneisses, phyllite and slate, together with a range of igneous rocks. It has been suggested that the beds were laid down under shallow water conditions near to a land mass, but the presence of graded bedding and sole structures in the interbedded turbidite sandstones indicates deeper waters, with possible slumping and sliding of gravel down slopes. A similar manner of deposition may also account for the Eocene clay pebble breccio-conglomerates of Ecuador.

The lenticular beds are more than 200 m thick locally and consist of poorly sorted pebbles among which clay varieties are dominant. The clay pebbles are sub-angular, have polished surfaces and, occasionally, have diameters of a metre or more. It is significant that the beds immediately overlying these conglomerates are of greywacke facies.

In Congo, Precambrian conglomerates which were originally thought to be boulder clays are now reinterpreted as melange-type sediments deposited from extensive mud slides and flows on the margins of a basin. The beds have been named tilloids because of their resemblance to tills or boulder clays. Breccias, excluding the volcanic varieties, are commonly associated with submarine topographic features, such as reefs, or with sliding processes active at the edge of platforms and escarpments. Weakly stratified limestone reef breccias can often extend as talus slopes to considerable depths and for many hundreds of metres in front of the parent reef core structures.

One of the most spectacular slide and mass flow breccias is the 310m thick Cambro-Ordovician Cow Head Breccia in Newfoundland, which is characterised by angular algal-rich, oolitoid and shelly limestone boulders up to 150m across. It appears to have originated as a result of the physical disintegration of a carbonate platform.

In many such instances the movement of the very coarse material is akin to the fluidised flows seen in long tongues of debris emanating from the foot of major terrestrial land slips. Water-saturated finer particles act as the lubricant. An indistinct type of graded bedding may be formed during final emplacement, and the resultant deposit is known as a debrite. The upper parts of submarine fans are often characterised by debrites.

Boulder Clays and Glacial Gravels

The most characteristic of all products of glaciation is boulder clay or till, which in its most typical form is a stiff clay with a varying proportion of rock boulders. The latter are commonly angular and have often been faceted or striated by attrition on rock surfaces, or by mutual pressure. Boulder clay sometimes contains intercalated beds of sand and gravel, and thus grades into fluvioglacial deposits.

To find true modern boulder clay we must turn to Greenland, Spitzbergen or the Antarctic, where it seems to be a characteristic product of glaciation by continental ice sheets. Pleistocene boulder clay is largely developed in Britain north of a line joining London and Bristol, and it is also very abundant in north Germany, Holland and the United States.

As British examples the boulder clays of the Yorkshire coast may be quoted; these consist of an exceedingly stiff and tenacious clay, in which are embedded numerous angular, sub-angular and rounded boulders whose derivation from Scandinavia, Scotland and the Lake District is easily recognisable. These boulders are frequently of one or two tonnes in weight.

Perhaps the most abundant of all glacial deposits, especially in mountain regions, are those to which the general term moraine is applied. Moraines consist of irregular accumulations of fragmental material, a marked feature being the indiscriminate admixture of blocks of all sizes; this is due to the fact that there can have been no sorting action as in water deposition. The blocks are commonly angular and often striated; in fact, moraine material resembles the coarser constituents of boulder clay, without the argillaceous matrix.

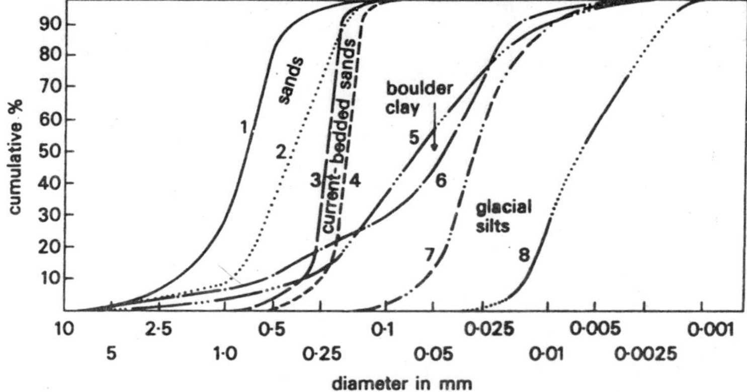

Figure 7.6 : Mechanical composition of glacial and fluvioglacial deposits expressed by cumulative curves based on sieving of the disaggregated particles. The quality of sorting (which reflects natural sorting processes) is indicated by the size range and slope of the curves. Narrow ranges and high slopes mean good sorting.

Very closely related to moraines are fluvioglacial deposits, which consist of moraine material sorted and redeposited by running water. Here there has usually been a certain amount of grading according to size, and sometimes a rough stratification may be observed, although this is more characteristic of the finer material. Fluvioglacial deposits are with difficulty distinguishable from ordinary torrential boulder deposits; but the inclusion of far-travelled blocks sometimes afford a useful index to their origin.

It is perhaps worth noticing here that the so-called anchor ice of northern rivers often transports large boulders, and mixes them with other deposits lower down the river or even in the sea. A similar process is the carrying of large quantities of material over the sea by floating icebergs derived from the land. This certainly accounts for many of the large boulders found in marine sediments; and in the Arctic and Antarctic regions it gives rise to a peculiar type of oceanic deposit.

The total amount of material carried south by icebergs during Pleistocene times and dropped on the Dogger and Newfoundland Banks, for example, must have been enormous. Instances are known in the far north where ice laden with boulders is driven on to the shore, or forced through a narrow channel, and both of these causes may result in the formation of boulder terraces along the coast. Pre-Pleistocene boulder clays are known by the name of tillite, and have been recognised in many parts of the world. They contain irregularly shaped or angular boulders of fresh, *unweathered rock*, with all sizes of pebbles mixed *indiscriminately* together in a fine-grained matrix of angular chips. The Permo-Carboniferous tillites of the southern hemisphere are perhaps the bestknown examples (e.g. the Dwyka Tillite of South Africa). The Talchir Boulder Bed of India is of similar age. All were deposited during the glaciation of Gondwanaland before the drifting apart of the continents.

The Dwyka Tillite forms the base of the Karroo System north of latitude 33°, and rests upon a striated pavement of older rocks. The boulders, which are often faceted and scratched, include an immense variety of rocks derived from pre-Karroo formations, such as basic lavas, granites, gneisses, grits, jaspers, slates, sandstones and quartzites. These are set in a tough, bluish-grey matrix of angular sand mixed with argillaceous material.

This deposit reaches a thickness of over 300m, and for the most part has all the characters of an indurated boulder clay. There are, however, at some localities beds of conglomerate with a calcareous cement, which appear to represent fluvioglacial gravels, and bedded shales are also found, occasionally showing a varve-like banding characteristic of clays laid down in glacial lakes.

Tillites of late Precambrian age have also been reported from many parts of the world, such as Canada (*Huronian Tillites*) and South Africa (Government Reef Series). The Dalradian Port Askaig, Kilcherenan and Lock na Cille conglomerates of Ireland and the western Highlands of Scotland may be tillites. The South Pole in Ordovician times was located in the northern Sahara and is expressed in that region by tillites and other glacial deposits.

Intraformational Conglomerates and Breccias

The conglomerates which have already been discussed in this chapter all contain pebbles or similar bodies derived from older consolidated rocks, and there can be a considerable difference in age between such pebbles and the matrix which encloses them.

There are, however, other rudaceous rocks in which this age distinction is slight or negligible; deposits of this kind are termed intraformational conglomerates or breccias, and in certain environments are formed in the normal course of continuous sedimentation, without the intervention of earth movements. Thus their stratigraphical significance is entirely different from that of most other rudaceous rocks.

Streams flowing over an alluvial plain tend to shift their channels laterally. In doing this, the streams erode the recently deposited sediments of which their banks consist, and large masses of coherent mud and silt, released by undercutting, fall into the channel and are rolled along its bed. Transported fragments of this kind come to rest on the inner, slack-water, side of the meander, and become buried in sand and mud to be preserved eventually as conglomerates and breccias.

In times of heavy flood, vigorous erosion of the banks takes place, and the debris may be broken up into small, semi-plastic pellets which are distributed by the flood water over the surface of the plain. Deposits of this kind, known as clay-gall conglomerates (with sandy matrix) and shale conglomerates (with argillaceous matrix), are of frequent occurrence in the Carboniferous Coal Measures and Cretaceous Wealden deposits of Britain, and are also conspicuous in a somewhat similar facies of the Middle Jurassic sediments on the Yorkshire coast.

Mud-flake conglomerates and breccias are formed as a result of the exposure of large areas of newly deposited mud to the atmosphere with consequent desiccation of the surface layers. They are found in a wide range of environments. In semi-arid and desert countries, for example, large pools or lakes, often no more than a few centimetres in depth, are formed by seasonal rains.

When the water disappears again and the freshly deposited sediment is exposed to the atmosphere, the surface layers shrink on drying, and a system of polygonal desiccation cracks appear. These polygonal masses of mud are sometimes picked up by the wind, and accumulate elsewhere to form an intraformational breccia, or they may remain, becoming hardened by further desiccation and baking in the sun, until they are buried under later sediments or removed by flowing water during a later flood.

8

Arenaceous Deposits

Sediments of arenaceous or sandy texture are characterised by grain sizes which fall on the Wentworth scale between 1/16 and 2mm. For the rocks described in this chapter and the next the term *siliciclastic* is also appropriately descriptive.

Quartz is the commonest sand-forming mineral, and may be regarded as the principal constituent of most arenaceous rocks. Many sands consist almost entirely of quartz, but frequently appreciable proportions of feldspar, white mica and clay minerals are present. Other components are subordinate and consist of such resistant minerals as were able to survive the weathering processes that effected the destruction of the rocks in which they originally occurred.

These are known as accessory or heavy minerals, some of the most common of them being ilmenite, rutile, garnet, tourmaline, zircon, staurolite and kyanite. The identification of all these constituents of sands and sandstones is now highly sophisticated and the modern techniques allow a qualitative and quantitative accuracy beyond that achieved by the *petrographic microscope*. X-ray diffraction (XRD) is widely used to identify detrital and *authigenic minerals*, especially the clay mineral component.

The scanning electron microscope (SEM), with magnifications up to several thousand times that of the normal microscope, is very useful for studying the morphology and surface features of grains, detecting dissolution effects or *authigenic outgrowths*, observing the nature of the pore lining or pore throat blockage by clays, and distinguishing between authigenic and detrital clay minerals. *Electron-microprobe* analysis allows quantitative chemical analysis of individual minerals even of the smallest sizes. *Cathodoluminescence microscopy* (CL) appears to be a key technique in detecting dissolution, *reprecipitation* and replacement processes affecting quartz grains. Repeated zonation in quartz overgro-wths, reflecting a complex cyclical diagenetic history, is readily recognisable. Fortunately, the petrographic microscope is still capable of providing a broad indication of the nature of most sandstones and the textural relationships between the constituent particles. Quartz is an extremely durable mineral and individual grains are passed on from one sandy deposit to another through long periods of geological time.

During this process, the external shape of the grains undergoes repeated modification, and the original characters are progressively obscured and *obliterated*. At the same time, the less stable minerals, originally associated with the quartz grains, are constantly being eliminated, so that sands of ancient derivation which have not received additions of more recent material gradually approach the condition of being pure quartz deposits.

In studying any particular arenaceous rock it is important to bear in mind the possibility of the grains having been derived from an older sandstone, and to distinguish between the age of the grains and the age of the deposit. Quartz grains derived from granites usually present an irregularly angular but roughly equidimensional *monocrystalline* form. When the mineral is liberated by chemical decay of the surrounding *feldspars*, the grains have a frosted and corroded surface, due to the solvent action of alkalis liberated from the decomposing feldspars.

A similar effect can be seen in grains loosened from a *calcite* (*alkaline*) cement. Quartz liberated by mechanical destruction of undecayed *granite* is *sharply angular*, with clean, glassy fractures. A study of the inclusions in detrital quartz grains may often give an indication of the source of the material; the inclusions in the quartz of granites most commonly consist of small crystals of other minerals, such as prisms of *tourmaline* and needles of rutile.

The quartz of acid lavas are more apt to contain glassy inclusions. In contrast, studies of fluid inclusions within authigenic overgrowths on quartz grains do not indicate provenance, but rather illuminate the timing and *physicochemical* conditions prevailing during the new growth. Detrital quartz, using cathodoluminescence, is blue, red, brown and orange depending on the presence or absence of impurities such as *titanium* and *ferric iron*.

The blue or blue-violet types indicate original igneous or fast-cooling high-grade *metamorphic sources* (>570°C), whereas the brown-orange reflect slow-cooling high-grade or low-grade metamorphic sources (<57°C). Authigenic quartz (mainly as overgrowths) is usually precipitated at temperatures below 300°C and is non-luminescent. However, if subject to later mild metamorphism it too can develop an orange luminescence.

Mica-schists and thinly foliated gneisses, when newly broken up, are liable to produce quartz grains which are flattened in the plane of foliation, and which may consist of a group of crystals rather than a single large individual. The grains are referred to as *polycrystalline*. Coarsely banded gneisses, on the other hand, give quartz grains very much like those derived from granites.

Gneisses derived from intensely metamorphosed sediments not uncommonly supply quartz grains characterised by inclusions of *sillimanite* in slender needles with a felted or plume-like arrangement. Polycrystalline grains of quartz derived from fine-textured siliceous rocks, such as flint or chert, are readily recognised between crossed nicols, when the minute constituent crystals extinguish individually to give a highly characteristic *minutely speckled* appearance.

Great care has to be taken to avoid provenance *misinterpretation* because quartz grains may be recycled, extensively reworked in the basin of deposition or transported over long distances. During all these processes the polycrystalline grains appear to be selectively disintegrated by mechanical and chemical agencies.

Undulatory or strain-shadow extinction, a common feature of quartz grains, is of doubtful value in provenance studies as many igneous and metamorphic rocks contain quartz of that type.

The rough, irregularly angular grains derived from weathered crystalline rocks, provide the raw material for most arenaceous deposits. On being picked up by rivers, this material is to some extent sorted according to grain size, but retains its sub-angular character, even after long transportation.

Beach sand, unless derived from older sandstones, is also sub-angular, but the constant to-and-to movement under the action of waves very slowly rounds off the sharper edges of the grains. However, the grains rarely become well rounded in these situations. Some of the best rounded of modern sands are found in deserts, such as those of North Africa, and are believed to owe their shape and smoothness to abrasion during long-continued transportation by wind.

The grains affected are usually larger than 0.1 mm in diameter. Below that size a marked degree of angularity may be retained. Similar well-rounded sands are found in the Fermo-Trias beds of north-west Europe. A common surface feature of desert grains is frosting, a micro-pitting caused either by wind abrasion or the precipitation of a vitreous film of silica from capillary waters.

It has been suggested that chemical corrosion by alkaline capillary solutions can create the same effect. On the other hand, in environments where water is more freely available and alkaline in nature, secondary chemical frosting can occur during cementation. This affects grains of all shapes and degrees of roundness. Marine calcareous sandstones show it and so do calcrete- and caliche-type sandy soils.

Electron microscope studies on unlithified sand deposits show that particular styles of transport and deposition impress certain types of pitting on grain surfaces. The distinction between and evaluation of new and inherited textures is difficult. In ancient sandstones the environmental diagnosis value of these studies is diminished by cementation and dissolution processes, which can destroy depositional surface textures.

CLASSIFICATION

The arenaceous deposits include ordinary unconsolidated sands, such as those found in offshore basins, on sea-beaches and in rivers, and their lithified equivalents, the sandstones. The range of dimensions of grains to which the term sand is commonly applied will be seen by reference to Table elsewhere in this chpater Sand grains usually consist of individual mineral fragments, though composite or lithic particles, derived from rocks, are not uncommon.

In most unconsolidated sands, the pore space between the grains is occupied by air or water, but in environments where sorting processes are not active, the sand grains may be embedded in a matrix of silty or argillaceous material. The grains may be welded together under pressure, without addition of cement from outside, and the pore space becomes reduced or disappears in the process. Open sands may become lithified by subsequent chemical deposition of some substance between the grains, this secondary binding material being termed a cement.

In the case of sands with a matrix already present between the grains, induration of this material may serve to lithify the deposit without the introduction of a cement. Many schemes of classification for the arenaceous rocks have been drawn up, some based upon theoretical principles, some emphasizing the mineral composition, and others devised empirically for convenience in field and laboratory description.

Current tendencies are to subdivide sandstones on the basis of their

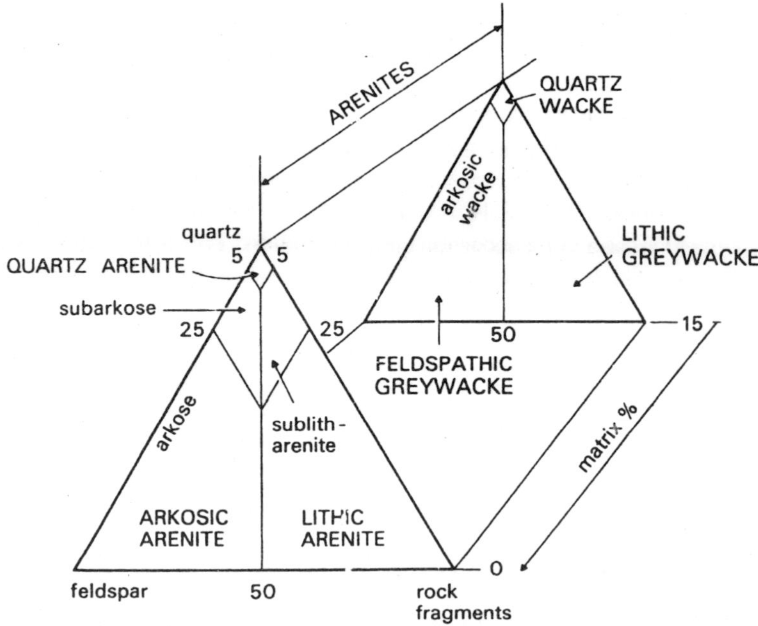

Figure 8.1 : Classification of sandstones. All wackes carry 15% or more matrix. Cements are not taken into account when establishing the variety of sandstone.

mineral content, or mineral and textural attributes. None of the proposed schemes is used universally, so that care is needed when interpreting the nomenclature used by various authors.

Some of the attributes of sandstones may be easily recognisable in the field, but almost invariably, require confirmation by petrographic and other means. Cements play no role in sophisticated modern classifications, though matrix minerals, which equally well may have been introduced or be authigenic in origin, are frequently taken into account. As most sand and sandstones contain more than quartz, feldspars, rock fragments and matrix it is usually necessary to recalculate the composition before allocation to a given category in modern classifications.

In this text the classification adopted is that of Dott modified by Pettijohn, Potter & Siever. Two main groups of sandstone, namely arenites and wackes, are recognised and separated on the basis of their matrix ($<30\ \mu m$) content. Arenites, which have less than 15% matrix, and wackes, which have more, are then subdivided on the grounds of various mineral and rock particle attributes.

Unfortunately, one of the consequences of adhering to any particular scheme is the inevitable disappearance of some well-established names. One such is *orthoquartzite* which, in the proposed classification, falls into the quartz arenite category. Another is *subgreywacke*, now replaced by *sublitharenite*.

As might be anticipated the mineral and textural parameters used in classification are reflected by the chemistry of sandstones. Quartz arenites have a very high silica content and low *alumina* content, arkoses have a comparatively higher potash and alumina content, and so on. Whether looked at from a chemical or mineral point of view it is necessary to appreciate that the rock as eventually constituted reflects the composition of the source rocks, the nature and maturity of the weathering processes, the quantity and quality of *diagenetic* changes and the presence or absence of *biochemical* or other contaminants.

VARIETIES OF SANDSTONE

Quartz Arenites

These are sandstones generally typifying slow sedimentation on broad shelves and have a quartz content greater than 95% with very little fine-grained matrix or cement. A few pebbles may be present. A small amount of the quartz is commonly in the form of overgrowths in optical continuity with the original clastic cores and, in some ancient beds, low grade *metamorphism* mildly enhances this effect.

The sand grains become firmly interlocked and the ultimate product is a very hard rock, but one susceptible to brittle fracture when stresses are applied. The rocks have also been called *orthoquartzites* and quartzites, the latter being a long-established term perpetuated *stratigraphically* by such names as Holyhead Quartzite (Precambrian of North Wales), Hartshill Quartzite (Cambrian of central England) and Eriboll Quartzite (Cambrian of northwest Scotland).

The high detrital quartz percentage almost certainly reflects a recycled origin from a deeply weathered quartz-rich terrain, plus an additional amount of winnowing-out and attrition of more labile minerals during contemporary transport and deposition. Sorting is very variable. It is hardly surprising that the bulk of these sediments accumulate in quantity on shelf areas.

Under conditions of prolonged marine transgression, quartz arenite sands may be smeared in extensive sheet-like and lensoid bodies above the plane of marine erosion. In Britain good examples are found in the Lower Palaeozoic strata of the Welsh Borderland and include the Wrekin Quartzite (*Cambrian*) and Stiperstones Quartzite (*Ordovician*). Fresh bedding surfaces of these rocks have a light grey to white lustre. Ripple marks and bioturbation structures are common at some levels.

When warm water currents prevail and soluble calcareous skeletal matter proliferates, the transgressive sand bodies may become rich in introduced calcareous cement as lithification proceeds. The proportion of cement might reach 40-60% so that the textural characteristics become atypical of quartz arenites. The name calcarenaceous sand (or sandstone) has been suggested for this special variety.

Some modern quartz arenites are recorded in moderately deep basinal waters and have either been swept from shelves and deposited by sheet flow or are in-place, *current-winnowed* relict deposits of earlier periods of lower sea level. There is also the possibility of deposition from *melting icebergs* and ice sheets in circumpolar regions.

Terrestrial varieties of quartz arenite include many of windblown origin, such as those forming major and minor dune fields in present inland desert, river flank and coastal areas throughout the world. A noteworthy feature in deserts is the conspicuous rounding of the larger grains, producing what is known as millet-seed texture, and the general absence of friable mica flakes, which cannot withstand the predominantly rolling movements in a highly abrasive situation.

Organic matter is usually absent or very rapidly disintegrates. Fossil examples are particularly evident in red-bed successions, such as the Penrith Sandstone (Permian) of north-west England. *Ganister* is an old English name for a terrestrial quartz arenite with good refractory qualities. It is found in coal-bearing successions and probably originated as a gley podzol (organic sandy waterlogged soil) in poorly drained ground adjacent to rivers. Traces of roots are present and plant debris occasionally abundant. The originally deposited sand has been subject to leaching and the feldspar and detrital mica content is low compared with associated sandstones.

Pore fillings of *authigenic kaolinite* are a consequence of the breakdown of these minerals. Quartz overgrowths and patchy impregnation by limonite are also characteristic.

Quartz-lage is another special type of very fine-grained quartz arenite, with angular grains, preserved within Carboniferous and other *coal-bearing* successions. The light brown-grey layers, usually interbedded within coal seams, are several centimetres thick and can be traced over moderate distances. The typical internal structure is a laminar aggregation of drusy authigenic quartz crystals lining and filling thin *lensoid cavities*. The origin of quartz-lage is uncertain but could be due to the decomposition of *Equisetalestype* plant debris included within a windblown loessic deposit. Certain modern varieties of these plants, including horsetails, have silica distributed in their tissue, and ancient equivalents, if of similar composition, could have been the source of some of the drusy quartz.

Some types of duricrust, known as silcretes, are pebbly quartz arenitic and are produced by prolonged surface weathering under warm humid to subhumid climatic conditions. Extensive upstanding land surfaces in parts of South Australia and Queensland are mantled by these deposits and though mainly of Tertiary age they range from the late Mesozoic through to the Pleistocene. The youngest deposits carry an opaline and chalcedonic silica cement, which in the older *recrystallises* into granular quartz or is redistributed as quartz overgrowths. Similar red, buff and brown silcretes of *post-Cretaceous* age are known in Britain, America and South Africa.

Lithic Arenites and Sublitharenites

These sands and sandstones constitute a major group and are common in terrestrial, shelf and shelf basin successions, expecially along passive margins.

The quartz content in lithic arenites (or *litharenites*) varies up to 75%, rock fragments are dominant over feldspar grains, and the matrix, which is mainly clay of mixed detrital and authigenic origin, forms less than 15% of the rock. Sublitharenites are a subgroup straddling the boundary between lithic arenites and quartz arenites.

Introduced cements commonly include secondary quartz, calcite, siderite and iron oxides and hydroxides. In most instances the quartz, calcite and siderite have been precipitated early during diagenesis, though recrystallisation and redistribution invariably occur during the course of deep burial, compaction and acute tectonism. The carbonates often show corrosive relationships to the clastic grains they are partly enclosing, this being detected by micro-embayments at the periphery of the quartzes and cleavage infiltration in feldspars and micas.

Limonite is widespread as cement and is usually attributed to secondary superficial changes, especially of pre-existing iron minerals. The detailed make-up of the clastic grains is very variable depending on the constitution of the provenance and the environment of accumulation. Modern alluvial and estuarine sands in south-east England contain *quartz*, *chert*, *flint*, *vein quartz*, *quartzite*, *feldspar*, *glauconite*, *barytes* and *clay* minerals derived from adjacent Pleistocene and Tertiary outcrops. This material is admixed with present-day silt and shell fragments and set in a mud matrix. The labile constituents such as the shells are progressively lost during lithification.

Terrestrial lithic arenites of Lower Old Red Sandstone age in the Welsh Borderland have Lower Palaeozoic volcanic rocks, argillaceous limestones and siltstones as their main lithic constituents. Some Tertiary terrestrial sandstones of the northern Mediterranean average 25% quartz, 10% feldspar, 25% limestone grains and variable amounts of chert and volcanic fragments. In fact, terrestrial successions the world over comprise lithic arenites to a large degree. In part this is a reflection of the type of weathering in the provenance areas, whereby temperate and semi-arid climates are more likely to see a preservation of more unstable constituents, tropical climates less so.

Also there is a more irregular pattern of physical and chemical destruction of fragments during transit on land compared with shelf sea areas. Many cross-bedded and lenticular sandstones, up to tens of metres thick and occurring as channel fills (*wash-outs*), are lithic arenites and sublitharenites. Good examples occur in the coal-bearing fluviodeltaic successions of Upper Carboniferous age in the Northern Hemisphere, Permian age in India and Australia, and *Palaeocene age* in *Dakota*.

Sometimes clastic mica, mainly muscovite, is concentrated along closely spaced bedding planes, easily splitting for flagstone production. The overall acidic nature of river and associated ground waters in swampy vegetated areas ensures that very little calcite is precipitated as cement within channel sands, though siderite is. Clay minerals, some clastic, others authigenic, frequently fully occupy or line primary pores. Kaolinite aggregates, which tend to be stable under mildly acidic pore-water conditions, show marked evidence of plasticity and distortion as compaction proceeds.

In addition to freshwater and sea-margin sequences, lithic arenites are reported as occurring frequently in post-Jurassic marine basin successions, where the sediment has been laid down and buried relatively quickly. An inherent property of all lithic sandstones, irrespective of environment of accumulation, is their high potential for provenance determination.

Rock fragments are more rewarding than individual crystals when inferring the constitution of source areas. Schist, slate and tuff fragments, for instance, are much more informative than monocrystalline quartz grains. Moreover, if the rock fragments are weak and susceptible to rapid physical disintegration, it is reasonably certain that they have not passed through several cycles of erosion and deposition.

Arkosic Arenites and Related Rocks

Sands freshly derived from undecomposed crystalline rocks, such as acid gneisses or granite, contain notable quantities of feldspar. If the feldspathic sediment is deposited almost at once, and sealed off from circulating ground water, the feldspar is preserved, even though weathering, erosion and deposition might have taken place in a humid climate. Molasse facies laid down in tectonically active and rifted converging continental plate situations commonly comprise a high proportion of arkoses and subarkoses.

The deposits are distributed in the form of fans and sheets, and channel and ephemeral lake fills. Interdigitation with marine sediments occurs in some basins. Work on the petrology of the Californian Coast Ranges has shown that the arenaceous rocks of all ages from the Jurassic onwards to the present day are prevalently feldspathic rather than quartzose. Orthoclase, often pink, and acid plagioclase frequently make up half or more of the sediment by weight, and sometimes are present in considerably greater abundance.

These rocks, having a feldspar content greatly in excess of 25% and matrix less than 15%, are arkoses in the strict sense. In fact, the clay matrix in the Tertiary beds rarely exceeds 4%, though carbonate cement, which is discounted for classification purposes, ranges up to 40%. The quartz of the Cretaceous arkoses contains fluid inclusions suggestive of derivation from granite, and the accessory minerals, green hornblende, magnetite, zircon, sphene, apatite and others, provide further evidence of this origin.

In parts of the Tertiary, additional minerals such as *glaucophane* and brown hornblende indicate that much of the sediment was supplied by rivers draining an area of Mesozoic Franciscan metamorphic rocks, a view which is confirmed by a study of the accompanying conglomerates. We thus have an imposing sequence of Mesozoic and Tertiary arkoses, derived directly from mildly chemically weathered *crystalline rocks*. The sands were rapidly transported and those with a high carbonate content were probably deposited in lacustrine and marine environments.

In the arenaceous rocks of western Europe, feldspar is not found to be so persistently abundant as in the region discussed above, but important deposits of arkose appear at certain horizons. In parts of the Torridon Sandstone (*late Precambrian*) of Scotland, feldspar is almost as abundant as quartz, reflecting the composition of the gneisses and granulites of the underlying older Lewisian source rocks. In spite of the great age of the sediment, the dominant feldspars, microcline and microcline-microperthite, are remarkably fresh and angular. The matrix of these sandstones is usually a mixture of clastic and authigenic clay minerals into which red iron oxides have infiltrated. The overall impression is of very limited transport and minimal sorting by moving water in a range of sub-aerial fan, fluvial and lacustrine environments. Cross-lamination, cross-bedding and channelling structures are widespread and there are occasional thin stromatolite lenses. Terrestrial red-bed deposition is implied.

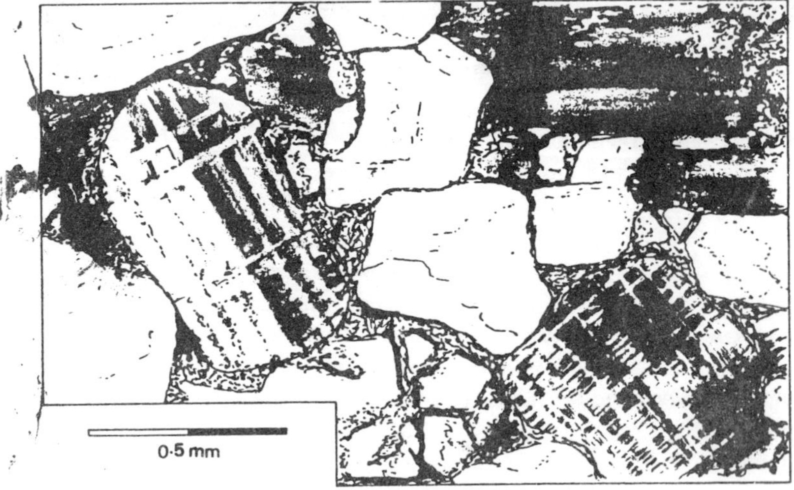

Figure 8.2: Arkose, Torridonian, north-west Scotland. A poorly sorted coarse sandstone consisting of a sub-angular to rounded quartz, vein quartz and metaquartzite grains, with a relatively high proportion of fresh microcline, microperthite and acid plagioclases.

Fresh microcline is well represented in the Millstone Grit (Namurian) sandstones of northern England, deposits laid down in the deltas of a river system draining a distant northern land area. Certain beds are genuine arkoses, but more have a 5-25% feldspar content and are more appropriately referred to as subarkoses. In contrast to the *Torridon beds*, these arkosic and *subarkosic sandstones* carry little clay matrix and have a relatively greater proportion of secondarily introduced chemical cement.

Greywacke Sandstones

Greywackes are prominent in fore-arc, inter-arc and marginal continental flysch-type basinal sequences. These poorly sorted, matrix-supported sandstones, which are various shades of grey in colour, are more difficult to define satisfactorily than any other group of sandstones. This is because their *mineralogy* tends to be complex with angular to

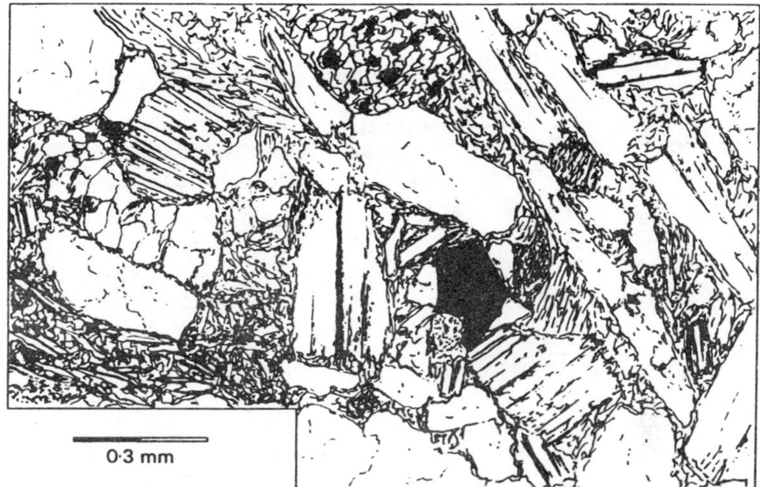

Figure 8.3: Feldspathic greywacke, Devonian, Germany. A classic example of a mediumgrained, matrix-supported greywacke consisting of a mixture of sub-angular to subrounded grains of altered and fresh feldspar, quartz and a range of rock fragments including metaquartzite, vein quartz, slate, tuff and trachyte. Marginal intergrowth of the clastic matrix of illite, chlorite and quartz with the larger grains is usual.

subrounded quartz, *plagioclase* and *rock particles* set in a finer-grained matrix. The matrix is difficult to resolve under the microscope, but generally consists of a mixture of silt- and clay-sized quartz, feldspar, illite, *montmorillonite*, *chlorite* and mixed-layer clay minerals. These are *intermingled* occasionally with epidote, pyrite and calcite of diagenetic origin.

The clay minerals are partly derived from the breakdown of unstable silicates in the source areas, and are clastic. The rest are almost certainly diagenetic, forming by recrystallisation after the deposition and burial of the sand. Some are redistributed as pore linings. Loading and subsequent tectonism, and mild metamorphism, accelerate the post-depositional processes of change of the unstable grains.

The progressive consequences of such change are particularly evident in many pre-Tertiary greywackes, where grain boundaries become diffuse because of intergrowth between the larger clastic grains and matrix minerals. Clay minerals recrystallise and alter in composition. Feldspars are replaced by secondary clay minerals, zeolites and epidote and, at a certain stage, the clastic plagioclases begin to convert into Na-rich albite.

The Na_2O content of older greywackes frequently exceeds 3%. The proportion of matrix, therefore, apparently increases with increased

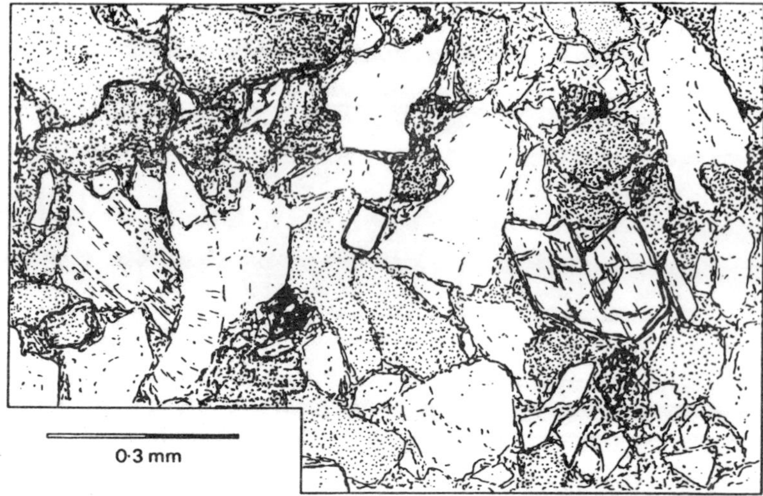

Figure 8.4 : Lithic greywacke, Silurian, southern Scotland. Angular highly degraded rock fragments, sericitized feldspar, quartz and occasional pyroxene (right centre) set in a relatively small quantity of matrix. Small grains of epidote (centre) are present in small amounts.

burial diagenesis, such that the larger, less labile grains become even more markedly matrix-supported than in their originally deposited state.

The classifications that take into account the matrix indicate that it must form more than 15% of the rock. When feldspars are dominant over lithic fragments the rock is called a feldspathic greywacke; when rock particles, such as volcanic ash, are dominant over feldspars the rock is called a lithic greywacke. Many lithic greywackes are rich in volcanic detritus and seem to have been laid down in troughs adjacent to active volcanic island arcs near plate margins.

The Tertiary Waitemata Group greywackes in northern New Zealand are of this type and carry abundant andesitic and basaltic fragments. In these situations quartz can be totally absent from the sediment. On the whole, greywackes carrying such rock fragments are more informative about the nature of the provenance than feldspathic equivalents.

Greywackes characteristically form the lower parts of repetitive, graded beds (Bouma turbidite sequences) found in thick successions. Individual beds are often several metres in thickness. The poorly sorted nature of these sandstones is in part a reflection of mass gravity transportation by turbidity currents, though the sorting is clearly exaggerated by post-depositional recrystallisation and compaction effects, especially in regions of high heat flow (island arcs up to 180°C).

Quartz wackes are not normally interbedded with greywackes in ancient successions but more often with finer-grained sublitharenitic and quartz arenitic sandstones and siltstones. They are much more common than is credited and constitute significant proportions of deposits laid down in littoral and shallow marine environments where suspended fine mud can be abundant. The mud mixes with the sand and silt grains, particularly during storm conditions.

DIAGENESIS

Compaction

The compaction or volume reduction of sands is predominantly caused by overburden loading, though the degree to which it occurs is a function of many small- and large-scale interdependent factors, such as pore pressures, grain shape and size, sorting, packing, mineral dissolution and tectonism. In general, the greater the loading and grain-to-grain stresses are, the greater is the compaction. Porosity is reduced.

On deposition, some well-sorted coarse sands may exhibit a very open, primary packing (or spatial arrangement) of the grains, with porosities verging on 40-50%. After mild loading and readjustment of the grains to a tighter packing, the sand might then have its porosity reduced to 25-30%. If smaller grains were originally introduced into the sediment, so as to occupy interstitial positions between the larger grains, then the loading might reduce the porosity to some 15% or less.

In the top few metres of newly deposited, highly porous and well-sorted sands the pore pressures may build up to exceed the loading pressures until a point is reached when any type of small vibration triggers off mass movement. The sand then flows or liquifies, fluid is dispersed and the grains readjust to a tighter configuration.

Several types of plastic deformation and de-watering structures are produced in this way, such as contorted bedding in channel sands, intrusive sand dykes and sand volcanoes. Sometimes, because of high rates of deposition in rapidly subsiding basins, sand is buried so quickly that pore fluids do not escape readily.

The sands then remain overpressured at depth, with relatively high pore pressures, and never compact to the expected amount. The oil-bearing, fluviodeltaic Brent Sandstone Formation (Middle Jurassic) of the North Sea Basin, even though now buried at depths of 2.5-3 km, has a relatively loose and friable nature as a consequence of being highly overpressured.

Under more usual conditions, compaction proceeds by individual grains being pushed closer together. Whereas an originally deposited sand may have grains which have small or tangential point contacts without interpenetration, a buried loaded sand may have a much higher proportion of closer, long, concavo-convex and microstylolitic (sutured) contacts. It has to be appreciated, however, that the original shape of grains determines to some extent the nature of the contacts, irrespective of subsequent loading adjustments.

At greater depths of burial or during phases of marked tectonism, quartz grains begin to slide over each other, fracture and split at contact points, and hence adopt a closer packing. Undulose extinction may develop adjacent to grain contacts. Feldspars break even more easily, whereas micas, clay minerals and lithic fragments bend and slide acting, together,with small amounts of pore fluid, as lubricants.

Arkosic sands (15-25% feldspar) are much more compactible than quartz arenites (5% feldspar). Clay-rich and lithic sands, e.g. greywackes, may achieve up to 40% compaction. By the time that plastic deformation, extensive shearing and granulation occurs, without necessarily increasing compaction, the processes pass into the realms of metamorphism.

Cementation, Dissolution and Authigenesis

Cementation and compaction are usually synchronous processes which reduce the primary porosity in sandstones. Cementation is a two-way process in which episodes of cementation are frequently interspersed with episodes of cement dissolution, the latter increasing the secondary porosity. The distinction between cements which are chemically precipitated and bond clastic fragments together and matrix, which also bonds clastic fragments, can sometimes only be resolved using SEM techniques.

Matrix constituents should be clastic but, where clay minerals are concerned, there is ample evidence that they occur both as detrital and as authigenic grains. Sands with open pore spaces are normally lithified by the introduction of a cement, deposited from circulating solutions, or formed by the redistribution of some original constituent such as calcium carbonate or colloidal silica.

The proportion of cement required to lithify a sand is comparatively small (5-10%), and the quantitites involved could in many cases be supplied by redistributive processes acting within the sand itself, or by the solutions expelled from adjacent argillaceous or calcareous rocks during compaction. The conversion at burial temperatures of 100°C or greater of hydrous smectites and kaolinites into illites can release large amounts of water. For expelled solutions to be effective agents of cementation the water must be in large quantity, free-flowing and maintain a reasonable degree of chemical uniformity for a long period of time.

The role of water, the usual fluid expelled, needs some emphasis as there is a tendency to underplay its importance in many diagenetic studies. Whether fresh or marine, meteoric, connate or juvenile the water is an active and changing entity, which has a prominent, if not dominant, effect on the course of diagenesis. Reactions and inter-actions are constantly taking place between the water and the materials with which it comes into contact and, as a consequence, the chemistry of the water changes, albeit slowly.

There may be a desalting of the water or, conversely, a concentration of salts, or a preferential increase or decrease of some dissolved constituents. In fact, the changes may be such that the present chemistry of the interstitial waters, if determinable, may bear little relationship to their original chemistry. The pore waters in marine strata, presumably originally derived from the sea, show a marked deficiency in Mg and marked increase in Ca compared with sea water. There are differences also in the Na/K and Ca/Na ratios.

Certain of these differences may be caused by ionic filtration processes which seem to be particularly effective when interbedded *argillaceous sediments* are compacted and a proportion of the salts are retained by the clay. Other changes probably involve adsorption, ionic exchange, pH-Eh relationships and microbial activities.

Some authorities have concluded that cementation of sands can only take place above the water table basing the argument upon the assumption that active circulation of ground water (i.e. meteoric water) is essential for the operation of this process.

Diagenesis does not bring about the conversion of quartz into other minerals, but it is very common to find opal and chalcedony changing into quartz. The change often goes hand in hand with cementation, the opal and chalcedony altering progressively into authigenic quartz overgrowths marginal to the original quartz grains.

The quartz cement of sandstones is usually deposited in optical continuity with the detrital grains, but may also form fringes of minute crystals growing radially outwards from such grains or the original pore space of the rock may be filled up by an irregular mosaic of minute crystals. In the deposition of secondary quartz around original nuclei of sand grains, the crystalline continuity of the new and old quartz can be demonstrated by their simultaneous extinction when a thin section is rotated between crossed polars. This is syntaxial overgrowth.

Sandstone showing this texture may be met with in deposits of all ages, though it is most pronounced in quartz arenites and least seen in immature sandstones, such as arkoses and greywackes. Quartz cement is unimportant, for example, in arc basin sandstones. Early Permian aeolian sandstones of Scotland and the North Sea Basin, which locally consist of well-rounded quartz grains formed by wind action (the so-called millet-seed sand), show a secondary quartz growth in crystalline continuity round original grains coated with hydrated ferric oxide.

Where the cementation has been continued sufficiently long, the cemented grains fit so closely together that they mutually interfere and prevent the development of crystal faces.

The problem of the origin of secondary or authigenic quartz cements has been discussed at great length but there is little general agreement. Some suggest that the new quartz can be produced quite adequately by the pressure solution of the original clastic quartz. The clastic quartzes dissolve at their points of contact and the silica in solution is immediately precipitated around the grains in positions of lower pressure.

In contrast, others suggest that the solution of interstitial grains of quartz (less than 0·02mm in size) may create 'siliceous fluids' within the loose sand and from these fluids secondary quartz cement is precipitated. Most cementation does appear to be early diagenetic occurring when burial has not proceeded beyond a depth of a few hundred metres.

Devitrification of volcanic glass in sands almost certainly contributes additional silica for quartz cementation; likewise the leaching of silica from clay minerals distributed in the sand. An important source for marine sands is siliceous organisms such as diatoms, radiolaria and sponges. On death their chemically unstable opaline silica skeletons dissolve, enriching the pore waters in silica and this becomes immediately available for precipitation as secondary quartz.

Sandstones are often cemented by fibrous, drusy and granular calcite; ferroan calcite and aragonite are rare. The quartz grains sometimes develop secondary embayments due to reaction with the carbonate-rich pore waters, so that in thin section the calcite appears to be corroding the quartz. This process releases silica into solution.

Precipitation of calcite cement depends largely on increasing the carbonate-bicarbonate ratio in the interstitial waters. This is accomplished either by increasing temperatures by burial or by increasing pH, both of which decrease the solubility of calcite.

The proportion of calcite to detrital quartz varies considerably, as also does the arrangement of crystals in the cement. In lightly cemented sands, such as some of the Pleistocene sands of Britain, there is merely a fringe of minute calcite crystals coating the quartz grains, cementing the rock near the original points of contact, but leaving large voids elsewhere. There is every gradation between this type and its completely cemented derivative in which the pore spaces have become completely filled by granular calcite.

In a calcareous sand, consisting of quartz grains and shell-chips, the pressure due to the weight of overlying sediment is carried at the points of contact between adjacent grains. The solubility of calcium carbonate is increased at these points, and the material of the shell-chips goes into solution, to be redeposited as crystalline calcite in the open pore spaces.

Many calcareous sandstones appear to have become lithified in this way. In some cases the cement in each interspace is a single crystalline unit, the crystals in adjoining spaces being independently oriented. The most remarkable structure is found in those rocks where the cement is in conspicuously large crystals, each enclosing a considerable quantity of detrital quartz sand.

On broken surfaces of the rock, the cleavage surfaces of the cement give a characteristic lustre, and each crystal unit independently gives a more or less brilliant reflection when held in an appropriate position. This phenomenon is known as lustre mottling. The large cementing crystals have irregular, wavy or slightly interlocking outlines, but in a few cases the cement occurs in large euhedral crystals which can be readily isolated from the less perfectly cemented sand in which they are embedded. The best-known European occurrence is that in the Oligocene Fontainebleau Sands of the Paris Basin.

The calcite here is in the form of rhombohedra, embedded in comparatively loose sand. Thin sections show that the detrital quartz gains within the cement are not in contact with each other. This 'floating' texture implies that the cementation was early diagenetic and occurred prior to any significant compaction of the sand.

On this recurrent line of evidence many authorities are of the view that the bulk of calcite cements originate early at very shallow depths. At moderate depths, in contrast, there can be a dissolution of calcite by mildly acidic ground waters, creating secondary porosity in the sands. Suggestions have been made that the acidity of the waters could be caused by the thermal breakdown of kerogenous material incorporated in interbedded mudrocks.

Ferruginous cements (mainly *haematite*, *goethite* and *limonite*) are common in red-bed successions. The iron oxide is often present as an investing pellicle around each quartz grain, and where silicification has subsequently taken place the iron coating often becomes enclosed by the new crystal growth. On the other hand, the secondary quartz itself sometimes becomes coated with iron oxide, and successive periods of ferruginous and siliceous cementation may thus be distinguished.

The haematite which appears to be the main red colouring agent in the Penrith and St Bees Sandstones (Permo-Trias) of north-west England is diagenetic and formed around clastic grains in the basins of deposition. In these, as in many other instances, the sediments were not red when deposited but have acquired that colour over tens of thousands of years as a consequence of the breakdown of detrital and authigenic iron-bearing clay minerals and ferromagnesian minerals.

Clay minerals readily adsorb iron on to their lattice and in some, such as chlorite and certain smectites, iron is an integral part of their composition. Under humid circumstances the iron from all these minerals is released by hydrolysis into the capillary pore waters and, depending on the Eh being positive, is progressively oxidised and precipitated, initially as goethite which progressively alters into red haematite.

The Fe_20_3 content of the sediment rarely reaches more than 5% and can be as low as 1%. Once precipitated, the haematite persists indefinitely if the chemical environment is favourable.

The term '*red beds*' is normally used in describing thick, reddened terrestrial clastic successions, often of molasse-type. Some of the oldest occur in the Precambrian Torridonian succession of north-west Scotland, the Sparagmite Formation of Scandinavia and the Keweenawan beds of the Lake Superior region. All the geological systems from Cambrian to Tertiary are known to have major 'red-bed' occurrences. Probably the best known in the northern hemisphere are of Devonian (Old Red Sandstone facies) and PermoTriassic (New Red Sandstone facies) age.

Parts of south-west Arizona, the Sahara and central Australia are mantled by Tertiary to Recent desert sediments which progressively deepen in red hues with age. Siderite (chalybite) is probably much more often present in the cement of sandstones than is commonly realised. In some cases there can be little doubt that the mineral has replaced original calcite.

An important series of sandstones in which siderite is the sole cement has been described from the Lower Lias of Sweden. The rocks consist of fine-grained, clean sand, enclosing unaltered marine fossils, and the sideritic cement is in the form of minute yellowish crystals. On weathering, the siderite is principally converted to red-brown limonite. In the British Isles barytes has been described as a cement in red-bed Triassic sandstones in the neighbourhood of Nottingham.

The barytes is present as a microcrystalline aggregate either uniformly distributed or aggregated in streaks and patches. In places it amounts to as much as 50% of the rock, occurring in patches which preserve their optical continuity over large areas, and thus produce a kind of '*lustre mottling*'. It is probable that in most cases the original cementing material was not a sulphate, but a carbonate (witherite) deposited from water in which it was dissolved as a bicarbonate, and that the sulphate has been formed by a metasomatic interchange with soluble sulphates.

Instances are known of the formation of isolated crystals of barytes enclosing 40-60% of quartz sand. Such crystals and groups of crystals are found in the mainly continental Nubian sandstones (*Carboniferous Cretaceous*) of Egypt.

Gypseous sandstone, in which the cementing material is gypsum, is usually characteristic of cementation under arid and semi-arid conditions. Such deposits have a wide distribution on the modern sabkha areas of Middle East countries and have also been described from the Russian steppes and from Bolivia. In all these cases the gypsum tends to form crystalline aggregates, sometimes with definite crystal shapes, enclosing as much as 60% of sand. Some of the aggregates are called 'desert roses' because of their distinctive petaloid morphology.

All cements affect porosity and permeability to some extent, but it is clay mineral cements in sandstones that have attracted the greatest attention over the last few decades. Pore lining and rim cementation by clay particles reduce porosity and permeability very little, whereas pore filling reduces them considerably. Early diagenetic kandite aggregates, sometimes with a vermiform or book-like habit and up to 1 mm in size, are present in some Carboniferous and Jurassic subarkoses and sublitharenites in Britain and the North Sea Basin, where they effectively play a role in blocking the migration of pore fluids. In the same rocks late diagenetic kandites tend to be blocky in habit due to recrystallisation during deep burial.

In marine sandstones, *authigenic illite* cement commonly consists of hairy fibres or filaments lining pores between quartz grains, whereas in non-marine sandstones the illites tend to occur as plates, laths and blades intimately associated with the breakdown of clastic and authigenic muscovite, feldspar, kaolinite and smectites. Some of the latter changes in thick basinal sequences are at temperatures exceeding 300°C and at depths exceeding 10 km. Illite in all its habits has the capacity to fill pores, bridge pores and, by physical displacement, effectively block pore throats, hence altering the *pore geometry* and reducing *intergranular porosity*.

Chlorite cement, formed from the breakdown of biotite and other ferromagnesian minerals, is usually present in marine and brackish water sandstones. The grains with their platy or blocky habit commonly occur as grain coatings, less frequently as pore fills.

Smectites and poorly crystalline mixed-layer illite-smectites and illitechlorites are found typically as patchy pore linings. Their ability to absorb water and expand is another factor in modifying the original primary porosity of sand and restricting the movement of fluids within the deposit.

ACCESSORY MINERALS

In addition to such minerals as quartz, feldspar and mica, sands contain small quantities of other minerals, some of which have formed in place (authigenic) and others of which are detrital.

Glauconite

Glauconite is particularly widespread in marine sandstones of all ages and, occasionally, is present in sufficient quantities to impart a green tinge to the fresh rock. The Lower Cretaceous greensands of south-east England, which are predominantly calcarenaceous sandstones, are prime examples. At weathered outcrops the green tinge is usually obscured by brown-red staining partly caused by the breakdown of the glauconite to limonite.

Glauconite comprises a group of green minerals all of which are potassium iron silicates. Glauconite proper is an iron-rich clay mineral with a wellordered, high potassium, mica-type lattice with less than 10% expandable layers. A disordered, non-swelling, mica-type lattice and a disordered, swelling, montmorillonite-type lattice are characteristic low potassium '*glauconites*', whereas other 'glauconites' consist of an admixture of two or more clay minerals. In swelling varieties the expandable layers may be over 50%.

Greensands in modern oceans originate near the edge of the continental shelf, though the mineral glauconite may be transported after formation into both deeper and shallower water. The principal deposits are situated in warm waters (15-20°C) along the south and east coasts of Africa with an especially strong development on the Agulhas

Bank; along the south and west coasts of Australia; along the west coast of Portugal; off the Californian coast, and along the edge of the continental shelf off the east coast of North America between Cape Hatteras and the Bahamas Banks.

The tops of off-shore Californian banks, between depths of 40 and 500m, are particularly rich in glauconite with percentages rising to 20. The pore waters in these areas have a pH of 7-8, and it is probable that the glauconitisation processes occur in mildly reducing situations (Eli 0 to -200m V) in the top few centimetres of the sediment. Formation is facilitated by the presence of decaying organic matter.

Glauconite may be formed from pre-existing clay minerals, especially if those minerals have a degraded layer silicate lattice. Potassium and iron appear to be progressively adsorbed on to the lattice, which probably has a low lattice charge, with the production of glauconite. Other suggestions are that it forms by the breakdown of iron-bearing silicates such as amphiboles, pyroxenes and biotite mica.

There is now substantial evidence that much of the glauconite present in *Cretaceous greensands* of north-west Europe originated by the breakdown of basic volcaniclastic glass shards and pumice fragments. For example, fragments from *contemporaneous* basic submarine eruptions appear to have been fundamental in the formation of the Hibernian Greensand of Northern Ireland.

Both ancient and modern glauconitic deposits sometimes contain reworked phosphate peloids and pebbles which suggests very slow sedimentation rates.

Heavy Minerals

Accessory constituents are usually scattered throughout the rock, but can be concentrated for study. The majority of these minerals have specific gravities greater than that of quartz, and are commonly separated by allowing them to sink in appropriate liquids while quartz and feldspar are floated off. The resulting concentrate is commonly spoken of as a heavy mineral residue.

The hard and stable minerals of regional *metamorphism*, such as garnet, kyanite and staurolite, are common accessory minerals, as also are some of the minerals of pneumatolysis, such as tourmaline and topaz. The accessory minerals of granite, for example *zircon*, *sphene* and *monazite*, also have a wide distribution. The minerals of purely thermal metamorphism appear to be less stable as detrital grains than are those of large-scale *thermodynamic metamorphism*, and such minerals as *andalusite*, *sillimanite* and cordierite have a distinctly restricted distribution. Other common detrital minerals are rutile, apatite and ilmenite; somewhat less common, though widely distributed are corundum, anatase, brookite, chloritoid and the *spinels*, *pyroxenes* and *amphiboles*.

When they are newly derived from crystalline rocks, the grains of these minerals tend to be comparatively unworn and consist of cleavage forms, angular fragments, or more or less complete crystals. If the material becomes involved in a second cycle of sedimentation, the content of accessory minerals usually suffers some change, even though no new material is added.

Unstable minerals tend to be eliminated, or to be replaced by new alteration products. Such minerals as olivine, hypersthene, augite and brown hornblende are particularly susceptible to alteration under sedimentary conditions, and would not be expected to survive more than one cycle of sedimentation. At the other extreme, hard and stable minerals such as zircon, tourmaline and rutile are almost indestructible, and are passed on from one sediment to another through many cycles of sedimentation.

Such grains in time become smoothed and worn, but the process is very slow, and the well-rounded grains found in some sedimentary rocks indicate that the individuals in question are of very great geological age as detrital grains. The preponderance of zircon, tourmaline and rutile in the heavy residue of a sand almost certainly indicates derivation from pre-existing arenaceous rocks, especially if the grains of these minerals show appreciable signs of abrasion.

Since the accessory minerals of a sand are principally survivals from the parent crystalline rocks from which the sediment was derived, and the character of the assemblage of these minerals depends upon the history of the sediment, it is often possible to learn much of this history by a careful study of the minerals present.

In particular, the accessory minerals usually yield evidence of the source (or provenance) of the sediment, either by indicating directly some specific terrain, or by ruling out those areas which could not have supplied the assemblages found, and thus narrowing down the possibilities. By using this method of investigation it is often possible to obtain information which is of considerable service in reconstructing the geography of the areas which supplied the basin of deposition.

It is necessary to understand that suitable allowances have to be made in all deductions for sizesorting and specific gravity-sorting of the heavy minerals during their transportation. Moreover, post-depositional processes may alter the constitution of heavy mineral assemblages under certain solution and weathering conditions. It is known that the effects of intrastratal solution depend on whether the process occurs at depth, with enhanced pore fluid temperatures, or at the surface.

Certain minerals, such as apatite, sphene and garnet, persist during deep burial, whereas they are much less stable and tend to dissolve during acid surface weathering. Andalusite, kyanite and sillimanite, in contrast, are more stable near to the surface than at depth. However, the infiltration of fluid hydrocarbons inhibits and sometimes stops all these changes.

As some ancient sandstones have passed through more than one cycle of deep burial and surface weathering, the quantity and diversity of their heavy mineral suite compared with that originally deposited can be markedly affected. Nonetheless, it is held that such suites, when studied over a large region and in conjunction with the overall sedimentary data, can give a reasonable assessment of provenance.

Work on the New Red Sandstone (*Permo-Triassic*) of the west of England illustrates the use of accessory minerals in tracing the source of arenaceous sediments. The heavy mineral assemblage in these beds was found to be large and varied, but two important sets of minerals could be distinguished, one derived from a region of granitic intrusions, and the other from an area of regional metamorphism.

Blue tourmaline, *topaz*, *garnet*, *cassiterite*, *fluorite*, *rutile* and *brookite* suggest derivation from the contact-metamorphic and pneumatolytic rocks and the minor intrusions associated with the granite masses of Devon and Cornwall. The minerals of regional metamorphism, such as staurolite and kyanite, on the other hand, could not be derived from the same area, and the metamorphic rocks of Brittany provide the nearest available source. Thus we may picture sands and torrential gravels being swept into this Permo-Triassic desert basin from a massif of metamorphic rocks lying to the south, and also from the region of the granite intrusions lying to the west.

The assemblages of accessory minerals present in the sands of a persistent basin of deposition usually vary not only within each bed when traced horizontally along its outcrop, but also vertically from one formation to another. If the horizontal variations are slight when compared with vertical changes in mineral composition, the latter may be used as a basis of correlation over restricted areas. This method has been found to be of considerable service in recognising horizons in thick series of unfossilferous sands found in some oilfields, and thus providing a means of correlation between the sections in different wells. Correlation by this means is not always possible and even where it is practicable, considerable caution should be used in interpreting the results.

The causes of local variation in the heavy mineral content of a sandy deposit are several. In the first place, sediments from different sources may be brought independently into the basin of deposition, for example by rivers, by glaciers, or by coast erosion with or without littoral drift. Even in basins supplied by a single drainage system, the distribution of accessory minerals is not, as a rule, quite uniform, since some minerals are less easily transported than others, and thus tend to remain concentrated near to the source of supply.

The differences found between the mineral contents of contemporaneous sands in closely neighbouring basins are sometimes very great indeed. For example, the marine Middle Jurassic sands of the English Midlands contain garnet, staurolite, kyanite, glaucophane, chlorite, chloritoid and topaz, and have quite evidently been supplied by rivers draining a region of metamorphic rocks. Further north in Yorkshire, the Middle Jurassic sands contain an extraordinary abundance of the titanium minerals rutile, ilmenite, anatase, and locally brookite, the only other important minerals being *zircon* and *tourmaline*, often in much-rounded grains. An assemblage of this kind strongly suggests derivation from older sediments, and not from metamorphic or igneous rocks.

Vertical variations in the heavy mineral content of a series of superimposed sand formations may arise from changes in the direction of littoral drift, from the arrival of sediment supplied by a different drainage system, or from changes in the existing system, such as the exposure of deeper-seated rocks as denudation progresses.

Variations of this kind are strikingly displayed by the Upper Palaeozoic rocks of Britain. The earliest Carboniferous deltaic sandstones of eastern Scotland contain a mature assemblage of subrounded heavy minerals (zircon, topaz, rutile, tourmaline, anatase, and monazite) suggestive of derivation from pre-existing sediments. These sediments were probably Lower Palaeozoic and Precambrian in age and mantled a complex core of metamorphosed rocks which was only exposed by deep erosion in later Carboniferous times. The evidence for this latter conclusion rests with the presence of a more variable heavy mineral assemblage (garnet, magnetite, ilmenite, zircon, topaz, rutile, xenotime, monazite, tourmaline and brookite) in the later and more widespread deltaic sandstones.

There is also a wide range of metamorphic and igneous pebbles in some of the later beds. The source of the deltaic materials throughout was a land area to the north and north-east of the Midland Valley of Scotland.

Beach Concentrates and Placers

Concentrates of dark sand on beaches often consist of heavy mineral grains of high specific gravity, selectively concentrated by wave action. Their composition naturally varies from place to place. Beach concentrates or beach placers of ilmenite, rutile and zircon in Florida have been extensively worked as a plentiful source of zirconium and titanium. Monazite sands are of economic importance since this mineral is a source of the rare earths, especially thorium.

Monazite is a widely spread accessory constituent of granitic rocks, and in certain localities, especially Brazil and southern India, it has accumulated in large quantities as beach placers. Around the shores of the Pacific Ocean large deposits of 'black sand' consist chiefly of magnetite and ilmenite and other heavy and stable minerals including workable quantities of gold. They are extensively developed in Alaska, Washington State, Oregon, California and New Zealand.

Alluvial Placers

Alluvial placer deposits containing precious accessory minerals are of considerable commercial importance in many parts of the world. Alluvial sands and gravels accumulating in the valleys of rivers and in lakes, to which they have been transported by running water, at one time yielded the bulk of the world's output of gold. The gold is concentrated in the coarse gravels and among the boulders at the bottom of the placers, the most valuable accumulations being often actually on the bedrock itself.

If the latter happens to consist of steeply dipping schists or slates, the upturned edges act as natural riffles or bars to catch and retain the gold particles. The character of the gold is very variable: it occurs in flat scales and flakes, in rounded particles, and as irregularly shaped grains and nuggets bearing evidence of much attrition. Crystallised gold also occurs. Some of the richest gravels are formed by a secondary mechanical concentration or re-sorting of earlier auriferous gravels, which today are found as terraces tens of metres above the present river beds, as in California.

The platiniferous placers of the Iss and other rivers draining the eastern slopes of the Ural Mountains in Russia deserve special mention. In these deposits the platinum is associated with chromite and magnetite; and their origin has been traced to intrusive masses of peridotite.

A large proportion of the tin ore of the world is obtained from alluvial and other deposits. Cassiterite (SnO_2), the only important ore-mineral, is very hard and heavy, and highly resistant to weathering; it therefore tends to concentrate naturally in all superficial deposits derived from tin-bearing granites, in which it occurs as a primary mineral, and to an even larger extent from the associated pneumatolytic and metamorphic rocks.

Cassiterite may be concentrated in workable quantities in residual deposits left by differential weathering at the outcrops of tin-bearing rocks, but more commonly it has been transported some distance from the original source. As in the case of the gold placers, the concentrates tend to accumulate at the base of the gravel, and to rest upon the underlying floor or bedrock. The early tin output of south-west England, which goes back almost to prehistoric times, was almost entirely from alluvial deposits.

Much alluvial tin has also been obtained from Nigeria and eastern Australia. Gem-gravels and sands in which, among other precious stones, diamond is found are also of economic importance. Diamantiferous gravels in which the diamonds are sufficiently numerous for profitable extraction are not common. Those of the Vaal River in South Africa are perhaps the best known. The diamonds are found in residual and alluvial gravels derived from lavas and Carboniferous Dwyka conglomerates.

9

ARGILLACEOUS DEPOSITS

The fine-grained clastic sediments, sometimes collectively referred to as *mudrocks*, include *clays*, *shales*, *mudstones*, *marls* and *loess*. Shales are finely laminated rocks in which the individual parallel laminae represent periodic phases of slow sedimentation in a low energy environment. They are fissile, splitting easily along the planes of lamination. Mudstones are blocky, massive and non-fissile in character, with a general absence of laminae. The constituent particles seem to have been deposited at a faster rate than in shales.

Although the bulk of the particles in argillaceous rocks are less than 0.004 mm in size, it is common experience to find an ample proportion of siltsized particles. Most of the rocks contain a mixture of siliciclastic materials, silt, sand and clay minerals, hence hybrid names such as silty shale (25-45% silt particles) and sandy shales (25-45% sand particles). Many glacial clays and *aeolian dusts* consist principally of particles of quartz and feldspar; this material is known as rock flour and, except in its grain size, differs little from the material of sands. Clay is commonly used to describe mudrocks which have been subject to less compaction and cementation than shales and mudstones.

In addition to these materials chemical precipitates, carbonaceous and bituminous matter and colloids are found in varying proportions as accessory constituents. Calcareous shales (20-35% $CaCO_3$) and marls (35-65% $CaCO_3$) are mixtures of clay and calcite; the latter either being of organic origin, as in some pelagic Cretaceous-Palaeogene deposits of the *Mediterranean* and northwestern European regions, or in the form of a biochemical/chemical precipitate, as in certain windblown terrestrial Triassic deposits of the Northern Hemisphere. *Calcitic cocco-*

lithophores (algal *nannoplankton*) are now being subscribed to siliclastic muds in increasing amounts in the Dutch and German intertidal zones (waddens) and in the deep Norwegian Trough. Extensive blooms develop in the surface waters (*photic zone*) and appear to be triggered off by enhanced nutrient input, probably agricultural nitrate, into the North Sea. Consequently, the muds in some places are beginning to take on the calcareous characteristics of marls.

In deep and shallow quiet waters the acquisition of siliceous organic debris and volcanic ash often leads to partial diagenetic chertification of muds, producing siliceous shales with up to 75% silica content.

COMPOSITION

Rock Flour

The mechanically formed rock flour which is present in various proportions usually consists of quartz, feldspar, muscovite and biotite, but granules of other rock-forming minerals may be present. These minerals are reduced to a fine powder either by abrasion during transportation, or by crushing. In addition, minute crystals of stable minerals such as *zircon*, *tourmaline* and *rutile* are released from igneous rocks by the decay of the less stable minerals in which they are embedded, and ultimately find their way into the argillaceous deposits.

Clay Minerals

Only a few of the crystalline phyllosilicate clay minerals can be considered here. Of the two-layer types, with a sheet structure composed of alternate layers of silica tetrahedra and alumina octahedra, the kandite group is probably the best known. Kandites have an approximate composition of $Al_2O_3.2SiO_2.2H_2O$ in which ferrous and magnesium ions often partially substitute for aluminium ions. Individual members include kaolinite *sensu stricto* and *dickite*. *Halloysite* is a related form.

Of the three-layer types, with sheet structures composed of two layers of silica tetrahedra interleaved with alumina di- and trioctahedra, the best known are the smectite and illite (*hydromica*) groups. The smectites include montmorillonite and nontronite and are characterised by an ionic lattice capable of expansion and contraction. This physical change is brought about by the adsorption or loss of water molecules. The illite group by contrast consists of non-expanding minerals which have a complex chemical composition resembling that of muscovite and sericitic mica.

Apart from the variety illite, the group contains glauconite (in the strict sense). Closely related in composition are members of the chlorite group in which ferrous ions are prominent. The chlorites have a mixed-layer type of structure with alternate two-layer (*kandite*) and three-layer (smectite) type arrangements of ions. Many clay minerals other than the chlorites appear to be mixed-layer type and are commonly referred to as chlorite-illites, kaolinite-illites, and so on, instead of devising new specific names for each mixture.

When certain clay minerals, notably montmorillonite, are in equilibrium with the pore fluids present in a slightly calcareous sediment or soil, they absorb calcium ions, and in this condition the colloidal clay-complex is known as a calcium clay. If a solution of sodium chloride is allowed to percolate through a calcium clay, calcium ions are taken into solution, and are replaced by sodium ions, thus forming a sodium clay.

This phenomenon is known as base exchange. Reaction with acid solutions, such as rain water containing carbon dioxide, or with acid humus, tends to remove exchangeable bases, and replace them by hydrogen, the result being a hydrogen clay. Such clays are commonly developed in sour or acid soils, from which the calcium has been leached by acid rain water. Other exchangeable bases are magnesium and, in cultivated land, potassium and ammonium.

From the sedimentary point of view, the importance of base-exchange lies principally in the effects of different replaceable bases upon the permeability and degree of flocculation of clays. Hydrogen clays are highly dispersed, and are impermeable. Sodium clays, such as those deposited from salt water, are flocculated and allow a slow passage to gases and liquids. Calcium clays, which may be deposited from hard fresh water, are strongly flocculated, and are much more readily permeable than either hydrogen or sodium clays.

If any of these clays are deposited from water containing electropositive colloids, such as ferric hydrosol, or certain organic colloids, the charges on the electronegative clay particles are neutralised, and a highly flocculated precipitate is formed.

Accessory Constituents

Siderite, *limonite*, *calcite* and a range of other carbonates, such as ankerite, occur frequently in mudrocks as early diagenetic cement, concretions and nodules. The first three minerals are not co-precipitated because of their differing geochemical requirements.

Authigenic pyrite in the form of tiny cubic crystals (*framboids*) or nodules owes its origin to an in-place reduction of iron salts by organic compounds or bacteria.

Organic matter, in the form of humic colloids, or as solid animal or plant debris, is often deposited in considerable quantity with argillaceous sediments, and gives rise to various diagenetic products. Organic radicals also attach themselves loosely to the clay mineral lattices. *Fissile pyritic* black shales carry laminae comparatively rich in bituminous and carbonaceous matter, though the total percentage is relatively small. They are deposited slowly in anoxic fresh and marine waters. Laminated *lacustrine* and *marine oil-shales* are much richer in bituminous matter.

Ferric oxides, usually in a hydrated condition, are another frequent constituent of mudrocks. Some are of detrital origin, especially in regions of lateric weathering, but others are diagenetic from the alteration of detrital iron-rich minerals. Low organic productivity and rapid oxidation of any organic matter are essential conditions during these alteration (and *reddening*) processes.

DIAGENESIS

Chemical and mineralogical changes are liable to begin almost as soon as the mud is deposited. These earliest changes are often made strikingly apparent by an alteration in the colour of the sediment, which in some environments becomes perceptible very shortly after deposition.

The clay mineral components of argillaceous rocks show evidence of change starting with mild ionic substitution in lattices and sometimes ending with complete conversion, such as montmorillonite into illite. Montmorillonite forms 60% of the clay mineral content in some Gulf Coast sediments at a depth of 1800 m below the present surface.

At 4500 m depth where higher temperatures prevail the *montmorillonite* content is reduced to 20%, probably due to conversion into illite and mixed-layer minerals. However, the assumption made in these assessments is that the change is entirely diagenetic and unrelated to progressive changes in provenance. Large amounts of interlayer water may be released at intervals from the clay minerals if the changes are truly diagenetic, thus creating abnormally high pore pressures in the original and adjacent sediments. Some of these waters may be capable of mobilising oil and bitumens and transporting them into adjacent, more permeable beds.

In muds containing iron pyrites, oxidation of the mineral forms sulphuric acid which then attacks granular kaolinite and forms, in its place, vermicular kaolinite and dickite. Montmorillonite can form from halloysite, and kaolinite from feldspar, by the action of siliceous waters. Small illites sometimes recrystallise to give large hexagonal grains 2

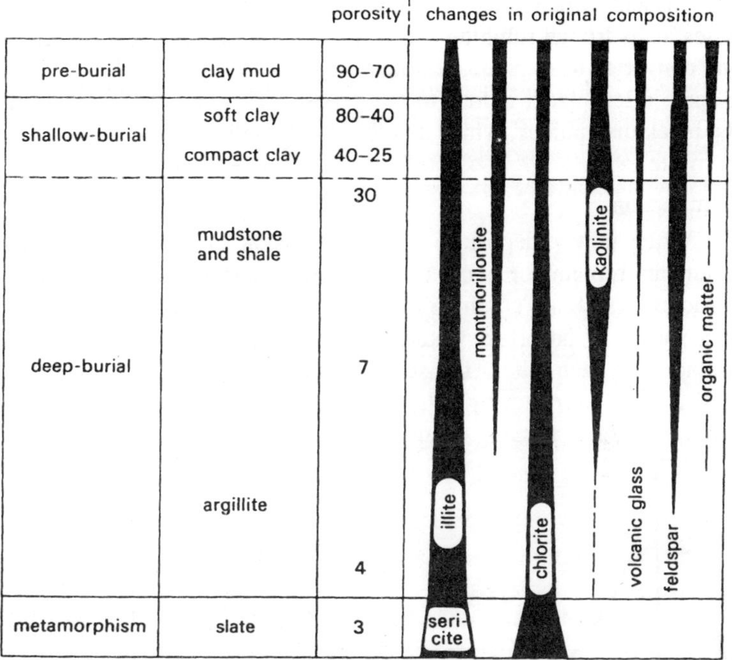

Figure 9.1 : Diagenetic changes in the common constituents of argillaceous rocks, excluding quartz. Significant changes in mineralogy occur only with deep burial or with a rise in the geothermal gradient. Clay minerals in sandstones are affected in a similar fashion.

mm across. Clay minerals almost certainly migrate in colloidal solution at early compactive stages while porosity is still relatively high. The smectite group minerals are usually most mobile and often replace the calcite in shell debris. The kandites are less mobile except in acid water, plant-rich environments.

The muds of temperate and tropical rivers, lakes, estuaries and seas usually contain an appreciable amount of organic matter. This material is the main source of hydrocarbons now present in oil and gas reservoirs. The surface layer of mud, having access to the dissolved air of the overlying water, tends to remain in an oxidised condition, and supports a population of aerobic bacteria.

The iron in this layer is usually in the form of limonite which gives a brown or yellowish colour to the sediment. At a slight depth, sometimes merely a fraction of a centimetre, circulation of oxidising solutions is sufficiently restricted to allow reducing conditions to prevail, the reducing environment being intensified by an active population of anaerobic bacteria. The sulphates are reduced, and organic substances such as proteins and carbohydrates are broken down, with the production of such gases as hydrogen sulphide and methane. Iron compounds are reduced and converted to sulphide, colloidal $FeS.H_2O$ being first formed, giving a dark, often black, colour to the deposit. This becomes converted to the black disulphides, which at a late stage of diagenesis slowly change to pyrite.

Compaction

When newly deposited from suspension in water, argillaceous sediments remain for some time in a fluid condition; the component particles, which are commonly minute flaky crystals, are each encased in a sheath of adsorbed water, and at this stage are not in physical contact with each other. The sediment in bulk contains only from 10% to 30% of solid matter by volume, and can have a porosity of 70-90%, much of the space being occupied by mechanically enmeshed water.

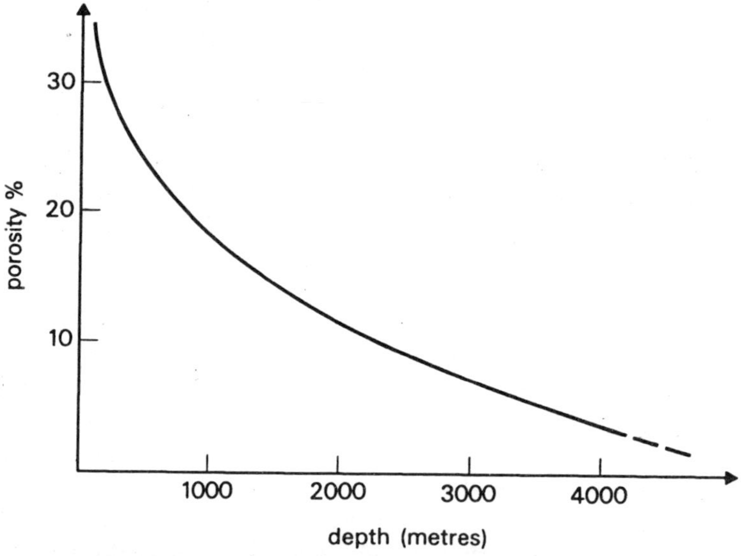

Figure 9.2 : Porosity in shales. This idealised graph illustrates the progressive decrease in porosity with increase in depth of burial for some Liassic clays of north-west Germany.

Muds containing much finely divided organic matter may have initial porosities exceeding 90%, and flocculated deposits in general have a greater initial porosity than those which settle in a dispersed condition. The clay particles begin to pack together to form a less open network, especially if more sediment continues to accumulate above. Free water is expelled as the larger interspaces become reduced, but the sediment remains fluid until the pore space has diminished to 75%

or less. Under an increasing load of strata, further expulsion of free water takes place, and the aqueous sheaths of adjacent grains are brought nearer together until they are in close contact.

The expulsion of water becomes increasingly difficult as the pore space diminishes, since the frictional resistance offered to the passage of water becomes correspondingly greater. The weight of overlying strata is now principally carried by the aqueous envelopes of the particles, but with still increasing loads the solid grains break through the adsorbed films at the points of greatest pressure, and the mineral particles themselves come into contact.

An aqueous sheath still encloses the remainder of the surface of each grain. It is believed that the porosity at this stage is usually between 30% and 35%. So far, the compaction of the sediment has been effected principally by expulsion of free water, accompanied by mechanical adjustments of the solid particles such that many of the clay minerals begin to assume a marked preferred orientation parallel to the bedding surfaces. A considerable proportion of the free water has now been removed and further consolidation involves expulsion of part of the adsorbed and interlayer water, deformation of the mineral particles themselves and precipitation of chemical cements.

Diagenetic cementation by carbonate, silica, iron and phosphate minerals is usually initiated early and can continue through several subsequent episodes of dissolution and reprecipitation.

In the late stages of compaction the formational temperatures at depth are probably more important than the loading pressures. The thermally induced movement of the interlayer water can lead to additional *overpressurisation* within the sediment but, if the residual pores are breached, then further *dewatering* occurs. The original volume of the deposits can be reduced another 10% or so.

The physical strength of clays is basically a function of decrease in pore space, this usually being a consequence of overburden pressure exerted by the sediment itself. In deep water some pelagic clays are often stronger than would be anticipated as a consequence of simple pressure and are referred to as being overconsolidated.

In all probability the overconsolidation of these sediments reflects slow rates of deposition and enhanced rates of chemical cementation. Strictly, the soil mechanics term '*overconsolidated*' should only be applied to sediment which has been subject to overburden pressures greater than those presently effective at the depositional surface. But the usage of the term has been extended to include pore space reduction effected by desiccation, cementation, unusually strong interparticle forces and other factors. For example, the upper layers of alluvial and supratidal mud during dry climatic phases become much more rigid than the immediately underlying sediment and can be said to be overconsolidated.

That is, they have hardened at a faster rate than would have happened if the sediment had been subject to 'normal' compactive processes. Subsequent deposition can then lead to the incorporation of these overconsolidated layers into the pile of softer sediments. These layers are best seen in recent partially lithified sediments; they are very difficult to detect in a fully lithified succession.

On the other hand, induration of this kind is often reversible, and the sediment breaks down again if allowed to remain sufficiently long in contact with an excess of water. Rapidly deposited shallow water clays may be less strong at depth than anticipated and may be termed underconsolidated. In these cases the overburden stress is largely borne by the overpressured pore water rather than the solid particles.

The marine clays being deposited adjacent to the actively prograding Balize delta lobe of the Mississippi River are generally underconsolidated. During phases of very fast deposition and loading these clays often distort and flow. Fracturing also occurs though the planes of fracture tend to reseal without leaving slickensided surfaces.

Rheotropy, Thixotropy and Dilatancy

The role played by pore water in the flow and consolidation of mud has attracted much attention though its effects in many instances are often difficult to confirm. The science of the flow of materials is called rheotropy and the study of natural muds is simply one small field of endeavour. None the less it is known from laboratory evidence that if a firm, but unconsolidated, mud is subject to vibration it will either become partially mobile or completely mobile.

These two states are referred to as false body and sol respectively and seem to be created by changes in the packing of the grains. This results in pore water being released and immediately makes the mud more mobile. In the true sol state the mud flows under the slightest shearing force. When the vibration ceases the liquified mud once more becomes firm and readsorbs most of the water released earlier; this is called the gel state.

In this state, flow will only occur when vibration or a shearing force is reapplied and reaches a minimum critical value. This value is called the yield point and is naturally dependent on factors such as the mineralogy of the mud and its grain packing. Smectite-rich muds (bentonites) have a low yield point, rapidly becoming fluid when vibrated and equally rapidly returning to a gel or firm state when the vibration ceases. Illite-rich muds flow less easily and have a higher yield point.

Rheotropic changes can also be accelerated in the presence of suitable electrolytes and it is known that muds, in general, tend to have a lower yield point in marine waters than in fresh. Clays raised above sea-level and leached of sea salts are known to increase in yield value and decrease in plasticity by as much as 60%.

Sediments which have low yield points are usually called highly thixotropic. Most clays are of this type. The fact that many thick clay horizons (*Mesozoic Weald Clay*, Gault, Fuller's Earths of southern Britain) show poor or little bedding may be one expression of rheotropic behaviour, the original bedding planes being destroyed by *penecontemporaneous mobilisation*.

Whereas highly thixotropic muds increase in mobility with increase in shear there are others which decrease in mobility with increase in shear. The latter in are said to have dilatant characteristics. Some soft silty and sandy muds, when compressed, vibrated or simply walked over, become relatively firm but rapidly return to a soft state when the stress has been removed.

Dilatancy, as applied to muds, has not been recognised as a significant process in fossil sedimentation. This is probably due to lack of suitably preserved criteria in the rocks.

CLAY MINERALS AND ENVIRONMENT

Although clays, mudstones, shales and marls consist of a variable admixture of clay minerals and other clastic particles, it is appropriate to consider at this point the distribution of clay minerals in sediments.

The majority of clay minerals which eventually find their way into the complex of water-covered environments are eroded from a mixed terrain of exposed rocks. In these areas of sub-aerial weathering and soil formation the physical and chemical environments are probably as important as the composition of the parent rocks in determining the type of clay mineral produced. The formation of kandite group minerals in soils is favoured if rainfall exceeds evaporation and leaching is intense. Strong leaching removes Ca, Mg, Na and K ions if they should be present in the parent rocks and stabilizes any iron by converting it into oxide or sulphide form. Excessive silica ions also need to be removed so as to maintain a high Al: Si ratio in soils. Under these conditions the kandites will form irrespective of whether the parent rocks are granites, gabbros or volcanic ashes.

The smectites are generally formed under conditions which are the reverse of those needed for kandite production. Evaporation should exceed precipitation and leaching processes should be negligible and alkaline conditions prevail so that a low Al: Si ratio is maintained.

Non-acid, potassium-rich conditions favour illite creation. Rainfall and consequent leaching should be only moderate and intermittent. A lack of sufficient potassium in the soils usually means that a degraded variety of potash-deficient illite forms but conversion to normal illite is soon accomplished when the illite is transported into an alkaline marine basin.

Some petrologists consider that clay minerals alter little, if at all, during fluvial transport but become very susceptible to change once they reach open bodies of water such as large lakes, seas and oceans. It has been suggested that large changes in composition occur in river-transported clay minerals deposited in saline waters, with the preferential formation of smectites, illites and chlorites and loss of kandites. If the clay minerals are not buried quickly and the ionic composition of the water is appropriate then the changes may occur relatively rapidly.

Figure elsewhere in this chapter illustrates changes in clay mineralogy detected in bottom samples over a distance of 160 km extending from the Guadalupe River in Texas via Aransas Bay into the Gulf of Mexico. The increase in salinity and chlorinity of the waters is believed to be reflected by the increasing conversion of montmorillonite into chlorite.

In contrast, the clay minerals laid down at the mouths of the Mississippi delta do not show any significant differences from those being carried downstream in the upper reaches. Only along the outermost seaward edge of the delta is there any suggestion that conversion of river-borne montmorillonite to illite is occurring. Even here, the evidence is indirect and is simply based on the fact that illite is more common in these muds than in those nearer to the delta proper. The variations may be accounted for by the reworking of relict clay materials, variations in source or the differential distribution of incoming clastic clay grains.

In the Gulf of Paria off the north coast of South America the distribution of clay minerals shows the influence of flocculation processes. These processes are undoubtedly very important in clay deposition. Upon contact with marine water, clay particles suspended in fresh water tend to flocculate and form larger composite particles. In the Gulf, illite and kaolinite appear to be flocculating quickly, usually as relatively coarse particles which are then rapidly deposited. *Montmor-*

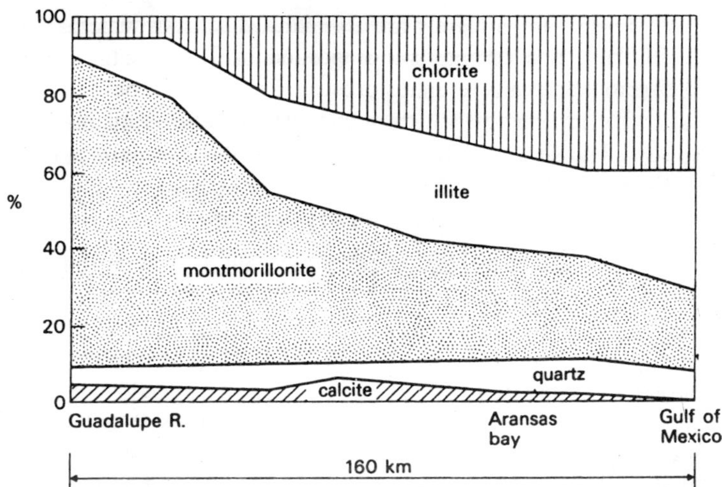

Figure 9.3: Clay mineral distribution in modern sediments near Rockport, south Texas. The diagram shows the mineral distribution in the <1 μm size fraction of the sediment being transported and deposited in the lower reaches of the river and adjacent sea.

illonite is flocculating more slowly as small particles which take longer to be deposited. Consequently, there is a progressive sorting of the clay particles such that the illite- and kaolinite-rich zones correspond with the fresh and brackish water marginal deltaic areas and the montmorillonite-rich zones with the more open marine areas. Similar effects probably account for the changes during the transition from nonmarine (kaolinite) to marine (illite/mixed-layer) shales in Namurian strata of northern England.

Changes in the climate of a source area may be ultimately reflected in the clastic clay mineral content of derived sediments, though the changes in the sediments are likely to be gradual and spread over at least a moderate thickness of strata. On the occasions when a sharp change in clay mineral composition is detected at a precise level within a sequence other considerations arise. For instance, in conformable beds the marked change may represent a hiatus of considerable duration, during which a significant change in the nature of the provenance or a total change of provenance occurred.

Although clastic clay minerals commonly reflect the nature of their provenance, it is recognised that they are susceptible to alteration in varying degrees during burial diagenesis in deep basins. In some circumstances this can mean that the original source for the clay cannot be traced.

Authigenic clay minerals are also formed in environments of clay accumulation, thus diluting the proportion of detrital clay grains. In some Triassic playa-type mudstones of England, detrital illites and chlorites are intermingled with authigenic sepiolite, palygorskite and corrensite. The smectites in marine bentonitic clays are also *predominantly authigenic*.

MARINE DEPOSITS

The influence of marine processes on mud deposition is felt from immediately adjacent to coastlines, outwards across continental shelves, and into the deeper oceanic basins. The distribution of muds follows no regular pattern, being controlled by coastal physiography, bottom topography, the pattern of tidal currents and the location of the clay sources. Off the mouths of major deltas, such as the Mississippi and Orinoco, the clays laid down in shallow marine waters are little different in lithology from those deposited well away from any deltaic influence.

During river flood conditions, plumes of suspended sediment can be transported far off shore, some eventually finding its way into deep oceanic basins. Loose mud thus deposited on the bottom slopes or perched on the edge of more pronounced topographic features is susceptible to mass displacement as turbidity currents. Fine-grained mud turbidites are centimetre-thin units with graded silt passing upwards into rippled, convoluted and mildly bioturbated mud laminae.

Other muds are redistributed by bottom currents moving nearly parallel to the slopes and are redeposited in thin layers, with a mild degree of internal lamination, pockets of coarser silt and sharp contacts. Bioturbation can be intense. These deposits are muddy contourites and are distinguished from hemi-pelagic muds (hemipelagites) on the basis that the latter are more homogeneous and thicker bedded, indicative of very little current activity. Blue or slate-grey mud of terrestrial origin is widespread in modern seas and oceans. Green muds, such as those in the basins off shore from California, seem to be much more common than originally realised.

The green colour is caused by green illite and montmorillonite; chlorite and fine-grained glauconite are very subordinate in amount. A high organic matter content (about 7%) is a typical feature of green basinal muds. Foraminiferal tests, which show little evidence of solution, often form an important proportion of the deposits. Green shales and mudstones are common in the Lower Palaeozoic rocks of Wales. Marine argillaceous sediments derived from lateritic land areas tend to contain conspicuous quantities of red ferric oxide. The red muds of the western

Atlantic are believed to owe their ferric oxide pigment to lateritic material transported by the Amazon and Orinoco Rivers.

Black Muds and Shales

In a few parts of modern seas, black muds containing considerable quantities of undecomposed organic matter, and large concentrations of carbon dioxide together with varying amounts of hydrogen sulphide, are developing where there is a strong oxygen deficiency in the water immediately overlying the sea floor. Depth of water, in itself, has little or no controlling influence upon the formation of such sediments, the principal requirement being lack of circulation.

Thus black muds may be deposited at almost any depth, varying from the shallowest pools down to profound depressions such as that of the Gulf of California, Black Sea Basin and deeper Norwegian fjords, provided, that the water in contact with the sea floor is foul and stagnant. The oxygen deficiency in such an environment inhibits benthonic animal life, and any remains which may be found are those of planktonic or nektonic creatures, whose dead bodies have sunk from the better-aerated surface waters.

The rarity of benthonic life in such an evironment appears to be primarily due to this lack of an adequate supply of oxygen, rather than to the toxic effects of such substances as hydrogen sulphide or carbon dioxide. Examples of ancient black shales which were probably formed under similar conditions are abundant. The black graptolitic shales of the Lower Palaeozoic are pigmented partly by finely disseminated iron sulphide, and partly by carbon, which amounts to as much as 6% in some beds.

Much smaller quantities will suffice to give a sooty black colour to the rock, and many analyses of black shales show about 0·5% of carbon, and between 1% and 2% of iron sulphide. The absence or rarity of any remains of bottomliving organisms in these deposits, while planktonic forms are often abundantly preserved, suggests that whereas the surface waters were wholesome, the layer in contact with the sea bed was unfavourable to animal life. Black or dark coloured shales of basinal facies are found at various horizons in the Palaeozoic and Mesozoic.

In north-west Europe the goniatite shales of the Devonian and Carboniferous, and some bituminous shales of the Jurassic, present a remarkable uniformity both in their lithology and in the character of their fauna. The rocks often contain pyritic calcareous concretions and frequently are slightly oily. Their tendency to weather into thin laminae gives rise to the term '*paper shales*'. Bottom-living forms are characteristically sparse and the benthic fauna principally epifaunal. There is little or no bioturbation. The dark marine Kimmeridge Clay (Upper Jurassic) of north-west Europe is so highly bituminous at some levels that it warrants the name oil-shale. The organic carbon ranges up to 35%.

Five main oil-shale bands are recognised in some areas and within each there are many partial or complete mudstone-bituminous shale-oil-shale cycles. The organic constitution of the clay is of some importance as it may well be one of the main sources of the oil present in the North Sea Basin successions. Where black shales commonly differ from other shales is in containing higher concentrations of minor elements such as uranium, arsenic, copper, molybdenum, lead, vanadium and zinc.

Clearly the geochemical circumstances in these specialised environments favours the precipitation of these elements. Adsorption by organic complexes and clay minerals is probably an important retention process. Syngenetic metallic sulphides, sometimes in economic quantities, are also concentrated in the shales.

The Permian Kupferschiefer-Marl Slate extending from Poland to northern England has yielded locally up to 3·6% copper and 1% zinc from sulphides. Up to 10% copper, mainly as sulphide, has been recorded in the two black shale horizons of the late Keeweenawan (Precambrian) Nonesuch Shale of Michigan.

Brown (Red) Clay

By far the most widely spread and most characteristic of all the abyssal deposits is that to which the name of brown clay is applied. It is formed to some extent by the disintegration of rocks and minerals *in situ,* but to a greater extent by fine terrestrial detritus being transported from shelf areas, partly via the nepheloid layer, and redistributed by deep-sea currents.

The older name, red clay, is now considered obsolete because the deposit typically has a chocolate-brown colour. It occurs in all the deepest parts of the oceans, and in the Pacific it is estimated to cover an area of nearly 70000000 sq. km; it is also abundant in the Indian Ocean and in the deeper parts of the Atlantic, and, in fact, it is found in almost all regions where the depth of the water exceeds 4.5 km.

It is plastic when wet, but dries to a hard mass, showing many of the characteristics of clay. Its carbonate of lime content is low, not exceeding 4%, reflecting deposition below the calcite compensation depth (CCD). Siliceous organisms are often abundant, and brown clay grades into radiolarian ooze. Sponge-spicules are generally present, and sometimes diatoms. The mineral components of the brown clay show a good deal of variation according to the locality: in the South Pacific the most abundant constituents are minute fragments of basic volcanic rocks; in the North Pacific pumice prevails; while in the South Atlantic quartz is dominant, with some feldspar and clay minerals.

These particles are partly derived from the decomposition and decay of fragments ejected by volcanic eruptions, some of which may have been submarine. The fact must be strongly emphasized that brown clay has no definite composition, but varies according to the source from which the material is derived; it is believed that the eruption of Krakatoa in 1883 strongly influenced the composition of the abyssal brown clay deposits of the Indian Ocean.

In some parts of the Atlantic are found grains of sand which must have been carried by the Harmattan winds from the deserts of Africa. Another substance common in samples of brown clay is phillipsite, which is often very prominent and together with related zeolites can form up to 50% of the sediment.

Diagenetic rounded and botryoidal ferromanganese nodules and encrustations are abundantly formed just at and below the surface of brown clays, although they also occur in association with other substrates, such as carbonate and siliceous oozes and terrigene deposits. They are a mixture of MnO_2 and goethite with variable trace amounts of diagenetic Ni, Cu, Zn, Co and Pb. The nodules are scattered over the floor of oceans in deep basins (4-6 km deep) and on sea mounts and plateaux (170 m-4 km deep), and it has been estimated that at least 10% of the Pacific area is covered by them.

The nodules vary in size from pellets a few microns in diameter to slabs several metres wide. Laminar internal structure is usual which may be connected with the activities of encrusting arenaceous foraminifera and fungi around the periphery of the growing nodules. The nodules are often associated with vast numbers of sharks' teeth and the ear-bones of whales.

The abundance of these organic remains indicates that at the greatest depths brown clay forms with extraordinary slowness, since remains of extinct species are often dredged up along with those of existing forms, showing that since Tertiary times the thickness deposited has been insufficient to bury them completely. Apart from these larger constituents, the brown clay is summed up as a decomposition product of aluminous silicates derived partly from the rock fragments spread over the oceans by various subaerial, submarine and volcanic agencies, and partly from 'weathering' in place of the rocks forming the sea floor: from one point of view it can be regarded as a residual deposit from which most of the carbonate of lime has been removed by solution below the calcite compensation depth (CCD).

Marine Marls and Marlstones

These are impure calcareous deposits, such as the pelagic Cretaceous Chalk Marl of southeastern England, which owe their grey nature to an abundance of fragmented calcareous coccoliths and foraminifera admixed with clastic and authigenic clay minerals, dolomite, clastic quartz and authigenic pyrite. Technically, marls should contain between 35% and 65% of carbonate minerals but, in common usage, the percentage is often much greater than that.

To confuse further, some Mesozoic bedded iron-formations are known as marlstones, for example the Marlstone (Middle Lias) of central England. These iron-rich sediments are not true maristones but calcareous sideritic chamosite grainstones and mudstones. Younger marls are friable, porous soft rocks, whereas older equivalents are more indurated and hence are marlstones. Mesozoic and Tertiary pelagic marls are widespread in the Mediterranean-Alpine region where it is not unusual to find them cyclically interbedded with more tightly cemented chalks.

The Ammonitico Rosso (Jurassic) of the region is a peculiar variety of red nodular marly limestone in which lime-rich diagenetic nodules are set in a red marl matrix of clay minerals, iron oxides and organic siliceous fragments. The sediment appears to have accumulated slowly and quietly at depths of 200 m to 3 km.

FRESHWATER AND TERRESTRIAL DEPOSITS

Freshwater Marls and Marlstones

Freshwater marls and marlstones include every gradation between calcareous clays and argillaceous limestones. The degree of lithification is very variable and not necessarily a function of age, so the point at which it is preferable to use marlstone instead of marl is quite subjective. Many of the red Permo-Triassic examples in the Northern Hemisphere are still comparatively soft and easily weathered so the name marl is appropriate, whereas the more indurated Old Red Sandstone (Devonian) equivalents are marlstones.

The beds are frequently massive and lenticular and appear to have been deposited rapidly on alluvial plains and in shallow lakes. The origin of the calcareous matter is not always clear. In some red-bed successions there is evidence for vadose diagenesis of calcite from ground waters, the calcium being derived in part from the breakdown of clastic calcium-bearing silicates within the original fine sediment.

Marls are sometimes formed in lakes by the activities of plants, especially calcareous algae. The modern green alga *Chara* deposits low-magnesium calcite within its tissues, and thus builds up a skeleton which is almost wholly calcareous. In most lake marls, the calcium carbonate formed in this way is mixed with argillaceous sediment, other plant debris and calcareous shells. Freshwater marls containing the remains of charophytes, together with *Limnaea, Viviparus* and *Planorbis,* are to be found in the Bembridge and Headon Beds (Tertiary) of southern England.

Clays of Glacial Lakes

Clays of glacial lakes consist principally of material removed mechanically from the parent rock and crushed to a fine powder or rock flour. In Sweden, where considerable glacial erosion of crystalline rocks has taken place in Pleistocene and recent times, rock flour is the principal constituent of many lacustrine deposits.

Argillaceous deposits laid down from suspension in the cold waters of glacial lakes commonly show well-marked lamination, as in the varve clays. Each double layer or varve in these deposits is believed to represent the sediment accumulated during one year, and observation on the rate of deposition in glacial lakes supports this belief. Sediment is carried into the lake only during the warm months when melting is in progress.

The coarse grains sink with comparative rapidity, but the finer material sinks extremely slowly, partly because no flocculation takes place in pure glacial water, and partly because the viscosity of water is much increased near freezing point. It is believed that sedimentation of fine clastic particles continues throughout the winter beneath the surface ice of lakes which do not freeze solid. The coarse (usually silty) deposits of the summer layer thus merge upwards into the extremely fine-grained winter layer, to be succeeded abruptly by the coarser material of the next summer layer.

Clays and Muds of Non-Glacial Lakes

Clays and muds of non-glacial lakes usually contain a high proportion of clay minerals, together with organic matter. Carbonaceous remains of plants are often abundant in the muds deposited in the marginal parts of lakes, while in the deeper central parts remains of planktonic algae may form an appreciable part of the sediments. Algae were probably the mother substance of the kerogenous constituents of many finely laminated, lacustrine oil-shales, such as those in the Green River Formation (Eocene) of the United States of America and in the Oil-Shale Group (Carboniferous) of Scotland.

The muds of some lakes are highly siliceous owing to the abundance of diatomaceous material. Such diatomaceous deposits are not uncommon in the postglacial lakes of Britain, especially where the supply of terrestrial debris was small.

Fine laminations are characteristic of quieter, often deeper water areas in lakes. The laminae frequently represent annual or seasonal deposition, but may also represent spasmodic and rapid, non-seasonal influxes of sediment, possibly by small-scale turbidity currents.

While most clays probably contain a mixture of clay minerals, deposits consisting principally of kaolinite are known, as for example in the ballclays of southwestern England. The Oligocene lake deposits of Bovey Tracey and Petrockstowe consist of the debris of kaolinised country rocks and plant debris, washed down by streams draining the highlands of Dartmoor. The quartz was deposited in extensive sandy deltas at the margins of the lake, the finer clay and plants being drifted further out to form beds of white pottery clay and lignite.

Lacustrine clays of bauxitic composition (i.e. containing appreciable quantitites of hydrated aluminium oxides) have been described from the Carboniferous rocks of Ayrshire in Scotland, where they appear to have been formed by the redeposition of bauxite eroded from lateritised basalt.

Loess

Fine-grained dust-like aeolian deposits of purely detrital origin are usually grouped together under the name of loess, and are extensively developed in China, North America and northern Europe. In constitution they differ strongly from most water-transported muds and clays, since they consist of chips of rock-forming minerals, mainly quartz, and the true clay minerals are either absent or present in subordinate quantities only.

The particles are small (20-50 μm) and sharply angular and probably the product of mechanical fracture in the source area, either by the grinding action of ice or by intermittently active soil-producing processes. Much of the material in the vast Pleistocene deposits of the Northern Hemisphere, often distributed regionally in ribbon-like zones, appears to have been picked up from the outwash of glaciers and ice sheets. On the other hand, there is considerable evidence that the finest detritus formed by wind abrasion in large deserts, such as those of central Asia, North Africa and western America, is blown out of the desert areas and accumulates in vegetated regions, and especially in grassland such as the steppes or the prairies, to form a deposit sometimes called warm loess.

The loess of China and of Europe is a yellowish fine calcareous silt or clay which is entirely unstratified and very uniform in texture. It is quite soft, and crumbles between the fingers. It resists denudation, however, in an extraordinary manner, probably on account of its homogeneity, and often stands up as vertical walls tens of metres high. This property is probably assisted by the presence of numerous fine tubes arranged vertically and lined with calcium carbonate; they are supposed to have been formed in the first place by fibrous rootlets.

The loess of central Europe is, for the most part aeolian, formed by redistribution of fine glacial mud originally laid down by water, and, after drying, carried by the wind often to considerable heights.

Loessic deposits mantling land surfaces are susceptible to reworking by processes such as solifluction and rain downwash, and can be partly leached of their calcareous components. In several areas in northern France, Belgium and southeastern England the product of these changes has been of commercial value over many centuries as a variety of brickearth. Adobe is a calcareous deposit similar to loess and is largely developed in the Mississippi valley and in semi-arid basins of the western United States.

10

LIMESTONES

The minerals which go to form the calcareous rocks are few in number, and the great variation in appearance and properties of different limestones arises principally from the almost endless variety of organic and other structures into which the crystals of these minerals may be aggregated. Calcite is the stable form of calcium carbonate at ordinary temperatures, and may be regarded as the principal mineral of limestones. Aragonite is the form which calcium carbonate normally adopts when precipitated from sea water.

Conditions favouring its precipitation in preference to calcite are warm water, high alkalinity, supersaturation and an abundance of sulphate ions in solution. These conditions are usually met in the warm current areas of seas and oceans. Water rich in sulphate ions and trapped in sediments may help to preserve aragonite much longer than normal. Aragonite is unstable (ormetastable) and under normal temperature and pressure conditions is liable to be converted into the more stable calcite.

The older limestones normally contain no aragonite, and any shells which originally consisted of this mineral are found to be represented by open moulds or by coarsely crystalline calcite. In most phyla, the shells consist either of calcite or of aragonite and, except in certain mollusca, the two minerals are not found together in the same shell.

Even in the complicated structures of molluscan shells, calcite and aragonite each build separate layers, and the crystals of the two minerals are never indiscriminately mixed. Magnesium only occurs to a very limited extent (less than 1%) in skeletons consisting purely of aragonite, while it is much more compatible with the calcite lattice. In magnesian calcite shells the lowest percentages of magnesium are found in species inhabiting cold water. At water temperatures near to freezing point the $MgCO_3$ content in a range of common invertebrate skeletons varies between 0% and 6%, whereas at water temperatures of 15°C and higher the $MgCO_3$ content is often greater than 11% and occasionally as much as 40%. Calcite with less than 5% $MgCO_3$ content is called lowmagnesium calcite; with more than 5%, high-magnesium calcite. Ferrous iron is also capable of ionic substitution in calcite lattices, giving rise to a group of minerals known as ferroan calcites. Dolomite grains, $CaMg(CO_3$ carry over 40% $MgCO_3$ in contrast to the 40% maximum in high-magnesium calcite grains.

However, the total percentage of $MgCo_3$ in any given limestone need not necessarily reach a high figure before dolomite appears. Within Palaeozoic and Mesozoic rocks dolomite rhombs occur when the $MgCO_3$ reaches 2-3%. There is a complete isomorphism between dolomite and ferro-dolomite, $CaFe(C0_3)_2$, with ionic substitution of magnesium ions by ferrous ions in all proportions. The ionic radii of magnesium ions and ferrous ions are very similar and this eases substitution. Ferro-dolomite is rare in limestones, but there are frequent occurrences of stable carbonates intermediate in composition between dolomite and ferro-dolomite. These are generally called ferroan dolomites or ankerites and have theoretical formulae which are expressed either as $Ca_2MgFe(CO_3)_4$ or $Ca_3(Mg_2Fe)\ (CO_3)_6$.

In general, ferroan calcites and ferroan dolomites are present as cements in limestones and tend to be formed late in any sequence of cementation. The similarity in optical properties of all these carbonate minerals makes them difficult to identify microscopically unless the thin sections have been artifically stained using alizarin red S, potassium ferricyanide or titan yellow. Even then care is needed in interpretation because the quality of stain absorption depends on grain size and crystallographic orientation.

The ARS/PF method, in common use, is such that non-ferroan calcite is stained pink, ferroan calcite is stained mauve, purple and royal blue, ferroan dolomite is stained sky blue to greenish-blue and dolomite remains totally unstained. Staining also has the advantage of emphasizing textural features of lime deposits and assisting in determining the evolution (paragenesis) of carbonate cements.

Cathodoluminescence microscopy is another valuable tool in paragenetic work, visibly emphasizing, as do staining techniques, the chemical differences between successive phases of cementation. The accessory constituents of lime deposits include a great variety of substances including detrital, diagenetic and authigenic phosphates, clay minerals, glauconite, quartz, albite and pyrites. Non-calcareous organic remains are common, important amongst these being siliceous sponge spicules and tests of radiolaria. Some of these organisms may be dissolved subsequently and reprecipitated as chert.

Pure limestones are white or very pale grey in colour, but even small traces of impurities will act as strong pigments, and may impart striking colours to the rocks. The most usual pigments are iron or manganese compounds, and finely divided carbonaceous matter.

CLASSIFICATION

Grain size is a useful basis for subdividing clastic limestones, as in Figure elsewhere in this chapter. The system adopted is based on the Wentworth scale for fragmental deposits with suitable modifications. A disadvantage of the scheme is that it is uninformative about the nature of the constituents, although appropriate descriptive names can be added, such as oolitic calcarenite and algal calcisiltite.

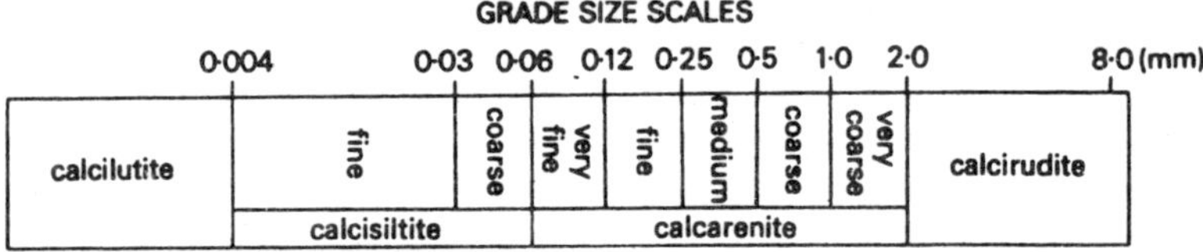

Figure 10.1: Clastic limestone classification, based on particle size.

The classifications of Folk and Dunham put more emphasis on constituents and texture than on grain size. They are more explicitly descriptive than some schemes and have the added advantage of indicating textural elements from which depositional and diagenetic processes can be deduced. A point worth emphasizing with these classifications is that the name adopted applies to the sediment as seen now and this appearance commonly post-dates a whole host of early and late diagenetic changes, which may have considerably modified the original deposit.

The basis of the Folk grouping is that most lime deposits consist of three identifiable components-allochems, micrite and sparite.

Allochem is a collective word for all varieties of discrete carbonate aggregate most of which have undergone transport at some stage in their history. The most important allochems are intraclasts, pellets (peloids, pelletoids), ooliths or ooids and shells (skeletal debris). Intraclasts are limestone or dolomite fragments of all sizes formed by *penecontemporaneous erosion* of either the adjacent sea floor or exposed carbonate mud flats, but do not include fragments (extraclasts) derived from older lithified limestone outcrops. They range from sand-size grains to pebbles and boulders. Some fragments become rounded, then subject to peripheral concentric algal growth.

They are called oncoliths or oncoids. Other small rounded grains, less than 0.15mm in size, may consist of homogeneously fine-grained limestone and are difficult to differentiate from pellets, which are micritic and have an entirely different origin. Pellets are ovoid to spherical grains, usually between 0.04 and 0.08 mm in size, which have no ordered internal structure unlike ooliths and most intraclasts. They are commonly of faecal or algal origin but, if the origin is indeterminate, the non-genetic term peloid is preferable.

Because of identification difficulties, some classifications based on Folk's scheme do not distinguish between pellets and intraclasts. Micrite is the name applied to inorganic and/or biochemical grains of aragonite and calcite less than 4μm in size though, because of this fact, individual grains cannot be studied with the normal petrological microscope. En masse they tend to have a uniform pale brown-grey appearance under polarized light. These grains are normally the dominant constituents in rocks referred to as *lithographic limestones*, *chinastones*, *calcite mudstones*, *cementstones* and, somewhat confusingly, micrites.

Calcimicrite is a better term for the last. However, in some of these rocks micrite can be markedly subordinate to silt-size particles. These silt particles are either detrital or possibly formed by recrystallization of micrite mud. The latter process produces grains sometimes referred to as microspar. The Devonian Griotte and Cephalopodenkalk pelagic limestones of southern France and Germany have been intensively converted into microsparite rocks.

To complicate matters even further there is a diagenetic process affecting lime deposits known as 'grain diminution' in which carbonate particles of all sizes are altered early diagenetically into new or neomorphic micrite. On the basis that detrital micrite matrix can be distinguished from micrite of diagenetic origins it is then assumed that the proportion of detrital micrite present reflects the hydraulic energy prevailing during deposition.

Detrital micrite-rich rocks imply low energies and low turbulence, whereas detrital micrite-deficient rocks indicate the converse. Sparite (spar) consists of chemically precipitated, coarse-grained aragonite and calcite grains more than 10 μm in size. The grains are a cement filling preexisting cavities and pores and if these are large the grains can reach a size well over a millimetre. When the sparite consists of radially arranged equant, bladed, dogtooth, acicular or fibrous crystals growing inwards from pore edges, it is described as drusy. If the drusy lining maintains an even thickness then the term isopachous cement is also used. In certain situations, commonly subaerial, the drusy cement precipitated from incompletely water-filled pores can even take on a dripstone or microstalactitic habit.

Sparite frequently infills shell cavities; sometimes the whole cavity, sometimes only the upper part, the lower being occupied by fine-grained micritic detritus. Sparite is also common in limestones formed dominantly of ooids or well-sorted, well-rounded shell debris. There is a tendency for the individual calcite crystals to increase in size and decrease in number away from the allochems they are cementing towards the centre of the original pores and cavities.

Sparite has to be carefully distinguished from recrystallized calcite, the grains of which may also be larger than 10μm. Recrystallized calcite is a secondary replacement of earlier carbonates and consequently tends to transect the boundaries of pre-existing textures and structures. In the absence of such obvious features identification becomes very difficult and dependent on a range of criteria, such as wavy grain boundaries, floating relics and variably patchy grain size, not all of which may be present or detectible.

Where replacement is extensive only 'ghost' traces of the original textures may remain as evidence of alteration. The grain size within recrystallized calcite patches is very variable, possibly as a consequence of pressure-solution effects. Differential pressures along the boundaries of original calcite grains give rise to solution at points of strain and reprecipitation at points of least pressure. In this way the original grains change shape and selectively enlarge until a patchy secondary mosaic of irregular shaped grains is created. One product of this type of alteration is pseudo-breccia. Lower Carboniferous pseudo-breccias from northern England and Wales exhibit '*fragments*' which are dark grey in colour and set in a pale grey micrite matrix.

Thin-section inspection shows that the matrix contains fossils whereas the 'fragments' consist predominantly of irregularly shaped patches of coarse calcite. These patches, in fact, are simply areas of more advanced recrystallization. Occasionally, crinoid plates '*floating*' within these calcite patches have enlarged by marginal growth. Under these circumstances, the growth of new calcite on the crinoid margins may be in structural and optical continuity-an effect known as syntaxial or rim overgrowth. The cementation ensuing from overgrowth is sometimes called 'rim cementation'. Multiple consecutive phases of rim cementation have been observed affecting shell fragments in certain limestones.

Figure elsewhere in this chapter illustrates how the three main components of limestones plus the variation in the nature of the allochems is used for subdivision. The rocks are named by combining, in abbreviated form, their major petrographic attributes. The first part of the hybrid name refers to the allochem component and the second part to the cementing or matrix materials, whichever is the dominant. For example, intrasparite consists of intraclasts cemented by sparry calcite. Biomicrudite is a further style of modification in which the grain size of the dominant allochem is taken into account. In this particular rock the average size of the shell fragments set in a micrite matrix is greater than 2 mm.

Dunham's classification is essentially textural and is most valuable when used in a purely descriptive way for lithified rocks. Textural maturity is implied in that the least mature varieties are richer in mud matrix than the more mature grainstone varieties. However, depositional deductions based on these textures alone need great care.

Table 10.1: Classification of Common Carbonate Rocks according to Depositional Texture.

Depositional texture recognisable					Not recognisable
Contains mud			Lacks mud and is grain-supported	Original components bound together	CRYSTALLINE CARBONATE
Mud-supported		Grain-supported			
Less than 10% grains	More than 10% grains				
MUDSTONE	WACKESTONE	PACKSTONE	GRAINSTONE	BOUNDSTONE	

The term mudstone is not quite synonymous with calcimicrite or calcilutite as it includes all grains up to 20μm in size. Wackestone consists of more than 10% of grains between 20μm and 2 mm in size 'floating' in a matrix of mud (floatstone if the particles are greater than 2 mm), and packstones are composed of grains in close contact with each other with interstitial mud cement. In analysing the grain size distribution of modern sediments it is found that the lutite and silt-grade materials are often so inextricably mixed that division into those above and below 20μm in size is very difficult. Lutite and silt may then be grouped together as 'mud' and the term 'grain' reserved for sand-sized and larger particles (>60 am). Grainstones are mud-free carbonate rocks indicative of relatively strong bottom currents. To make the

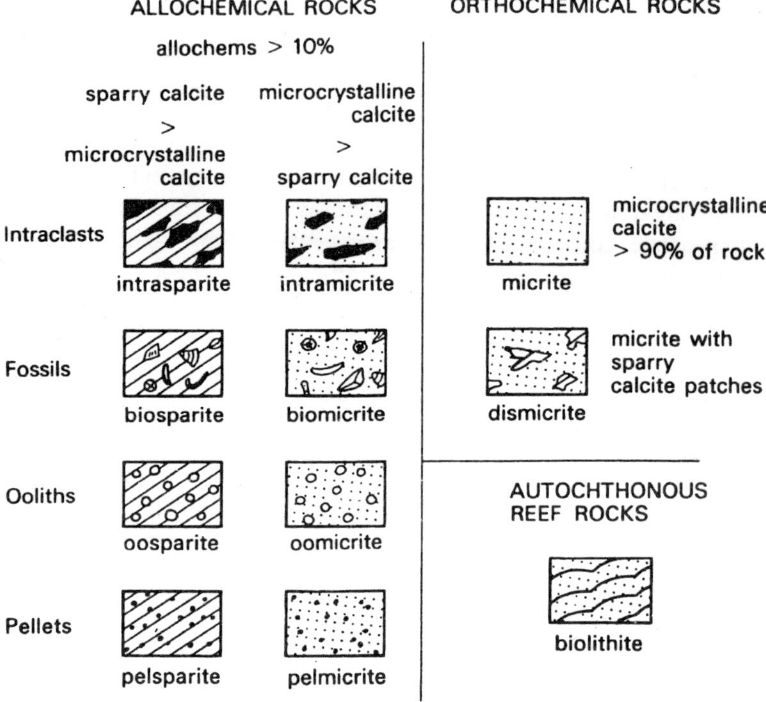

Figure 10.2: Folk's basic classification of limestones. Application of the terminology to the common varieties of limestone. The scheme is valid for both aragonite, which may be more abundant in modern deposits than calcite, and calcite.

classification comprehensive the term boundstone is used for rocks formed mainly of binding, net-like organisms, such as algal buildups. If the buildups are formed mainly of frame-building rigid skeletons, such as branching corals, then framestone is possibly more appropriate. Crystalline carbonates are those in which nearly all original texture is lacking, such as dolomites.

ALLOCHEMICAL LIMESTONES

Within this largest of all limestone groups occur a range of rocks which are predominantly formed of mechanically deposited carbonate particles (allochems).

Intraclast-Bearing Varieties

These comprise endogenetic rock fragments up to boulder size, which are set in a primary micritic matrix and/or secondary sparite cement. True intraclasts are intrabasinal products and are essentially reworked fragments of penecontemporaneous lithified, and partly lithified, lime deposits. It is sometimes difficult in the laboratory to distinguish between these products and those introduced from extrabasinal sources, but the field evidence is usually conclusive one way or the other.

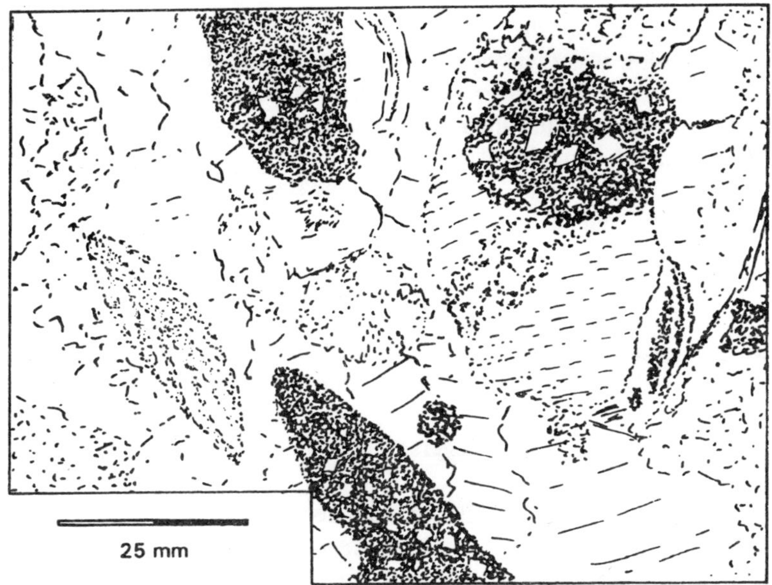

Figure 10.3: Intrasparite, Palaeozoic, Wyoming. Intraclasts of dense micritic limestone set in a coarse sparite cement. Small dolomite rhombs are scattered through the interclasts. Polarized light.

The roundness of intraclasts varies considerably, depending on the amount of transport. Some supratidal and lacustrine mud-flake deposits originating through desiccation contain highly angular fragments. So do some of the extensive submarine talus deposits associated with modern and ancient reefs and other buildups. In contrast, reworking in protected, shallow water environments can give good rounding.

Aeolian and storm-wave processes, effective in moving sand-sized particles for the littoral into the beach and dune zones, as in the Persian Gulf, also produce good rounding. The quality of sorting in intraclastic limestones is as variable as the particle roundness factor. In shelf areas with restricted current activity or in deeper waters where turbidity current emplacement has occurred, the sorting may be poor and the quantity of micritic matrix relatively high.

The textural characteristics are then of the wackestone-packstone type. Supratidal carbonates commonly have a wackestone texture. The slopes of deep basins and troughs bounding shallow water areas of lime accumulation are the loci for graded bed deposition, in which the basal packstone layers of each graded unit are sometimes pebbly, with genuine intraclasts and exotic extraclasts. Calciturbidite beds of this kind have been called allodapic limestones.

The cleaner washed, well-sorted intraclastic limestones generally typify shallow water, strong current and wave-affected environments. In these situations any pre-existing micrite is winnowed-out and sparry calcite precipitated in suitable interstices, as soon as the fragments are mechanically stabilized. The texture usually falls into the grainstone category. Ripple marks, cross-bedding and scour structures characterize both modern and ancient beds.

Grapestones are lumpy clusters of strongly micritized and algal-bored allochems, cemented mainly by sparry calcite. They used to be regarded as a variety of intraclast, but it is now thought that they form as a consequence of in-place cementation of grains immobilized in algal-rich mats. These subtidal mats commonly cover large areas of the sea bottom, as in the northern and southern parts of the Andros Bank, and are pale brown-green layers, a few millimetres thick and rich in micro-organisms.

Filamentous algae and diatoms are dominant and trap and bind the clastic particles into a gel-like coherence. The layers are moderately resistant to erosion and stabilize the sediment underneath to a much greater degree than would be possible otherwise.

Biosparites

Among the well-bedded marine limestones, the most widespread type consists of broken or disintegrated fragments of large calcareous skeletons, mixed with the complete shells of smaller organisms, and cemented together by comparatively clear sparite. Many grainstones are of this type.

Frequently, in the well-sorted varieties, it will be noticed that all the recognizable fragmental remains are those of calcite organisms. Possibly the carbonate of any original aragonitic organisms has been dissolved, redistributed in solution and reprecipitated in pores and cavities as early diagenetic sparite cement.

This mechanism for cement production, however, is likely to be too simple, as it does not take into account the probability that some of the coarse calcite is neomorphic sparite, formed by in-place replacement of original cement or matrix. Certain thin Liassic limestone beds in Britain and parts of the thicker, more massively bedded Lincolnshire Limestone (Middle Jurassic) of the Midlands of England have sparry calcite of mixed origin. In general, the greater the lithification and age of a biosparite the more it should be suspected that mixed origin sparite is the rule rather than the exception.

Although shell debris (and other allochems) in biosparites are frequently recognizable by their morphology or specific internal structure, it has to be appreciated that the fragments may have been modified on the sea bottom prior to final incorporation into the rock. This modification is commonly initiated by the activities of fungi and endolithic non-calcareous algae, which bore into the surface of the fragments and create a dense network of tubes, each tube being up to about 6μm in diameter.

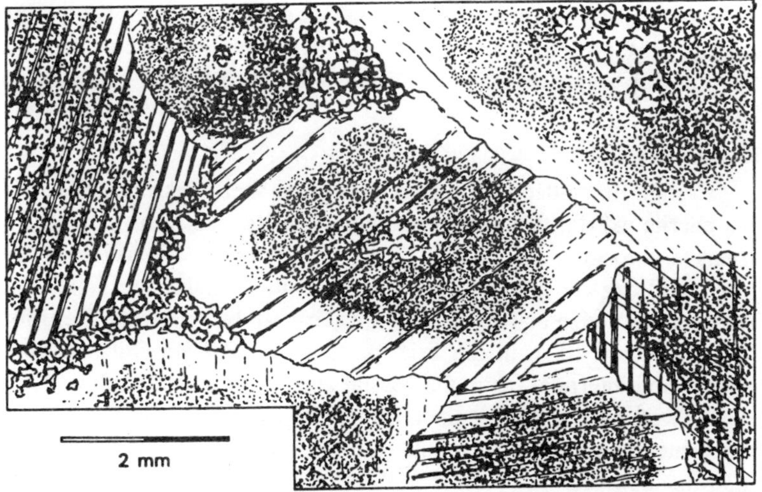

Figure 10.4: Crinoidal biosparite, Carboniferous, Forest of Dean, England. Dusty-looking crinoid ossicles enlarged by syntaxial (rim) cement. The outgrowths are in optical continuity with the calcite of the ossicles. There are small interstitial pockets of microspar cement.

Within the tubes, possibly aided by bacterial activities, conditions are established suitable for the precipitation of micrite. The process of micritization has the tendency, therefore, to destroy original organic internal structures. So, some rolled micritized shell fragments may begin to closely resemble pellets in shape and composition.

Crinoidal biosparites are rocks in which the ossicles of crinoid stems are conspicuous ingredients. Each ossicle consists of calcite which behaves as part of a single crystal, and breaks along cleavage planes. Syntaxial rim cementation is common. Consequently, a fractured surface of the rock may give a deceptive appearance of a coarsely recrystallized limestone, owing to the abundant large and conspicuous cleavage surfaces. This effect is particularly striking in some of the shelf crinoidal rocks associated with reefs in the Wenlock Limestone (Silurian) of the Welsh Borderland and the Lower Carboniferous successions of Wales and England.

Modern shell accumulations which might lithify into biosparites are found in various intertidal and subtidal situations. In the shallow waters on the southwestern flanks of the Persian Gulf there are extensive areas of clean washed mollusc, foraminiferal, algal and coral sands, where conditions appear appropriate for the precipitation of interstitial aragonite and calcite.

The rate of cementation depends on many obvious factors, such as the geochemistry, temperature and movement of the pore waters, but also on less obvious factors such as shell mobility. If the deposits stabilize for some reason then cementation is likely to occur much faster.

Biomicrites

Certain limey deposits consist of organic shelly detritus embedded in a micrite matrix; they are biomicrites or shelly wackestones. Deposits of this kind form in quiet water, where .fine-grained carbonate mud can settle, and organic remains are entombed in an unbroken condition, irrespective of size or fragility. The waters can be relatively deep, for example below 40 m in the axial parts of the Persian Gulf, or very shallow and intertidal, as in the coastal zone of the same region. Much of the shelly part of the Wenlock Limestone (Silurian) is shallow water biomicrite and probably accumulated in sheltered pools among the reefs of stromatoporoids and corals.

On the slopes leading from such current-swept shelf edges crinoidal and other organic debris is sometimes transported quickly and accumulates downslope as coarse turbidites. *Graded crinoidal calciturbidites* of Upper Palaeozoic age in the Sverdrup Basin of Ellesmere Island contain as much as 95% of disarticulated crinoids set in a micrite matrix.

Turbidity currents also account for some thinly bedded, laminated and graded micrites, mudstones and wackestones found within Palaeozoic slope and basin successions of Nevada, Appalachia and Britain. The beds are usually dark coloured, pyritic and carry few trace fossils. Resedimented fossils of shelf type and basal scour structures indicate downslope transport in a fashion probably akin to that producing distal turbidites in cyclical basinal sandstone and siltstone successions.

Algal Limestones

Calcareous algae secrete a hard lime skeleton. Two families, the Corallinaceae (red algae; *Rhodophyta*) and the Codiaceae (green algae; *Chlorophyta*), are more important than any others at the present day, and are confined to marine or brackish waters. The Corallinaceae require a hard foundation upon which to grow, and are principally found off rocky coasts or in coral reefs.

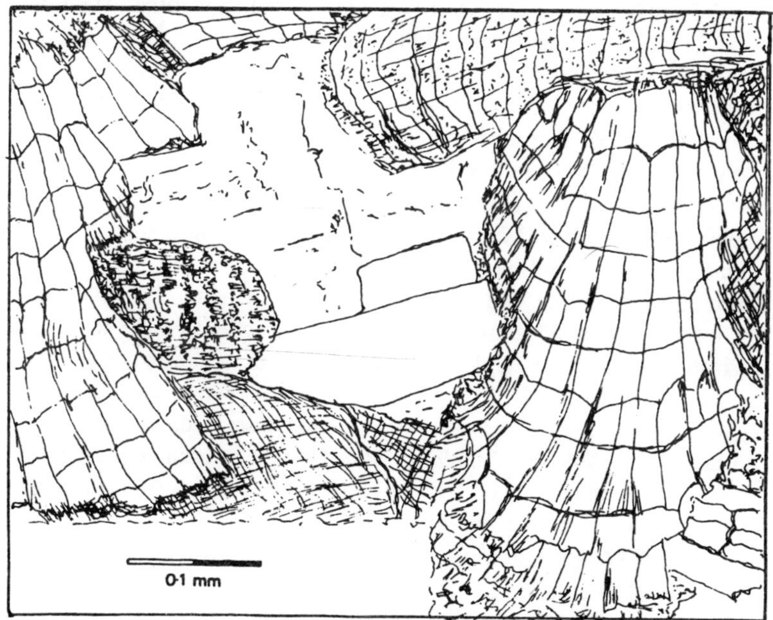

Figure 10.5 : Lithothamnium limestone, Eocene, Bavaria. Two species of the red alga Lithothamnium are present and set in a coarse, drusy sparite cement and micrite matrix.

The skeleton consists of closely packed, parallel calcified tubes, commonly formed of high-magnesium calcite in modern genera. These tubes are usually only discernible through a microscope; longitudinal sections show well-marked concentric growth surfaces. The family is important in modern seas and in Tertiary deposits, some of the best represented genera being *Lithothamnium, Archaeolithothamnium, Amphiroa and Lithophyllum*. The remains of these plants are sometimes found as rolled fragments and sometimes as encrusting or frondose masses preserved in the position of growth.

In calcareous rocks older than the *Cretaceous*, the *Corallinaceae* (*Carboniferous-Recent*) are less prominent than the closely related red algae Solenoporaceae (Cambrian-Miocene). *Solenopora* is an important rockformer at certain levels in Carboniferous limestones of northern England, and *Solenoporella,* although a locally restricted fossil in Britain, is closely associated with Jurassic coral reefs in much the same way as *Lithophyllum is* associated with modern reefs.

Lithophyllum and other red algae can occur as discrete spheroidal and ovoidal growths, on average 5 cm in diameter, on the sea bed of reef lagoons and flats. These growths are called rhodoliths. The living algae completely mantle any suitable nucleus, such as a pebble, and continuity of growth is accounted for by periodic disturbance and re-orientation of the rhodoliths by strong tidal currents. Rhodoliths are frequently mixed in reef habitats with similar sized spheroidal coral growths (coralliths) such as *Porites.*

The Codiaceae, such as *Halimeda, Penicillus* and *Udolea,* are important contributors to the deposits of modern tropical seas, but their remains are not found in any appreciable quantity in older limestones. These forms, unlike the Corallinaceae, are capable of colonizing unconsolidated muddy bottoms, and are consequently able to live in environments which would be impossible for the Corallinaceae. They are firmly established on the Bahamas Banks where *Halimeda,* on disintegration, contributes sand-sized, often micritized, grains to the sediment. In contrast, *Penicillus* and *Udotea,* which appear to be constructed of more weakly aggregated aragonite needles than *Halimeda,* tend to break down into fine-grained mud.

From the Cambrian to the Miocene the principal calcareous green algae belong to another family, the Dasycladaceae, which are poorly represented in British rocks, but make up a large part of the Alpine Mesozoic limestones in south-east Europe and Arabia. Dasyclads generally thrive under conditions of relatively strong current action, with little accumulation of micrite. Members of the class Chlorophyceae (?Precambrian-Recent) are less prominent than the Dasycladales but are widely distributed in many freshwater to fully marine environments.

The colonial variety *Botryococcus* flourishes and blooms in modern freshwater lakes and is rich in hydrocarbons. Fossil equivalents contribute significantly to freshwater bituminous limestones and oil-shales. Single-celled varieties, such as the small hollow *Calcisphaera,* are found in marine limestones and chalks from the Devonian onwards.

Some algae, especially filamentous cyanophytes like the present day *Schizothrix,* do not necessarily secrete a calcareous skeleton, but in certain circumstances become encrusted in calcium carbonate. If these forms grow under water they will only precipitate calcium carbonate if the water in contact with them is saturated with this substance. The algal nodules known as '*water biscuits*' or '*lake balls*' and found in modern freshwater lakes in many parts of the world are formed in this way. Similar structures also occur in the deposits of Tertiary freshwater lakes and lagoons.

The modern freshand marine-water genus *Scytonema* sometimes forms cushion-shaped tufts of roughly parallel tubes, which may become calcified, almost indistinguishable in thin section from colonies of the Palaeozoic genera.

Cyanophyte (blue-green) algae are often among the earliest colonizers of sediment newly deposited in intertidal and supratidal zones and in shallow fresh and brackish water ponds and lakes. It is well known that such algae exert a strongly stabilizing effect upon the sediment they colonize, largely because they bind the particles together with their growing mucilaginous filaments, and ultimately cover them with a mass of felted tubes.

On flats, where there is an abundance of drifting fine-grained sediment, the algae tend to collect it and progressively create a discontinuous, irregularly laminated and mat-like structure, known as a stromatolite. Bacteria also play an important role in the accumulation process by establishing conditions suitable for the precipitation of calcium carbonate. The laminations represent growth cycles, some possibly daily, others longer term.

These structures cover extensive areas of the high flats in the Bahamas, south Florida and the Persian Gulf where they exhibit mud-cracking, window-like cavities (*fenestral fabric*), and pockets and pseudomorphs of evaporite minerals, especially gypsum. In the more saline parts of Shark Bay in Western Australia stromatolites extend into the lower flat and subtidal zones. Freshwater equivalents are well developed around the periphery of many East African Rift lakes.

The individual laminae in all these situations are rarely more than a few millimetres thick and it would be unusual at the present day for a whole *stromatolite* to have a thickness greater than 1 m. As the accretion of successive laminae is usually uneven, because of exposure, light, salinity, substrate irregularity and tidal factors, the external morphology and internal structure vary considerably. Some bulbous stromatolites consist of laterally linked hemispheroid growths, others grow as discrete stacked hemispheroid columns or '*algal heads*'.

Structures comparable with modern stromatolites are well developed in ancient shallow water limestone successions, reaching an acme in diversity in late Precambrian to Upper Ordovician times. Ovoidal, dome-like stromatolites as much as l5m across and 1 m high are present in the Middle Magnesian Limestone (Permian) reef facies of north-east England.

The lamina frequency varies between 36 and 48 per centimetre thickness. The largest and most complex growth forms are associated with the reef crests and upper reef slopes, whereas the smaller, less complex varieties appear analogous to those forming modern reef flats and in adjacent lagoons.

Stromatoporoids (?Cambrian-Lower Carboniferous) exhibit structures remarkably similar to stromatolites. They are associated with buildups in many marine limestone successions, as in the Silurian of the Welsh 3orderland and Gotland in the Baltic Sea.

Some late Precambrian and Palaeozoic limestones (and iron-ormations) show traces of cellular or filamentous structure. The organic haracteristics, seen particularly in the close-linked hemispheroidal rowth forms of Collenia and discrete, vertically stacked hemispheroids f Cryptozoon, suggest the activities of algae similar to the modern ranophytes.

These algal structures are of special interest since they not only rm stromatolites but also build up reef-like masses. Particularly fine ecambrian examples are to be found in the Superior-type iron-formations f North America, the Algonkinian limestones of Montana and in the Transvaal Dolomite of South Africa.

Not all algal mats are stromatolitic or internally laminated. Within relatively deep and shallow seas a thin film of algal growth can form on the bottom without any vertical accretionary structure. This growth is a thrombolite and at present, for example, is forming on parts of the Bahamas Banks. The branched and tangled filaments of *Girvanella* (Cambrian-Recent) often dominate thrombolites in ancient shelf limestones.

The minute algae grouped together as the *Coccolithophyceae* (JurassicRecent) are mainly marine living in the upper 200m of oceanic waters and consist of a single, approximately spherical cell, up to 60μm in size, encased in a sheath of low-magnesium-calcite plates. The plates vary in size between 2 and 15μm and disaggregate on the death of the organisms. The coccolithophores have a wide distribution in the sediments of temperate and tropical oceans at depths above the calcite compensation depth (average 4.5 km depth) and are specially abundant in the globigerina ooze of the North Atlantic. The spinose rhabdoliths (another variety of coccolithophore) are, on the other hand, confined to warmer waters, such as the Mediterranean and the tropical parts of the open oceans.

Chalk

Chalks are fine-grained, finely porous, friable pelagic limestones in which coccolithic material is usually prominent. They are mainly biomicrites occurring in moderately thick, well-bedded and regionally extensive successions in many parts of the world, especially the late Mesozoic and Tertiary of the Northern Hemisphere.

Most chalks are marine, but a few freshwater varieties are thinly represented in modern Alpine lakes and in Neogene-Pleistocene lake deposits laid down on the site of the present Black Sea. The white pure chalks contain more than 90% calcium carbonate in the form of low-magnesium calcite. The white colour and high reflectivity to light indicate its purity and fine-grained constitution. Some 75-90% of the carbonate is in the form of organic particles less than 4μm in size.

The impurities in chalks are clay minerals of which smectites, often of volcanic origin, and illites are common. Glauconite is widespread in some chalks, but usually in minute quantities. It is not known whether all, or just some, of these minerals are authigenic. This contrasts with pyrite granules and nodules, and small alkali feldspars, which are undoubtedly authigenic. Authigenic francolite is sometimes concentrated as impregnations, grains and nodules at levels which are believed to have had very reduced or zero rates of deposition.

Coccolithophores form up to 80% of certain chalks. Preservation of the oval plates is occasionally good, but physical degradation and compaction generally break them down into smaller fragments. As lithification proceeds calcite overgrowths and recrystallization obscure their morphology. The effects of dissolution can become apparent during transit to the sea bottom. At depths below about 4km most plates are completely dissolved as the waters are undersaturated with respect to calcium carbonate.

The rest of the calcareous organic debris consists of whole or worn fragments of a wide range of invertebrates. Bivalves, often thin-shelled, echinoderms, polyzoa and foraminifera are the most common of these fossils. Planktonic foraminifera in some Cretaceous-Palaeogene chalks of the Mediterranean region are often equally as abundant as coccoliths. In addition to foraminifera, the chalks contain spinose spherical bodies, up to 500μm in diameter, which superficially resemble large isolated chambers of globigerina in appearance. The central area is usually occupied with sparite cement. A foraminiferal origin is nevertheless

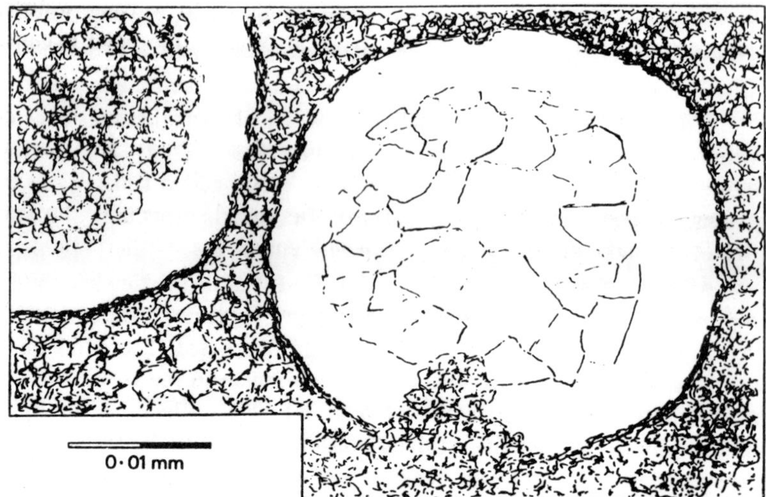

Figure 10.6 : Calcisphere in the Chalk, Upper Cretaceous, south-east England. These organic structures are readily visible using an ordinary microscope.

improbable, since they possess no aperture and appear to be imperforate. These calcispheres are especially abundant in certain hard or nodular bands. They are most probably the remains of some planktonic organism, possibly akin to modern chlorophycean algae.

Radiolaria, diatoms and other siliceous organisms are variously incorporated and are likely to be the main source of the silica eventually precipitated as replacement chert (or flint) nodules and nodular layers.

Chalks usually have a high porosity (36-65%) and it has been surmised that this partly reflects the original composition of the sediment. As most of the coccolith plates are formed of low-magnesium calcite there has been no subsequent conversion of aragonite into calcite as in other limestones, so that the original primary porosity has been retained to a high degree.

However, in Scotland, Antrim and northeastern England the Cretaceous chalk is moderately well cemented by secondary calcite infilling original pores and forms a comparatively hard limestone, whereas in southern England it is much softer and more friable. The density of the Antrim chalk varies between 2.60 and 2.64 and that of the soft English chalks between 1.70 and 1.95 because of the difference in cementation. The reasons for the peculiar geographical distribution of the hard and soft chalks is obscure but it has been suggested that the hard beds originally had a greater proportion of micrite deposited between the larger organic particles. This micrite may have recrystallized into syntaxial grain growths around the margins of the larger constituents of the chalk and so created a tougher rock.

On the other hand, pressure solution between the sediment constituents may have provided syntaxial cement without fine micritic particles being involved at all. In the Antrim chalk there is evidence of sedimentary structure deformation, fracture of fossils and interpenetration of shell fragments, which would suggest hardening by overburden weighting, possibly by the thick piles of Tertiary lava flows. The extended phase of Tertiary igneous activity centred on Antrim and western Scotland also implies a regional rise in the geothermal gradient, which may have increased the rate of diagenetic change in the chalk.

On a different, much smaller scale, hardening of beds in some chalks is accomplished by contemporaneous submarine cementation processes. Hardgrounds are layers a few centimetres thick, where the degree of aragonite, magnesian calcite and calcite cementation decreases downwards from the surface. At present, hardgrounds are forming on carbonatedominated shelves, as around the south-west margins of the Persian Gulf, and on the flanks of carbonate-dominated deep sea channels.

The Cretaceous equivalents are commonly bioturbated with organic encrustations, phosphatization and glauconitization typifying their upper surfaces. Very reduced rates of deposition over long periods seem probable, therefore, for the successful development of a hardground. The manner of deposition and depth of water in which the varieties of chalk were deposited are difficult to estimate from lithology alone. Intense bioturbation at many levels, often associated with flint and chert layers, implies prolonged phases of quiet deposition.

On the other hand, cross-bedded carbonate banks in the Normandy chalks, with amplitudes as much as 50m and up to 1·5 km long, are indicative of considerable bottom currents. Slump and graded bedding suggest slopes and instability of the bottom muds, and rapid emplacement by turbidity currents is recognized in many rhythmically bedded chalk successions wherever suitable basins and bottom slopes existed. Calciturbidite chalk beds are usually a few centimetres thick and often show partial Bouma divisions, as in the Palaeogene Lefkara chalks of Cyprus.

Faunal and floral considerations put a minimum depth limit for the finest chalks at about 100m. The bulk of the coccolith fragments show little '*corrosive etching*', which suggests deposition well above the calcite compensation depth, possibly at depths not much greater than 600m. The evidence of the molluscan fauna of the hardgrounds indicates depths as shallow as 50m.

A curious feature of the Cretaceous chalk beds of western Europe is the development of a distinctive brick-red staining at certain marly, nodular and bioturbated levels. In eastern England the zones of reddening are usually about 1 m thick, though they can reach 6m locally. There is evidence of the staining transecting primary bedding features but the distribution, in general, is concordant with bedding suggesting a contemporary or penecontemporaneous origin. The colour is caused by granular haematite, up to 40% in quantity, which is distributed unevenly throughout the zones. There is a possibility that some of the haematite is secondary after pyrite and glauconite, but how much remains conjectural.

The conditions giving rise to the reddening are also problematical. Suggestions have been made that the cause was the introduction of clastic lateritic mud into the Chalk Sea.

Globigerina Ooze

This deposit contains more than 30% globigerina and is one of the most widely distributed in modern oceans, since it is estimated to cover an areas of nearly 125 000 000 sq km. It attains its maximum development in the Atlantic, where equatorial sedimentation rates are of the order of 3·5cm per 1000 years, and also covers great areas in the Pacific and Indian Oceans; it is known to extend as far south as latitude 60°, and as far north as the Arctic Circle. The globigerina ooze occurs at all the medium depths of the ocean removed from continents and islands and is especially developed in those regions where the surface of the sea is occupied by warm currents and is organically very productive.

At depths greater than about 4500m (the calcite compensation depth), the concentration of the *planktonic globigerinacea*, which have tests composed of low-magnesium calcite, falls rapidly in the oozes to very low values. This is because the waters are markedly undersaturated with respect to calcium carbonate, so that solution of the tests of organisms occurs relatively quickly.

Globigerina ooze is a cream-coloured, pink or pale grey substance, muddy when wet, and powdery when dry; it consists for the most part of the tests of foraminifera, of which various species of *Globigerina* are the most abundant. Many other organisms also occur, especially pelagic molluscspteropods and heteropods - and coccolithophores. Siliceous organic remains are sometimes present in considerable quantity and consist chiefly of *diatoms*, *radiolaria*, *silicoflagellates*, and the spicules of sponges. Some samples dredged up contain relatively high proportions of celestobaryte grains-a solid solution of celestite ($SrSO_4$) and anglesite ($PbSO_4$) in barytes ($BaSO_4$). The probability is that the calcareous organisms extract these materials from the near-surface waters in the construction of their skeletons. On death, skeletal disintegration at depth leads to the accumulation of the discrete mineral grains of celestobarytes.

Fossil lithified equivalents of globigerina oozes are represented by pelagic, relatively deep water chalks and marls of Cretaceous and younger age, and are found in many parts of the world. Many of the thin glassy tests lose their lustre in these older deposits and show evidence of dissolution and fracturing. The dissolved calcite appears to be partly reprecipitated as interstitial cement and overgrowths on the remaining tests.

Ooid-bearing Lime Sediments

These include a number of carbonate rocks agreeing in the possession of a characteristic texture somewhat resembling the roe of a fish. A normal oolitic deposit is made up of an aggregate of spherical allochems called ooliths or ooids, usually 1 mm or less in diameter. When examined under the microscope, the grains are found to be built up of two or more concentric layers, usually around a nucleus, which may consist of a shell fragment, peloid or a grain of authigenically modified clastic quartz.

The material which makes up each of these concentric layers consists, in most modern deposits, of aragonite crystals laid roughly tangentially to the surface of the oolith. In many of the older oolitic limestones, on the other hand, the ooliths are found to be made up of little radiating prisms of calcite, created by internal recrystallization, set at right angles to the concentric rings.

The periphery and some of the inner rings of ooids often exhibit narrow zones of micrite, which have a dense grey-brown appearance in thin section. The causes of these zones of micritization are likely to be related either to encrusting algal perforation associated with direct precipitation of micrite or to fragmentation and dissolution of adherent aragonite crystals followed by renewed precipitation of micrite as cement in the micro-interstices of the ooid. In some instances the *micritization* is so thorough that it completely obscures the true original nature of the ooid, making it resemble a peloid.

The microporosity of ooids is commonly increased by micritization and, in ancient oosparites and grainstones, is known to effect a bimodal pore system within the rocks as a whole, one being intragranular (within the ooids) and the other intergranular (between the ooids).

All the important developments of oolitic limestones among Pleistocene and older rocks are undoubtedly marine or salt-water sediments. Modern occurrences are in the Persian Gulf, the Red Sea, on the shallow submarine platforms of Florida and the Bahamas, and on the shores of the salt lakes of the western United States, such as the Great Salt Lake of Utah. In Florida and the Bahamas ooids are growing where water saturated with calcium carbonate is drifted backwards and forwards in tidal channels and across adjacent subtidal flats.

The depth of water in the channels is never very great, usually not more than 6m, and the sea floor is so scoured by currents that all traces of micrite are commonly winnowed away. The grains which remain are kept in motion and become encased in concentric layers of aragonite. The mobility of ooids often means that they are current-swept en masse into shallower subtidal and intertidal situations to form migratory shoals, asymmetric banks, beaches and spits.

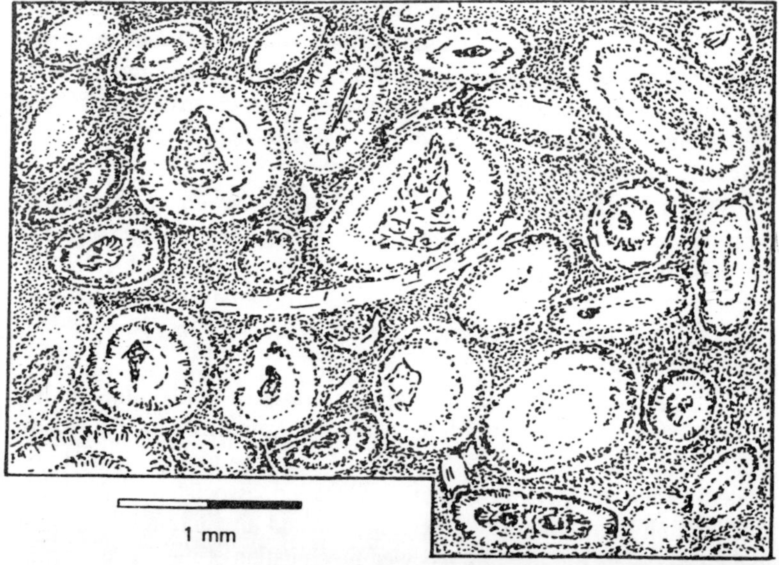

Figure 10.7: Oomicrite or wackestone, Lower Carboniferous, Morocco. Ooids and one-layer superficial ooids set in a micrite matrix.

As a consequence, individual ooids show marked peripheral abrasion. Under some circumstances, bottom currents transport ooids into less disturbed waters and they become incorporated into micritic mud. Alternatively, due to some change in current paths, mud may be introduced into the interstices between the ooids. In both instances the growth of individual ooids virtually ceases and the sediment is prospectively an oomicrite or oolitic wackestone.

The oolitic sands of the Gulf of Suez (Red Sea) form wide stretches of yellow, ripple-marked sand, exposed at low tide. When thus exposed to the tropical sun, part of the sand dries, and tends to be blown inland as small dunes, which advance towards the desert. Around the Persian Gulf, blown oolitic sands are recognizable some 40 km away from the shoreline. The name aeolianites is applied to these cross-bedded deposits. The nuclei most commonly consist of grains of quartz, feldspar or foraminiferal shells.

There is considerable evidence that ancient oolitic limestones were mostly formed in shallow, well-stirred waters. For example, the Middle Jurassic Great Oolite Series of southern England is characterized by cross-stratified, rippled and shelly oolitic limestones deposited in a marine intertidal flat and tidal channel situation. The bulk of the beds are oosparites or oolitic grainstones. Oosparites are also well developed within Lower Carboniferous marine shelf successions of South Wales and England.

Oncoids are rounded, sometimes irregular-shaped bodies which superficially resemble ooids, yet have algal origins. Modern oncoids usually consist internally of concentric layers of greyish, organic-rich micrite and clear aragonite. These layers are sometimes wrapped around nuclei of molluscs, algae and lithoclasts. Oncoids commonly form along the margins of subtidal channels and shoals on carbonate platforms, but can be rolled away from their point of origin. Good fossil examples are found in marine Carboniferous limestones of England and Scotland and in the Middle Jurassic oolitic beds of England.

Oncoids often closely resemble pisoliths, which are 2mm or more in size. Pisoliths are recorded from the sabkha zones around the Persian Gulf and seem to be of concretionary origin, the carbonate being precipitated by evaporation in the partly freshwater water- and air-filled pores around the water table (the vadose zone). In the same areas aragonitic crusty beach-rock, on the higher levels of the intertidal zone, is partly aggregated of pisolitic grains, formed by direct chemical precipitation from vadose water during low tide periods.

Pellet and Peloid-bearing Lime Sediments

The pellet-bearing carbonate sediments (e.g. pelsparites, pelmicrites, peloid packstones) generally owe their peculiar characteristics to organic agencies. When the pellets have no detectable organic affinities the term peloid is preferable. Pellets or peloids can vary considerably in size, but usually consist of ovoid or elongate bodies less than a millimetre in size, composed of micrite and aggregated to form a rock superficially resembling an oolitic limestone. They accumulate in quantity in low to moderate energy situations, such as on the lower parts of basin slopes, on wide tidal flats and in adjacent lagoons. Foraminiferal, algal and molluscan debris are frequent associates.

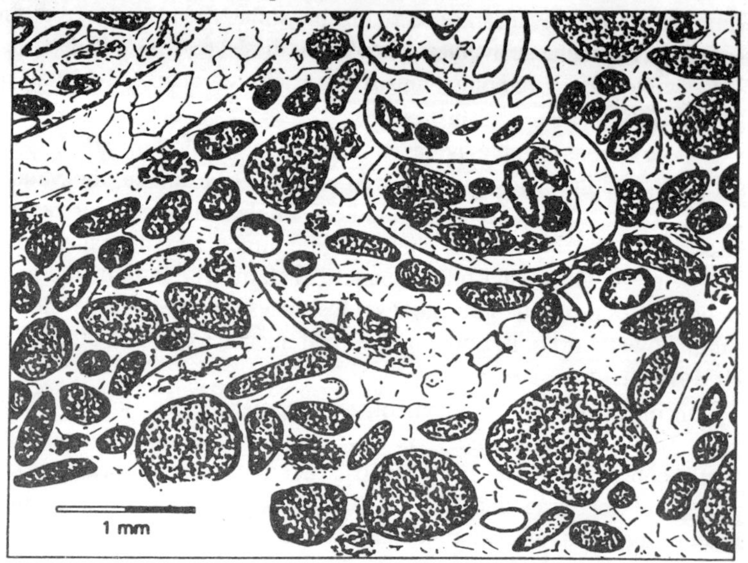

Figure 10.8 : Pelsparite, Middle Jurassic, England. The peldids are devoid of organized structure internally and are composed of micrite. They partly occupy the chambers of a gastropod shell (centre). The cement is sparite.

The majority of the pellets are faecal in origin and formed by the activities of gastropods, worms, molluscs and crabs. The various mud-feeding invertebrates pass the sediment through their bodies and leave the rejected material in the form of subspherical, ellipsoidal and rod-shaped grains. Within any given layer faecal pellets usually show a marked uniformity in size, ranging in different rocks from 0.03 to 0.16 mm. In some lagoons of the West Indies the rod-shaped pellets of the marine gastropod *Batillaria* locally form more than 90% of the bottom sediment.

They are composed of silt-sized particles embedded in organic mucus with traces of algae. Both modern and fossil pellets are generally rich in organic matter which gives them a brownish tinge.

Some shallow water muds on the Florida coast contain 50% faecal pellets, but a metre or so below the surface it has been noted that compaction has destroyed their individuality. The total destruction of pellets by compaction is normally prevented by interstitial aragonite tending to harden the muds relatively quickly. For instance, around Andros Island the faecal pellets, which form up to 30% of some sediments, are rapidly bound together quite firmly by aragonite cement.

Many fossil carbonate rocks show a pellet structure when examined in thin section and in most cases a faecal origin seems likely. In Britain this type of lithology is seen in the Lower Carboniferous beds of the classic Avon Gorge section at Bristol, also in Wales, the northern Pennines of England and in Ireland.

ORTHOCHEMICAL LIMESTONES

These are microcrystalline sediments characterized by a marked dominance of micrite or lime mud. Lithified equivalents are referred to as calcilutites, micrites, mudstones and wackestones (the last two in the strict sense of Dunham only). In modern deposits the individual aragonite particles can only be detected with the electron microscope whereas, in ancient rocks, the particles are often diagenetically enlarged, such that the granularity can be more readily observed with an ordinary microscope.

Allochems of all types form a low volumetric percentage or may be completely absent. Sparry calcite tends to occur in irregular patches only if the original ooze has been disturbed by bottom currents, escaping gas bubbles during desiccation or by the activities of organisms, especially algae. The structure created by any of these disturbances is known as 'bird's-eye' or fenestral and it is typical of intertidal and supratidal deposits. Within fenestrae geopetal structure is commonly developed. Folk describes these disturbed muds as dismicrites. Laminated and bioturbated dismicrites are sometimes given the name 'bird's-eye limestones'. When rhythmically interbedded with other shelf deposits, including reef buildups, they are also referred to as loferites and constitute part of repetitive, thin shallowing-upwards, and eventually sub-aerially exposed sequences known as loferite cycles. These cycles are probably controlled by variations in relative sea-level and are recognized in the Lower Carboniferous of Montana, Permian of New Mexico and Triassic of the Austrian Alps.

Fine-grained carbonate mud accumulation occurs in low energy environments, where there is a general lack of strong winnowing currents. Protected shallow water lagoons and shelves, and deeper offshore basins are common situations. Deeper water varieties of mud occupy depressions at water depths greater than 30 m in the Persian Gulf. In the Tongue

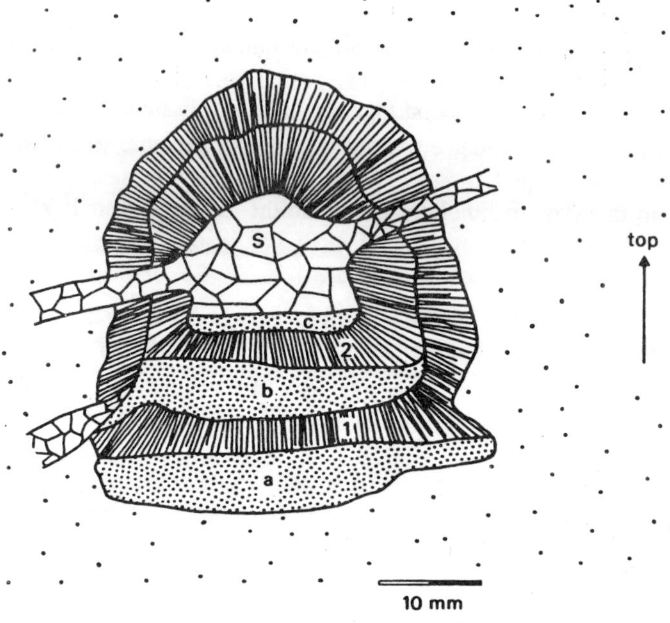

a, b, c 3 generations of internal sediment

1, 2 2 generations of fibrous sparite

S cavity fill of granular sparite

Figure 10.9 : Stromatactis structure with geopetal fabric. Complex chronology of infilling of a flat-bottomed cavity in an algal buildup of Devonian age in New South Wales. Three episodes of clastic micrite sedimentation in the cavity (a-c) are interpersed with two episodes of drusy fibrous calcite precipitation (1-2), followed by a final fill (S) of granular sparite. As the micrite sediment filters down to the bottom of the cavity it is possible to deduce the original orientation of the beds.

of the Ocean trough in the Bahamas similar muds are found, some of which are laminated and show small-scale graded bedding. They are essentially allodapic calciturbidites.

The most discussed lime mud deposits are those originating in shallow waters. Many petrologists suggest a physiochemical origin whereas others believe a biochemical or even biogenic origin is more likely. The problem is complicated by the fact that conditions which favour one origin also favour the others. These conditions are warm surface waters saturated with $CaCO_3$, restricted circulation of the waters and a hot climate.

Conditions which favour all the processes of accumulation are well illustrated on the Great Bahama Bank, where a considerable area of the sea floor west of Andros Island is covered with a white deposit of aragonite mud. The sea floor here stands only about 6 m below water level, and maintains this elevation more or less uniformly right up to the edge of the shoals, some 80 km offshore.

There is then a sudden drop, usually at a gradient of one in three, to depths varying between 600 and 1800 m. The extreme shallowness of the sea over these shoals effectively reduces circulation, and the water in the centre of the bank is virtually isolated from that of the open ocean. In the strong subtropical sun the water rises in temperature to as much as 37°C in summer, and increased evaporation takes place; since the rainfall is insufficient to make up for the loss by evaporation, water is drawn in from the ocean, and there is a concentration of salts ($>41‰$ salinity) in the centre of the bank.

The surface water of the open ocean in these latitudes is already saturated with respect to calcium carbonate, and any rise in temperature reduces the solubility of the salt; consequently, as the solution becomes warmed and concentrated in its passage across the shoals, the excess calcium carbonate crystallizes out. In quiet and sheltered parts of the bank, the precipitate takes the form of minute needles of aragonite, usually 1-5μm in length, forming a white, fine-textured sediment of almost pure calcium carbonate. Such a sediment is known as aragonite or micrite mud.

There is a variable percentage of skeletal fragments and pellets distributed through the mud, so the names biomicrite and pelmicrite are also appropriate. It is difficult on the bank to quantify the proportion of aragonite directly precipitated from sea water and estimates vary from virtually zero to substantial amounts, greater than 50%.

The contribution of identifiable skeletal debris to the Bahamas muds is about 10%, and the rest of the aragonite is either a chemical precipitate or unidentifiable organic debris, or a mixture of both. The proponents of biogenic sources for the mud favour calcareous algae, the argument used being the ease with which certain varieties *(Penicillus, Udotea, Rhipocephalus, Acetabularia)* disintegrate on death into single aragonite needles.

Theoretical calculations show that if the average density of such algal growth reaches two to three plants per square metre, over a sufficiently large area of sea floor, then the rate of needle accumulation can approach the known rates of sediment deposition. However, until some uncontroversial method of distinguishing between organic and inorganic needles becomes available, the arguments are likely to continue.

Aragonite needles are quickly re-suspended by bottom disturbance to form turbid clouds or patches, known as whitings. Whitings can be several hundreds of metres long and are carried along current paths until the velocity and turbulence diminish, at which point resedimentation occurs.

The drifting of needles inshore leads to a progressive accretion of micrite mud in the intertidal and supratidal marsh zones. Whether further accessions of chemically precipitated aragonite occur in the upper tidal zones as a consequence of in-place bacterial reactions is still conjectural.

Ancient micrite muds are most often found in shelf successions. In Britain, some of the best known examples are of Carboniferous age. The Cementstone Group limestones of Scotland and northern England are predominantly micrites or calcite mudstones, deposited under sabkha-lagoonal conditions, and later partly dolomitized.

The purer micrites are pale grey in colour, and have a smooth, almost conchoidal fracture, but small quantities of impurities may impart darker or more brilliant colours. Thin sections show that most of these rocks consist of grains ranging from 0·5 and 4μm in size. However, all silt-size grains (4-60μm) in patches and lenses may be present and occasionally dominant. Running through the rocks are irregularly branching little veins of sparry calcite; these are probably formed at the time of conversion from aragonite mud to a solid calcite rock.

In the conversion of the aragonite all traces of needle-shaped grains are lost and the new calcite grains rapidly become cemented into a rigid non-porous fabric by marginal syntaxial overgrowths which progressively infill the intergranular pores. Pressure solution at the contacts of the calcite grains probably plays an important role in the deposition of these overgrowths.

A particular variety of orthochemical micrite or mudstone, stratigraphically widespread, is that constituting diagenetic nodules. These mainly occur in relatively deep water successions where accumulation is slow. The nodules are disposed in layers, are of sand to boulder size, commonly ovoidal and usually have gradational contacts with the enclosing marl or limey argillaceous sediment.

They are light brown, grey or white, though in the Ammonitico Rosso facies (Jurassic) and Griotte facies (Devonian-Carboniferous) of southern Europe and North Africa red and green colours are very prominent. Evenly textured magnesian calcite is dominant with varying amounts of aragonite, calcite and dolomite. Skeletal matter, siliceous, calcareous and carbonaceous, is incorporated, larger fragments seemingly acting as nuclei around which the early diagenetic micrite is precipitated. The mechanism of carbonate segregation probably involves dissolution of calcareous skeletons and precipitation of carbonate from the interstitial waters at certain points where higher pH levels and concentrations exist.

TERRESTRIAL DEPOSITS

Almost all natural waters contain calcium carbonate in solution. The solubility of carbon dioxide in water is increased by pressure, so that springs issuing from a deep-seated source, often at a temperature above that of the air, immediately become saturated and deposit their excess of calcium carbonate in the solid form; thus many hot springs, especially in volcanic regions, give rise to great masses of calc-sinter, travertine or calc-tufa, as in Auvergne, and on a much larger scale in Italy, especially in Tuscany and near Rome.

The deposits show a remarkable development of concentric and botryoidal structures, resembling those of oolitic and pisolitic limestones; while in some cases nearly spherical concretionary masses have been formed with a diameter of 2 or 3m, and consist of innumerable, thin concentric layers of calcium carbonate.

Perhaps the best known of all deposits are dripstones, so extensively represented in limestone caves. The water percolating through the overlying limestone becomes charged with calcium carbonate, which is deposited when the carbon dioxide is given off in contact with the air. The process is assisted by the evaporation of the water.

A film of calcium carbonate forms over drops hanging from the roof, and, by gradual accretion long, pendent icicle-like bodies (stalactites) with radial or concentric structure are produced. Similarly, the drops falling on the floor build up masses which often take the form of pillars (stalagmites).

REEF ROCKS AND BUILDUPS

A well-known type of lime deposition is that associated with reefs, usually in shallow and warm waters. Reefs are a variety of carbonate buildup in modern parlance and are the product of the accumulation of organic constituents and their detritus. They stand up as definite topographic features above the immediate area of deposition and differ in nature from equivalent surrounding and, in the geological column, overlying deposits. On some occasions, reworked shell debris, for example crinoids and bivalves with no sediment-binding potential, constitute buildup frameworks, but the more important structures usually comprise appreciable amounts of calcareous and red algae, scleractinian (stony) corals, sponges, stromatoporoids and polyzoa.

These have sediment stabilizing properties, so that in reefs proper the framework develops a more marked rigidity and wave resistance than some other varieties of buildup, such as carbonate mud mounds, shell banks (biostromes) and lithoherms (deep and cold water carbonate banks). Buildup frameworks are often of a framestone, boundstone or biolithite type, but, whatever the composition, are all subject to wave and current erosion, even being totally destroyed on occasion. The products of disintegration are a range of lime deposits which are often much greater in volume than the framework itself.

On the seaward or fore-reef side of modern barrier and atoll reefs talus deposits (scree breccia) are usually thick, steeply dipping and grade into deeper water sediments. On the landward or back-reef side the reef detritus usually forms a thinner veneer and is closely intermingled with shallow water lagoonal deposits. The association of the reef proper and the sediments derived from its abrasion is commonly referred to as the reef complex.

If growth of reefs is extensive then their influence on sedimentation may be quite considerable. Circulation may be so restricted in the back-reef lagoonal areas that evaporites are precipitated. Terrigenous silts and muds may be prevented from migrating towards the deeper basins or may be channelled through gaps in the reef complex. In the latter case the materials may be transported via the mechanism of turbidity currents.

Modern reefs are mostly 4000-5000 years old and are generally confined to shallow water, though the majority stand at or near the edges of steep submarine slopes, so that their debris is often deposited in deep water. Reefs situated entirely in shallow water areas are referred to as patch reefs, for example the Alacran Reef on the Yucatan Shelf, and are of less importance at the present day than in the past. Reefs of the Palaeozoic and Mesozoic appear to be mainly of the patch type and grew principally on broad shelves in areas of epicontinental seas. They are frequently the sites of hydrocarbon accumulation, especially if dolomitized, as in the Devonian of Alberta.

Modern barrier, fringing, patch and atoll reefs do not consist by any means exclusively of frame-building coral - there are always present numerous shells and skeletons of marine invertebrates which live in the shelter of the reef. Encrusting algae are also important rock builders in such an environment. Nullipores (Corallinaceae, such as *Lithophyllum*) live in great abundance on the outermost part of the reef, where it is most fully exposed to wave action. Robust corals and molluscs also inhabit the more rigorous environments. *Halimeda, Penicillus* and other green algae colonize the more sheltered parts, such as the lagoon side.

The coarse calcirudite deposits associated with such reefs consist of large pieces of detached coral, sometimes encrusted with *Lithophyllum*. The finer fragmental deposits are shell, coral and foraminiferal calcarenites, of packstone and grainstone type, and calcilutites of wackestone type. The lagoon deposits of atolls contain large quantities of *Halimeda* segments, foraminiferal tests and coral fragments set in a micrite matrix. They are mainly mudstones (biomicrites), floatstones and wackestones.

The proportion of structured organic constituents in the framework of modern and ancient buildups varies from 5% to 90%, the remainder consisting of physically comminuted detritus plus cement, both of which reduce the original porosity. Penecontemporaneous marine aragonite and magnesian calcite in fibrous, bladed and micritic form are the dominant cements and are precipitated most readily where high wave energy is concentrated, forcing the sea water through the buildup interstices.

In Palaeozoic and Mesozoic reefs the calcite cement is commonly radiaxial, with a layer of outwardly diverging fibres and twinned crystals mantling allochems and in-place materials. This cement appears to be a direct precipitate following on the dissolution of aragonite.

Lowerings of sea-level, as during Pleistocene glacial epochs, also have a marked effect on buildups. Irregular karstic relief and caliche profiles develop on the exposed and inactive surfaces, and vadose cementation from freshwater pore fluids ensues. Isopachous and even blocky and drusy calcite is precipitated at these times.

An interesting feature of many varieties of buildup is stromatactis structure, well seen in some Devonian and Lower Carboniferous successions of northwest Europe. The structure is of uncertain origin, possibly due to the growth and decay of algae or de-watering, and can occupy as much as 70% by volume of some buildups. It is essentially a flat-bottomed cavity, up to 20cm wide, with an irregular upper surface and lined with drusy calcite. The cavity can show a geopetal fabric.

DIAGENESIS

Compaction, dissolution and cementation processes and effects are so closely interwoven in carbonate deposits that it is difficult to separate one from the other. The problems originate from the ease with which soluble carbonates are sedimented, dissolved and reprecipitated. Aragonite muds can become slightly lithified by sub-aerial desiccation and by the progressive conversion of aragonite into calcite, soon after burial.

Cement is commonly introduced early into deposits, so inducing further rigidification, as in hardgrounds and beach-rocks. Under these circumstances the original depositional porosities of 70% can be reduced quickly to less than 5%. If it happens that recrystallization and cementation occur at relatively slow rates, then a considerable amount of grain repacking and de-watering under load is feasible, with volume reduction and porosity reduction which can amount to 50% of the original porosity.

Shells show breakage and pellets, ooids and oncolites are deformed and reoriented. In some Miocene chalks and bioclastic limestones of Cyprus the presence of penecontemporaneous slumped beds, occasionally exceeding 12 m in thickness, suggests reduced diagenetic rates. The beds incorporate isoclinal, overturned and recumbent folds all generated during downslope movements of only partly lithified materials.

The development of stylolitic and microstylolitic junctions within limestones is indicative of compaction, though the depth at which the effect is generated or the reasons why some limestones are more susceptible to their formation than others are unknown. The sutured junctions, sometimes several centimetres in amplitude and tens of metres long, are caused by pressure-solution mechanisms.

Loading gives stress at grain contacts, solution ensues at the contacts and the carbonate taken into solution becomes available for reprecipitation as cement in adjacent pores. Considerable amounts of late diagenetic cement are probably produced in this way. Suggestions have been made that pressure-solution processes alone can reduce the thickness of partly cemented limestones by 30-40%, but these figures are probably excessive. On the other hand, there is little doubt that the process is the predominant porosity-reducing mechanism in burial diagenesis.

At present, cementation processes are active in a wide range of environments, extending from sub-aerial situations of dripstone and caliche accumulation, via intertidal zones where cementation is rife, to deep sea situations of calcilutite ooze deposition. In all these environments the bulk of the cementation is concurrent with deposition and, if not, mostly soon afterwards. It is unlikely, however, that cementation ceases once the sediment is removed from contact with the depositional waters, and further accessions of cement would be expected under high loading, thick overburden pressures.

Cementation is not always a continuous, steady effect. Discontinuities are recognizable by changes in the chemical composition and morphology of the carbonate grains, or by cements clearly pre-dating and post-dating certain events, such as a phase of fracturing of shell fragments by compaction or a phase of dissolution. *Cathodoluminescence* techniques effectively distinguish these discontinuities and, with care, even allow a cement stratigraphy to be established for given limestones.

The time intervals between cementation episodes are very variable, extending from a few days to hundreds of millions of years. When medium-term intervals of time are involved sea-level changes can be an important factor. For example, a sediment partly cemented by aragonite and magnesian calcite in an intertidal environment can become isolated from the sea by a fall in level and quickly become susceptible to a different style of calcite and dolomite cementation.

The distinction between the two phases of cementation can sometimes be confirmed by determining the isotope composition of the cements. The oxygen (δO^{18}) and carbon (δC^{13}) isotopes for the freshwater cements are lighter than for the marine cements.

Longer-term burial diagenetic cementation is little understood, but is likely to be very important in conjunction with compaction in altering the distribution, composition and texture of cements. Primary porosity in deeply buried limestones is reduced by loss of early cements and their replacement by coarse-grained calcite and dolomite, which often enclose pre-existing allochems (poikilitic texture). On the other hand, secondary porosity can be increased by the development of solution cavities. Overall, porosity is very unpredictable in these limestones.

The sources of primary cements are always a bone of contention, some authorities laying stress on direct precipitation from sea water, others on the solution of aragonitic organisms at higher levels within a carbonate layer and reprecipitation as calcite at lower levels. Pressure-solution mechanisms are considered important during the later stages of lithification. In some caliche and calcrete deposits in the Mediterranean region the source of the blocky and micro-stalactitic cements is usually either underlying chalks, marls and calcareous shales or calc-alkaline igneous rocks.

It has been claimed that the type of primary cementation can be broadly indicative of environment. In the meteoric water zone, micritic and microrhombic low-magnesium calcite and acicular aragonite are commonly precipitated near the ground surface and coarse sparry calcite at greater depths. In intertidal and subtidal zones, sometimes marked by great fluctuations in salinity, cementation is generally accomplished by acicular and radial fibrous micritic aragonite and magnesian calcite.

Whether the distinctions between the cements are retained as the rock progressively lithifies is another debatable issue. Primary cements are very susceptible to change, either by dissolution or by in-place alteration. One common type of in-place alteration is neomorphic aggradation, which produces grains larger than the original, and another type is neomorphic degradation, which is a grain diminution process.

Whenever these processes become operative it means that the fabric of the sediment, and often its mineralogy, alters; occasionally the alteration is drastic. Although there are some criteria which allow neomorphic changes to be recognized in thin section, such as transecting boundaries, it is an inescapable fact that the postdepositional changes introduce great subjectivity into the interpretation of the origin and subsequent evolution of a limestone. Perhaps it is advisable when examining limestones of all ages to assume that, unless there is strongly contradictory evidence, the rock fabric differs to some degree from that of the sediment originally laid down.

11

Magnesian Limestones and Dolomites

Most carbonate rocks contain magnesium carbonate in addition to calcium carbonate, and analyses show every gradation in composition from almost pure calcium carbonate rocks to those with more than 50% of magnesium carbonate. The nature of the magnesium-bearing minerals, and their distribution within the rock, varies to some extent in different types of magnesian deposits. In the types with approximately equal proportions of magnesium and calcium, the double carbonate, dolomite, $CaCO_3.MgCO_3$, accounts for most of the magnesium present. The rest is often in the form of an independent mineral, magnesite ($MgCO_3$).

Because of the complexities in mineralogy the classification of dolomite deposits on a mineralogical or chemical basis is difficult. The simplest criterion for dolomitized rocks is the relative percentage of calcite to dolomite.

Another method of tackling the classification problem is to adopt the petrographic approach of Folk. Partially dolomitized limestones are described using his shorthand nomenclature plus suitable qualifiers such as dolomitic or dolomitized. Under this scheme, a partly dolomitized oolitic limestone with sparry calcite cement could be called a dolomitic oosparite. More complete replacement by dolomite with 'ghost' or relict textures would lead to an oolitic dolomite, and a completely replaced limestone with no 'ghost' texture would be called a dolomite, with suitable grain size designation, for example coarse-grained dolomite,

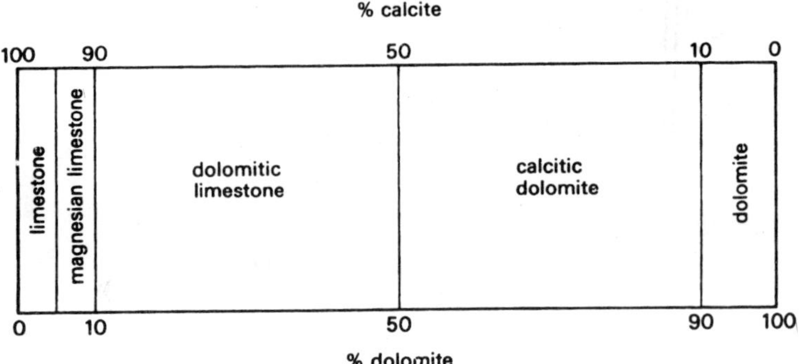

Figure 11.1: Limestone-dolomine classification.

dolarenite. At these high levels of alteration some authorities prefer to use the rock name dolostone instead of dolomite.

Dololithites

These are accumulations of dolomite crystals and dolomite rock fragments which have been derived from pre-existing dolomitic rocks. Subaerial weathering of mildly calcitic dolomites is capable of producing such rocks by the solution of interstitial calcite. The dolomite grains are progressively released and accumulate as sandy pockets and aprons mantling relict cores of dolomite rock. A very slight degree of recrystallisation is probably sufficient to destroy the clastic appearance of dololithites, and in the finer-textured types the changes involved in ordinary lithification would effectively remove all microscopic evidence of their origin.

DOLOMITIZATION

Most dolomitic rocks were originally laid down, as lime deposits, and have acquired their present composition as a result of early or late diagenetic processes. In most instances, the dominant process seems to have been replacement, with the dissolution of aragonite and calcite,

Table 11.1: Carbonate Reservoir Rocks, South-West Iran.

Dolomitization (%)	*Porosity (%)*
0	0-4
20	4-8
32	8-12
Over 58	Over 12

and precipitation of dolomite. However, a few modern instances have been described in which dolomite may have formed as an original

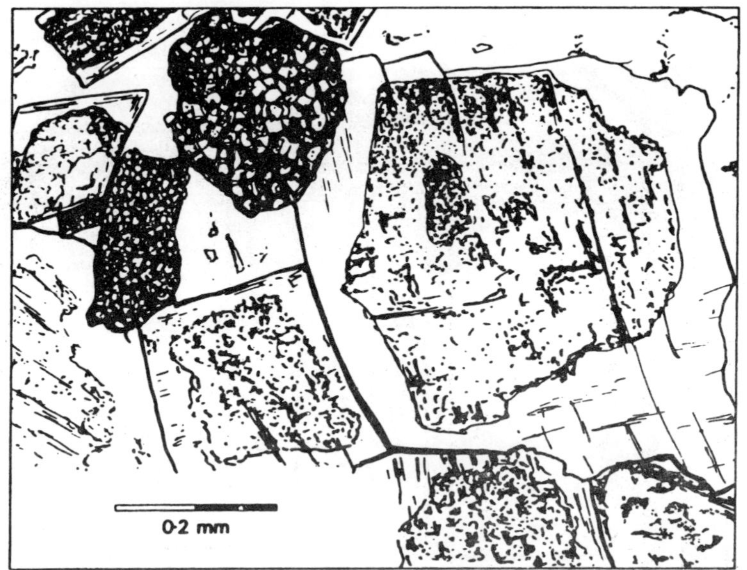

Figure 11.2: Dololithite, Cretaceous, Texas, Fragments of 'dusty' clastic dolomite grains and weathered finer-grained dolomitized rock (dark) set in a cement of predominantly syntaxial dolomite.

precipitate. These are invariably in desiccating shallow lagoonal or lake situations, such as The Coorong in South Australia, in which hypersaline pore waters appear to have the appropriate chemical qualities for direct primary precipitation. In many of these cases the dolomite is associated with evaporite minerals, such as gypsum and anhydrite.

Dolomite rocks preserve the structures and textures of their parents to varying degrees. In the earliest stages of dolomitization pre-existing textures and organic structures, like bioturbation and algal lamination, are clearly preserved but, as the intensity of change increases and the sediment takes on a more mottled appearance, there is a strong tendency towards their destruction. In some dolomite rocks which have undergone successive phases of shallow and deep burial alteration over a long period of time, the growth of coarsely crystalline dolomite effectively obliterates any primary fabrics.

Dolomitization is often seen to be selective in its attack upon the various constituents of limestones. The dolomitization of ooids often appears to proceed independently and in advance of the cement (Fig. 8.3). The fact that magnesium-rich calcareous algal remains are present in some ooids may be important in their speedier rate of alteration. In general, aragonite is more readily altered than calcite (even though the

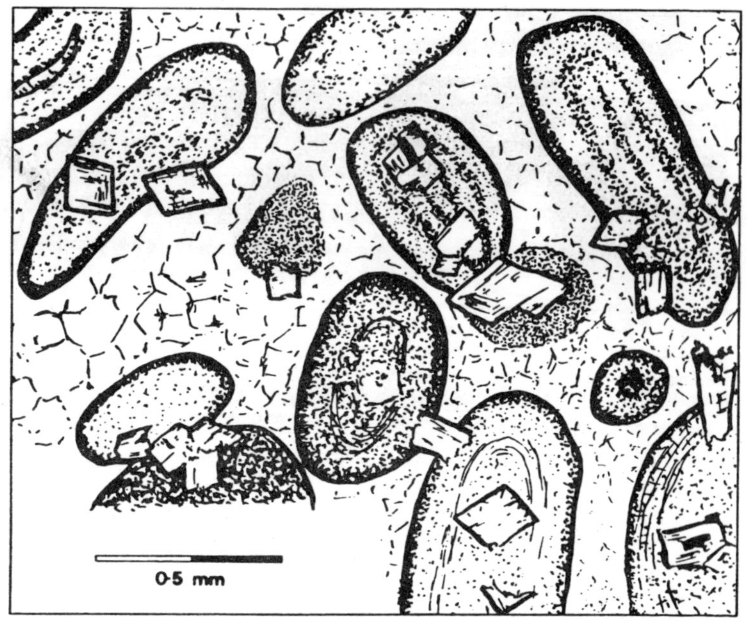

Figure 11.3 : Partially dolomitized oolitic limestone, Middle Devonian, South Morocco. A dolomitized oosparite. The rhombs of dolomite show preferential generation within the ooids, though transecting external boundaries of the allochems.

calcite may have a high magnesium content) and fine-grained compact material more readily than coarsely crystalline structures.

Aragonitic shells such as those of gastropods and cephalopods are the first to suffer dolomitization, and are often found to have been altered before the matrix. In contrast, stout calcitic fossils in calcarenites will often resist dolomitization even though the enclosing sparite cement and micrite matrix may be completely altered. In this way are formed dolomitic rocks enclosing more or less unaltered fossils, such' as crinoids, stout brachiopod shells or rugose corals.

The process of dolomitization itself may have far-reaching effects on porosity. The replacement of calcite by dolomite on a molecule-by-molecule basis involves a contraction in size of 12-13%. Under ideal circumstances this can mean that a limestone with a rigid fabric can have its porosity increased by 10% or more. In some Arabian Jurassic oil reservoir limestones porosities of 19% are recorded in the more highly dolomitized layers; permeability is also comparatively high.

The actual time which elapses between the deposition of the limestone and its subsequent dolomitization varies considerably in different cases. Many fine-grained carbonate deposits are altered very soon after deposition, while they are still lying in an unconsolidated condition. In other cases, there is evidence that dolomitization took place long after the original rock had been lithified, the magnesian solutions finding an entry through open joints and minute pores. It is thus important to try to distinguish between rocks affected by early diagenetic or penecontemporaneous dolomitization, and those which consolidated as limestones, and later were subjected to late diagenetic and subsequent dolomitization.

The final general point about dolomite is that weathering in the presence of pyrite or sulphate can bring about reconversion to calcite. This process of dedolomitization has been noted, for example, in Recent lime deposits of the Middle East, Carboniferous limestones of the Isle of Man, and in the Upper Jurassic Gigas Beds of Germany.

Early Diagenetic (penecontemporaneous) Dolomitization

Most bedded dolomites which preserve an approximately uniform character over a wide area are believed to have undergone dolomitization shortly after deposition. Rocks of this kind are sometimes of considerable thickness and form important stratigraphical units, such as the Sailmhor Formation (Cambrian) of north-west Scotland, the laminosa-Dolomite (Lower Carboniferous) of South Wales, the Magnesian Limestones (Permian) of northeast England and the Knox Dolomite (Ordovician) of the Appalachians of eastern North America.

On the other hand, the older the strata are the greater the care needed in interpretation and it is possible that certain of these units have been subject to one or more superimposed phases of dolomitization. Beds within the Sailmhor Formation have passed through one recognised phase subsequent to the initial alteration. The evidence for the penecontemporaneous date of alteration varies with the individual character and circumstances of each bed in question.

In many cases, stratigraphical evidence is sufficient, as in those examples which lie between unaltered limestones, showing that magnesian solutions cannot have been introduced subsequently. In South Wales, pebbles of dolomite may be found enclosed within the overlying conformable and unaltered limestone formation.

The conditions appropriate for penecontemporaneous dolomitization are a subject of endless debate. It is agreed generally that:

(a) almost all dolomites of this type are marine shallow water or supratidal in origin and currently are best developed in hot, relatively humid regions. Minor developments occur in salinas and coastal lagoons

(b) carbonate and bicarbonate concentrations need to be relatively high

(c) the original carbonate host-sediment needs to be very permeable to allow the free flow of ground water and the mass transfer of Mg^- and Ca^- ions

(d) time, probably 3000 years or more, is an important factor if significant dolomitization is to occur. During this period the Mg^{++} supply needs to be sustained.

It is unclear what role is played by organic catalysts or inhibitors in creating appropriate chemical conditions within the pore waters. Microbial sulphate reduction affecting sulphate ion content may be a promoter.

As many recent and ancient dolomites are interbedded with or pass laterally into evaporite beds it is hardly surprising that one of the widely supported hypotheses embraces the origin of both types of sediment. Supratidal flats, as in the Bahamas, West Indies and Persian Gulf (where they are known as coastal sabkhas or sebkhas), lie above normal high tide and are only flooded in their inner and central parts by highest spring and storm tides.

They are sub-aerially exposed for long periods between these highest tides. As a consequence, the laminated lime deposits are only subject to intermittent soaking or flood-recharge by normal sea water. Capillary evaporation then increases the concentration of salts in the pore waters above the water table until the point is reached where gypsum is precipitated in quantity. The fixation of calcium simultaneously removes the chemically inhibiting influence of calcium sulphate in solution on dolomite precipitation and increases the Mg/Ca ratio of the residual brine.

The flux of Mg-'ions into the high salinity brines is sustained by the upwards movement of pore water above the water table, replacing that lost by evaporation, a mechanism known as evaporative pumping. The conditions, especially in the intermediate sabkha zones, are then suitable for the replacement of carbonate in the original muds by dolomite. Aragonite is replaced in preference to calcite at this stage. Selective dolomitization of aragonite peloids and aragonite-filled algal borings in skeletal calcite fragments are cited as evidence.

The whole sequence of changes, initiated by tidal flooding and ending with dolomitization can be thought of as a single hydrological cycle. In practice, many thousands of such cycles, complete and partial, are needed to ensure any marked degree of dolomitization.

The proportion of original calcite in the upper layers of the sediment is usually diminished during dolomitization and is ascribed to contemporaneous leaching processes during flash-flooding or spasmodic rainfall. The carbonate ions taken into solution are probably immediately incorporated into the new dolomite grains.

The net result of all these complex processes is the production of very porous, light brown to grey crusts, patches and layers of weakly ordered or disordered calcian dolomite (protodolomite) in the top part of the lime muds. Protodolomite rhombs, 1-5 μ m in size, form 20-95% of these, and on the west side of Andros Island in the Bahamas thicknesses of crusts up to 150cm have formed in the last 5000 years.

Having a mildly imperfect structure, the protodolomite is unstable and eventually, usually long after burial, converts into 'ideal' dolomite with a well-ordered internal structure. Lagoonal dolomites of Jurassic age in north-west Scotland are still characterised by imperfectly ordered dolomite.

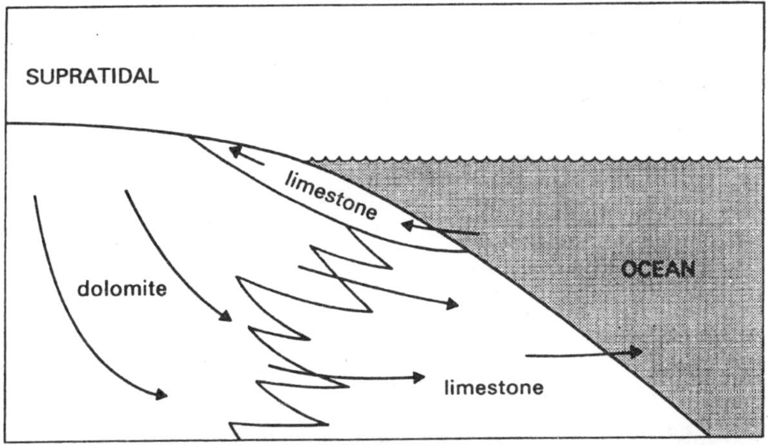

Figure 11.4 : Dolomitization and refluxion.

In coastal salt lakes undergoing desiccation and on sabkhas the high evaporation rates concentrate salts in the pore waters and progressively they become dense brines capable of seawards flow through the sediment, converting it into dolomite rock. This is a seepage-reflux mechanism. In The Coorong salt lakes of South Australia, where desiccation is seasonal, the periodic inflow of fresh water into the lakes appears to flush out the more soluble sulphate minerals precipitated in the sediments from refluxing brines, leaving dolomite and magnesite behind. This contrasts with sabkhas where the dolomite and other evaporite minerals persist in their association.

Seepage-reflux dolomitization is at its most effective when relatively high Mg/Ca ratios prevail in the ground waters. The Mg/Ca ratio can exceed 20 when recharge by the sea is frequent, though in intermediate sabkha zones where most dolomitization occurs and recharge is rarer, it is about 5-7. The latter ratio is little more than that of normal sea water; nonetheless, given an adequate length of time and adequate volumes of water, the ratio seems sufficient to promote the change.

On the other hand, it has been inferred on geochemical grounds that the Mg/Ca ratio need not be high and could be as low as 1 in some dolomitizing circumstances. The concepts embracing this view, known as the Dorag (from a Persian word meaning mixed blood or hybrid) and schizohaline models, are based on the mixing of ocean-derived brines (5-30%) and freshwater pore filling (phreatic) ground waters, such that a brackish water zone is created at a fluctuating interface in the sediment. The waters become supersaturated with respect to dolomite and undersaturated with respect to calcite, hence dolomite is deposited at the expense of calcite.

The advocates of these mixing models favour them because they explain much more satisfactorily than in the sabkha and related models the large and regionally extensive volumes of dolomites, especially those in late Precambrian and Palaeozoic successions. They also account for the absence of evaporites in many dolomite successions. The deficiencies of the proposals are that modern analogues are very rare, that dolomite grains often mould themselves around pre-existing aragonite and calcite cements without replacement, and that the geochemistry of the replacement is based on ideal well-ordered dolomite precipitation whereas, in fact, it is disordered calcian dolomite that is formed initially in the sediments.

Late Diagenetic (burial) and Epigenetic Dolomitization

Late and epigenetic dolomitization effects are often difficult to diagnose successfully because they are caused by a progressive series of events spread over what may be a considerable length of time. It is likely that many ancient dolomites have undergone unrecognised changes of this kind. Late diagenetic (burial) dolomitization by connate and ground waters occurs under more extreme loading and thermodynamic conditions than early diagenetic changes.

Some Mg-rich brines may migrate en masse from adjacent basinal muds undergoing compaction. However, some of the deeper-seated effects are probably contemporary with penecontemporaneous dolomitization effects at the surface. Some contemporary brines may seep downwards into underlying beds and create a whole host of cross-cutting dolomitic textures and structures. Primary bedding features may begin to be transgressed by patches of dolomite, and structures, such as ooids, fossils and sun-cracks, may become partially obliterated. Porosity also changes by contemporary solution of calcium carbonate through stylolitization, though this may be modified by later precipitation of new dolomite or calcite. Discrete dolomite grains coalesce and coarse, uneven sugary textures develop. Protodolomite converts into ordered dolomite.

Epigenetic changes by regional flow of heated Mg-rich ground waters are considered to be associated with periods of hydrothermal mineralisation, metamorphism, uplift and emergence of the rocks, and they are often found in the more highly folded and faulted regions. Shearing pressures seem to favour this late style of alteration and fractures often act as conduits for the mineralizing brines. Dolomites of this type are most easily recognised in formations which are not completely metasomatized, but still contain relics of unaltered limestone.

Lower Carboniferous limestones in Britain are affected by both penecontemporaneous and burial dolomitization. In some districts the dolomite grains are frequently well-formed rhombohedra zoned with haematite, and their presence in bedded formations underlying the PermoTriassic beds is taken as evidence that the rock was dolomitized in PermoTriassic times by infiltrating iron-rich ground waters.

The grey and yellow dolomites of neighbouring areas, on the other hand, contain no zonal inclusions of haematite and the alteration of these rocks is regarded as penecontemporaneous with deposition. Zoned dolomite euhedra represent episodic growth and fluctuations in pore water chemistry. The cleaner rims seen in such grains may be due to free growth in solution cavities.

It is recognised that late diagenetic and epigenetic dolomites may have passed through earlier phases of penecontemporaneous dolomitization. The overprinting of textures of one generation by those of another clearly makes difficult the interpretation of the nature of the original sediment. Many thick dolomites in Palaeozoic successions exhibit overprinted textures, and the uniformity and extent of their dolomitization implies that the burial and epigenetic changes were profound. Isotopic data support this inference.

MAGNESITIC DEPOSITS

Sedimentary rocks containing free magnesite as a principal constituent are much less abundant than calcitic or dolomitic limestones. Most

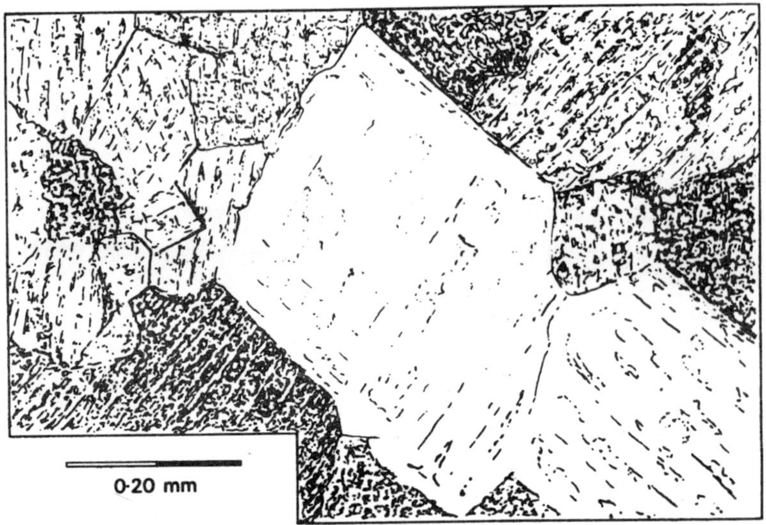

Figure 11.5: Limestone-dolomite classification.

occurrences of magnesite either result from the weathering of olivine-bearing ultrabasic rocks, or are formed by the action of magnesium-bearing solutions on other carbonates. At present the latter type of magnesite is forming in supratidal (sabkha) muds of the Persian Gulf and in coastal salina muds of South Australia.

Certain of these coastal lakes of the arid and semi-arid areas are characterised by the seasonal (summer) deposition of hydromagnesite and magnesite in association with dolomite. The influence of algae in producing high pH values in the water and concentrating magnesium in their skeletons may be a key factor in magnesite deposition, and it has been observed that certain major fossil deposits of magnesite are intimately associated with biohermal algal dolomites.

In Pitharagarh, northern India, lenticular magnesite beds, extending some 130 km at outcrop and having thicknesses up to 30 m, show this association. The beds contain about 42% MgO and 1·5% CaO. Other known examples have been formed, not in a marine environment, but in continental salt lakes. The deposits near Bissell, in California, appear to have accumulated in such a lake.

The magnesite forms a thin-bedded, compact white rock, interstratified with grey or green clays. Analyses indicate the presence of basic carbonates such as hydromagnesite, $3MgCO_3.Mg(OH)_2.\ 3H_2O$, in addition to normal magnesite. These deposits were evidently formed in a salt lake lying on the floor of a desert basin, and it has been suggested that the lake water contained sodium carbonate, which would react with the magnesium sulphate solutions brought in by spring water, precipitating hydromagnesite at the point of entry.

12

SILICEOUS DEPOSITS

These include purely organic deposits in which the organic remains are embedded in a matrix or cement of minutely crystalline silica, and purely inorganic deposits, such as multicoloured siliceous duricrusts which cap plateaux in dry regions of the southern continents. There is no sharp line of distinction between some of these deposits and some varieties of chert which bear the characteristics of diagenetic replacement of lime deposits.

Certain siliceous rocks have been claimed as chemical deposits derived from primary inorganic precipitates of colloidal silica. They include extensive bedded and nodular cherts of Precambrian and Cenozoic ages. Precambrian iron-formations partly consist of chert for which there is only meagre evidence of biological influence during deposition. In the alkaline, playa-like lake successions of the Magadi Basin of Kenya, Pleistocene cherts seem to have originated by the leaching and dehydration of a number of complex sodium silicates.

A few varieties of chert may have formed diagenetically from volcanic ash laid down in water. The terminology of the siliceous rocks requires a few words of explanation. The soft, entirely unconsolidated organic deposits are called oozes. Deposits which have remained unconsolidated although no longer in process of accumulation are termed earths, such as radiolarian earths and diatomaceous earths. The consolidated equivalents of these purely organic accumulations are termed radiolarites (radiolarian cherts) and diatomites (diatomaceous cherts).

The name chert is reserved for all siliceous deposits of a sedimentary nature whose main constituent is redistributed silica. Because of the length of time normally required for such redistribution, the bulk of true cherts are found in strata older than the late Cretaceous. They include rocks which are organic or inorganic in their immediate origin. In general, distinctive kinds of cherts are designated by the use of some qualifying descriptive term, but a few individual names, such as jasper, novaculite, lydite (lydian stone), and flint, are in common use for special varieties of cherty rocks.

Porcellanites (porcelanites), though not true cherts, are dull-looking, porous and highly siliceous sediments. They are widespread in deep sea successions of late Cretaceous to Pliocene age and are composed of low temperature opal-CT, mainly of biogenic origin.

FORMS OF SILICA

Several forms of silica are found in the siliceous deposits under consideration.

Opal-A is nearly amorphous hydrous silica, with a highly disordered crystalline structure, found in the tests and skeletons of phytoplankton, such as diatoms and silicoflagellates, zooplankton such as radiolaria, and other organisms including silicisponges and some gastropods and plants. Optically it is isotropic and has a lower refractive index than quartz. Its hardness and specific gravity (2.00-2.07) vary with the quantity of water present, usually between 2% and 13%.

Non-biogenic (inorganic) opal-A is much more restricted in distribution and is found in deposits adjacent to hot hydrothermal springs on land as siliceous sinter or geyserite, and sometimes at considerable depths in oceans. The flanks of oceanic spreading ridges are the site of such springs, the fluids at temperatures up to 100°C having become enriched in silica by the leaching activities of heated sea water passing through the crustal rocks. The passage of siliceous fluids through rocks also leads to secondary silicification. Opal-A occurs as a prominent cement in some sandstones.

Opal-C is inorganic in origin and is a hydrous variety of cristobalite found in lenses and pockets resting on and within altered submarine lavas and ashes.

Opal-CT is hydrous silica with a more ordered crystalline structure than opal-A and consists of layers of low temperature cristobalite and tridymite. Lussatite is a synonym. The opal occurs in the form of authigenic microspheres (lepispheres), blades and fibres only visible under the electron microscope. They are precipitated from the circulating pore waters during the early stages of burial diagenesis and represent a transitional state between that of biogenic opal-A and fibrous and microcrystalline quartz. Opal-CT is mostly detected in siliceous sediments of the porcellanite type of late Cretaceous age and younger.

Chalcedonic silica is minutely fibrous silica present in many types of chert as matrix and cavity fillings. It is more stable than opal and appears to have recrystallized from it. The surface usually has a spongy appearance due to the presence of minute water-filled pores and it is these pores which give a brown colour to the mineral in thin section.

Microquartz is a non-clastic anhydrous form of silica with a grain size usually less than 20μm and occurs as skeletal replacements, matrix and cavity fill. It is the common form of silica in all cherts older than the late Cretaceous and is regarded as a product of the slow conversion of opal-CT during burial. Occasionally, the conversion can be hastened in regions of high heat flow.

SILICA DEPOSITION

Most silica is transported into seas in true solution, probably as monosilicic acid (H_4SiO_4), and is very difficult to precipitate directly by electrolytic action. As the silica concentrations in rivers are generally higher by a factor of at least two or three than in sea water, there must be mechanisms accounting for the silica depletion. Changes in alkalinity have no effect on monosilicic acid. Possibly the silica in solution is adsorbed or coprecipitated onto colloidal and suspended solids within the river waters.

Where these waters meet the sea water, with its abundant electrolytes, almost all of the soluble silica might be removed, depositing it and the solids on the sea bottom. On the other hand, considerable doubt has been cast on primary inorganic precipitation hypotheses. There is a considerable body of opinion favouring the view that most transported silica is initially utilized by organisms (diatoms, radiolaria, sponges, etc.) in building their skeletons. On death, the siliceous skeletons accumulate and, if they do so in sufficient quantity, may recrystallize near to and just below the depositional interface to form, after prolonged diagenesis, cherty rocks.

In oceanic basins, at depths exceeding 2.5 km and often at about 4·5 km, a marked concentration of the skeletons is effected by the dissolution of associated aragonitic and calcitic tests, especially those of planktonic microorganisms. The waters at these depths are cold and undersaturated with respect to carbonates and are below what is known as the calcite compensation depth (CCD), so their solution ensues. The CCD is the level at which the rate of supply of organic carbonate detritus is matched or exceeded by its rate of dissolution. The level varies regionally in modern oceans and has also fluctuated through geological time.

Winnowing by intermittent bottom currents contributes to skeletal concentration by removing interstitial fine mud particles. In doing so, the currents also often effectively redistribute the skeletons down and along the bottom slopes forming a variety of common re-sedimentation structures. Turbidite- and contourite-like beds, with sharp contacts, sole structures and cross-lamination, are recognized in some bedded cherts.

The conversion of siliceous skeletons into chert is a drawn-out, complex affair which can take as long as 80 million years, especially in deep sea deposits. Most skeletons consist of opal-A which is partly soluble in alkaline interstitial waters. During the early stages of chertification within the soft bottom muds, such waters become enriched slowly in dissolved silica, mostly biogenic.

Degraded volcaniclastic detritus is an important secondary source. The eventual concentration of dissolved silica at lower levels in the muds, in part brought about by migration of siliceous fluids, can then approach the saturation level of amorphous silica, and opal-CT is authigenically precipitated in pores and as cement around detritals. Replacement of skeletal carbonate grains is initiated. This is the porcellanite phase of chertification.

As burial diagenesis proceeds, with higher pressures and temperatures, the opal-CT recrystallizes into fibrous chalcedonic silica and non-fibrous microquartz, with further and often extensive 'replacement of organic materials. The final product is chert in the strictest sense. A typical radiolarian chert at this stage could consist of 50-70% radiolarian tests in all phases of dissolution and replacement, 10-55% microquartz and 3-20% clay and iron minerals. The brittle nature of the cherts in late diagenesis is such that they readily brecciate as a consequence of loading and compaction. Chalcedonycemented chert breccias are common in some European Tertiary radiolarite successions.

BIOGENIC SILICEOUS SEDIMENTS

Diatomaceous Deposits

The diatoms are minute unicellular algae of the class Bacillariophyceae with cell walls composed of opaline silica. They form part of the plankton both in freshwater lakes and in the sea. Diatoms are most abundant in the photic layers of water, being found in the sea down to depths of 200m, according to the penetration of light. The siliceous frustules of the more robust species may be incorporated into marine sediments at any depth; however, they are not usually conspicuous in shallow water deposits, probably because they are masked by an abundance of other material. At the present day important deposits of diatomaceous ooze occur in the shelf and deep waters of the polar seas and in lower-latitude, cold-current areas bounding continental masses, for example the west coast of South America.

In these areas there is considerable exchange between surface waters and nutrient-rich, upwelling deep waters to produce high fertility all year round. The concentration of dissolved silica is several times greater than in surface waters elsewhere. In the Gulf of California, an almost land-locked deep basin, periodic diatom blooms contribute to the production of silica-rich and silicapoor laminae on the bottom. These sediments have a varve-like appearance.

In the Antarctic, diatomaceous ooze forms a continuous deep water belt encircling the zone of terrigenous deposits which fringes the edge of permanent ice. Consequently the ooze often contains a considerable admixture of terrigenous material carried by floating ice. The deposit is chiefly composed of the siliceous frustules of diatoms, together with some calcareous foraminifera; other constituents of organic origin are very rare.

When dried, the ooze is a white or yellowish powdery substance, which when examined under the microscope strongly resembles diatomaceous earth of freshwater origin (Tripoli powder or kieselguhr). Diatomaceous earths of late Pleistocene age are widespread in Scotland and northern England, usually in the form of lenticular beds laid down in proglacial lakes.

The diatomaceous earth of Kentmere, in the English Lake District, occupies an old lake basin of this kind. The deposit consists of diatom frustules mixed in varying proportions with clay and peaty matter, and is dark brown when fresh, but rapidly changes colour to deep olive green on exposure to the air. In its natural condition it is extremely porous.

Lithified diatomaceous deposits of Cretaceous and younger age are known from many parts of the world. In California marine deposits range from Cretaceous to Pliocene in age and some beds are estimated as carrying over 6 million frustules per cubic centimetre. Lenticular beds, as much as 30 m thick and devoid of internal stratification, were deposited extensively in the Rift Valley lakes of Kenya throughout Tertiary times. Some of the beds are very pure with silica contents exceeding 80%. The present-day alkaline playa-type lakes of the East

Africa Rifts, such as Lake Magadi and Lake Natron, are typified by high pH (> 9.5) and high dissolved silica content. The sources of the alkalies and silica are predominantly volcanic, commonly hydrothermal springs, but there is probably a contribution from diatoms. Complex chemical reactions ensue during which trona and hydrated sodium silicates are precipitated, often on a lake-wide scale. Magadiite ($NaSi_7O_{13}(OH)_3 \cdot 3H_2O$) is one of several such silicates constituting white, crystalline plastic beds up to 60 cm thick and which are contiguous with chert layers. Kenyaite ($NaSi_{11}O_{20.5}(OH)_4 \cdot 3H_2O$) is another and is commonly found in mantling chert platelets.

The chert appears to be formed by periodic leaching of the silicate beds by flashfloods during the rainy season, which lower the pH of the pore waters and remove sodium ions, alternating with more prolonged periods of exposure and desiccation. The cherts are marked by surface ridges and polygonal cracks suggestive of shrinkage which is estimated as being about 25% by volume of the original sediment. They also inherit deformation features from the original sediment. The conversion can be represented thus:

$$\text{Nasi}_7\text{O}_{13}(\text{OH})_3 .3\text{H}_2\text{O} + \text{H}^+ \rightarrow \underset{\text{(chert)}}{7\text{SiO}_2} + \text{Na}^+ + 5\text{H}_2\text{O}$$

Examples of magadi-type cherts are recognized in many ancient lacustrine successions, such as the Eocene Green River Formation of Wyoming.

Radiolarian Deposits

The bulk of radiolaria (Middle Cambrian-Recent) are confined to marine environments and tend to be abundant in tropical and subpolar waters. At the present day they accumulate to form radiolarian ooze in which the siliceous skeletons of radiolaria are a noteworthy constituent, ranging from 30% to 80%.

It is usually a reddish deposit, less plastic than the brown clay, owing to the smaller proportion of argillaceous material; apart from siliceous organisms the mineral components are much the same as in brown clay. Some specimens collected above the CCD contain appreciable amounts of calcium carbonate, chiefly in the form of foraminifera; diatoms and spicules of siliceous sponges are also present. The highest quality radiolarian oozes are confined to depths below the CCD and in the western Pacific and in the Indian Ocean cover a total area of about 4.5 million square kilometres.

Deposits containing abundant remains of radiolaria are known in the Cambrian and are found in most geological systems from then onwards to the Pleistocene. Many of the older radiolarian deposits are thought to have accumulated in shallow water; others, such as those associated with pillow lavas at spreading centres, were formed on the sea floor at depths which it is not always possible to evaluate, and a few examples, such as those of Barbados and Timor, may be true abyssal deposits, comparable with the modern radiolarian oozes although they are associated with globigerina marls.

In Greece, thinly bedded Mesozoic radiolarian cherts appear to have formed diagenetically within the basal radiolaria-rich division of silt and red clay graded units of presumed turbidite origin. Radiolaria are by no means confined to deposits such as these, but are widely distributed in fine-textured marine sediments. Because of their small size, they are usually inconspicuous in clastic deposits, but may be readily detected in the insoluble residues of many fine-grained limestones, such as the chalks. Radiolaria are also frequently found in phosphatic nodules, often in an excellent condition of preservation.

In some radiolarian deposits, such as those in the Oceanic Series (Upper Oligocene and Lower Miocene) of Barbados, the structural details of the radiolarian tests are perfectly preserved. Even in some Carboniferous deposits the preservation is sufficiently good to suggest the original sculpture of the shell. In most of the Palaeozoic cherts, however, only the shape of the radiolarian test is preserved in crystals of perfectly clear silica, surrounded by a turbid and darker matrix of minutely crystalline chalcedony.

In the Mediterranean Tethys the rifting of platforms in early Jurassic times created many elongated and narrow ribbon basins, at depths varying between 200 and 4000m, and suitable for rich radiolaria accumulations. The deeper water radiolarites, probably deposited near to the CCD, are often closely associated with ocean floor spreading ophiolites. They are evenly thinly bedded and rhythmically alternate with cherty radiolarian shales. Sharp basal contacts, sole structures and fining-upwards structure for some of the layers suggest episodic turbidity current styles of deposition.

The quieter phases, represented by the shales which carry a greater proportion of clay minerals and organic phosphates, also saw a variable amount of bioturbation by *Zoophycus-type,* medium-depth trace fossils. Radiolarian tests tend to be poorly preserved in the shales and their dissolution during burial probably subscribed to the chertification of the adjacent radiolarites.

The shallower water radiolarites in Tethys formed above the CCD and are thin, nodular and lenticular. They are interbedded with pelagic micritic limestones, chalks and marls, and occasional thin shales. Very thin 'filamentous' bivalves are present in quantity at some levels. The cherts exhibit both sharp and gradational contacts and, in many instances, there is evidence of replacement of carbonates.

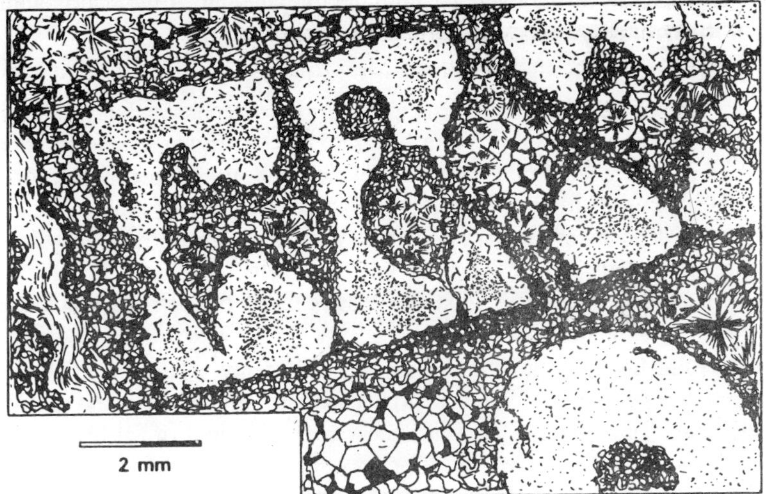

Figure 12.1 : Chert, Lower Carboniferous, northern England. Relict calcareous crinoid ossicles totally circumscribed and marginally replaced by fibrous chalcedonic silica and micro quartz.

Sponge Spicule Deposits

Although the siliceous sponges flourish principally in sea water, they are not entirely confined to such an environment, and spicules are substantial contributors to the diatomaceous deposits of a few freshwater lakes. Many dredgings taken from oceanic floors at present contain sponge spicules in noticeable quantities.

In Britain sponge remains play a variable role in the formation of cherts of Carboniferous, Jurassic and Cretaceous ages. The Carboniferous diagenetic cherts form nodular bands and sheets, and vary in colour from pale grey to brown and black, depending on the composition of the enclosing limestone or shale. They are replacive in origin and it is not uncommon to find a wide range of marine organisms showing all degrees of alteration. The source of the silica is probably the sponges.

The cherts in the Purbeck Beds (Jurassic) of southern England contain spicules of freshwater sponge *Spongilla,* and in places the chert is almost entirely made up of these spicules.

The Lower Greensand (Cretaceous) of southern England contains accumulations of spicules forming beds which vary from less than 1 cm to more than 1 m in thickness. The spicules are admixed with variable amounts of detrital material including glauconite. At many localities both the Lower and Upper Greensands contain spicule-bearing cherts, though it is far from clear as to the role the spicules have played in the formation of the chert. A considerable amount of the cherts with, in many cases, a loss of original internal texture.

Flint

The name flint is used principally for the siliceous nodules, nodular sheets and lenses in the White Chalk of western Europe, and in other regions where a similar facies is developed in the Upper Cretaceous. It is also used in describing the pebbly constituents of many younger detrital deposits of Tertiary and Quaternary age. The terrace gravels of the Quaternary River Thames and associated drainage systems in southeastern England, a region of extensive Chalk outcrop, are dominated by reworked, multicyclic flints.

Flint beds and nodular bands are commonly 10-20 cm thick and some can be breaks with a conchoidal fracture, giving clean and smooth surfaces. Thin splinters are translucent, and the colour of freshly broken nodules is usually brown, black or grey. Fresh flint has a surface patina of partly silicified chalk, usually chalky-white in appearance.

Flint consists of microquartz and chalcedony and is predominantly biogenic in origin. The silica is derived from the solution of opaline silica skeletons of sponges and radiolaria, the contribution of diatoms being uncertain. Solution is initiated early in the top few metres of the chalky ooze, and the silica is redistributed and locally concentrated and precipitated as cristobalitic porcellanite, especially in bioturbated and organically rich patches of the chalk.

The concentration of decaying organic matter in these patches may well have lowered the solubility of silica, allowing pores to be filled and fossils to be replaced. Many flints are casts of burrows, especially *Thalassinoides*. Further accessions of silica on to nuclei, and recrystallization into microquartz and chalcedonic silica, probably occurred during later burial diagenetic phases, with further replacement of the enveloping chalk.

Flint beds and nodular bands are commonly 10-20 cm thick and some can be traced over hundreds of kilometres, which makes them useful for correlation purposes. This widespread nature implies a periodic regional uniformity in the geochemistry of the interstitial waters of the chalk, probably most readily attained during pauses in sedimentation.

13

FERRUGINOUS DEPOSITS

Ferruginous deposits are characterized by anomalously high concentrations of iron compared with all other sediments. Whereas an iron-cemented or redbed type of sediment can contain as much as 10% iron, true ferruginous deposits frequently contain 15% and in some iron-formations more than 30%.

SEDIMENTARY IRON ORE MINERALS

The iron is represented mineralogically by a range of minerals. In ferruginous laterites formed during deep tropical weathering it is commonly present as redbrown hydroxides and oxides (limonite, $FeO(OH).nH_2O$; goethite, $FeO(OH)$; haematite, Fe_2O_3). These are generated under oxidizing conditions from many kinds of country rock. The ferruginous lateritic soils of Cuba, for example, rest on weathered serpentinites. The iron minerals and the soils form a large reservoir of iron feeding into water bodies, not only fresh water such as Lake Chad, but also estuaries, delta-fronts and oceans.

Freshwater bog ores, typifying certain present-day marshes and shallow lakes in Sweden and Finland, contain as much as 70% limonite in a disseminated and concretionary form. Manganese, which has a geochemical affinity with iron, can be precipitated with it in sufficient abundance to take the form of wad or bog-manganese (hydrous oxides of variable composition).

Residual and placer deposits developed by the selective concentration and mechanical and chemical winnowing of relatively heavy grains are often rich in iron minerals derived from nearby country rocks. Ilmenite

($FeTiO_3$) and magnetite (Fe_3O_4) are abundant in 'black sand' alluvial and marine placers.

In ancient iron-formations the dominant iron minerals are goethite, berthierine, chamosite, greenalite, stilpnomelane, siderite, pyrite, magnetite and haematite. Glauconite is comparatively rare.

Berthierine, an iron-rich chlorite, is widespread (in association with goethite) in Mesozoic and younger marine ores in the form of pale green peloids, ooids, intraclasts and matrix. At present, it is recognized in muds at depths of up to 150 m in the prodelta areas off the Niger and Orinoco, in the Malacca Straits and off Sarawak, where it is found as peloids, infillings of skeletal cavities and replacing calcareous shell fragments.

Berthierine is a diagenetic mineral precipitated from ferrous iron-rich pore waters under mildly alkaline, slightly reducing and bacterially active conditions associated with the decay of organisms. The degree of bacterial activity, which depends on the available organic matter, is important in bringing about a lowering of the Eh (Redox Potential) such that any ferric iron can be converted into ferrous iron.

Detrital and authigenic kaolinite already present in the soft sediment and carrying adsorbed ferric iron coatings probably plays an equally important role in making available the necessary alumina, silica and iron. With ageing and partial dehydration, at burial temperatures of about 145°C berthierine converts into chamosite, a very similar iron-rich chlorite.

This is the mineral generally found in older Phanerozoic iron-formations, such as the Ordovician Wabana and Silurian Clinton ores of eastern America, and the extensive Llandoverian to Upper Devonian ores of the Tindouf Basin in the Algerian Sahara.

Greenalite and stilpnomelane are complex ferrous silicates usually found in Palaeozoic and Precambrian ore bodies. Diagenetic greenalite has been identified in small quantities in recent deep coldwater muds but the bulk, as with stilpnomelane, appears to form during low-grade greenschist facies metamorphism of iron-rich sediments. The precursors are minerals such as chamosite and glauconite which alter and break down during metamorphism.

Stilpnomelane is also known to replace greenalite: Siderite ($FeCO_3$) and pyrite (FeS_2) are invariably diagenetic in ironformations, but are capable of being reworked by erosive bottom currents. The optimum chemical environment for siderite precipitation within muds is one with negative Eh (reducing condition), very high ferrous iron activity, excess

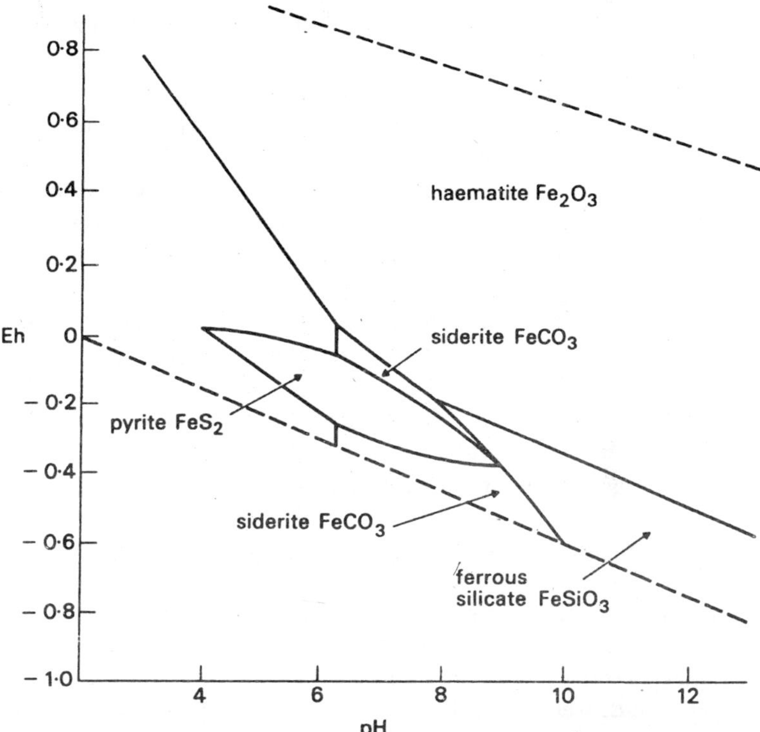

Figure 13.1 : Stability fields for iron minerals. Eh-pH relationships at 25°C and 1 atmosphere pressure, in the presence of water. The ferrous silicate field includes berthierine and greenalite.

dissolved CO_2 in the pore waters, and zero sulphide activity. Under these conditions the ferric iron in the sediment, originally introduced as oxides or adsorbed on clay lattices, is reduced and precipitated as carbonate. The carbonated pore waters are 'chemically aggressive' and it is a common observation that all pre-existing minerals and fossils within the sediments tend to show some replacement by siderite.

For pyrite to form in abundance a slower rate of sediment deposition and a greater quantity of slowly decaying organic matter is preferable than in the case of siderite. Ideal conditions are best established in relatively deep, stagnant marine basins, where the bottom and pore waters are oxygendeficient (Eh negative) and rich in sulphate ions and sulphate-reducing bacteria.

High hydrogen sulphide concentrations generated within the sediment react with the available iron and black monosulphides, such as hydrotroilite, are precipitated initially. The latter convert into pyrite, sometimes adopting a framboidal (microspheroids 4-8μm in size) habit. Magnetite as a diagenetic mineral is never abundant in Phanerozoic ironformations though it is present in quantity in some of Precambrian age. In the Upper Liassic Rosedale Magnetic Ore of north-east England the original berthierine ooids have a thick outer envelope of finely granular magnetite. The replacement of the berthierine may have occurred in the soft sediment under weakly oxidizing to mildly reducing, alkaline conditions, and the change may have been promoted by certain varieties of bacteria.

Most of the Upper Cambrian to mid-Ordovician Pisolitic Ironstones of North Wales are black in colour and contain finely divided magnetite replacing chamosite in concentric zones within the pisoliths and ooids. These rocks, however, have undergone low-grade metamor-phism and metasomatism, so that the origin of the magnetite and the timing of the replacement remains doubtful.

Glauconite is not a prominent mineral in iron-formations although it is known to reach concentrations of 20% or more in some modern and ancient marine shelf margin situations. The absence of glauconite, which is another diagenetic mineral, from the usual type of iron-formation clearly reflects different geochemical and environmental requirements from those needed for berthierine. Glauconite also seems to be formed via illite-smectite precursors, with the replacement of aluminium ions by ferric ions, whereas berthierine probably needs a kaolinite precursor.

TRANSPORTATION AND DEPOSITION OF IRON

Despite a broad understanding of the geochemistry of iron minerals in ironformations there remain many problems regarding how the iron is transported, concentrated and deposited. Iron is fourth in abundance among the elements in the Earth's crust so the concentration factor to produce an ore body is only 5 or 6.

Considerable concentrations of ferrous iron can be carried in neutral to acid waters. In subsurface waters, with a deficiency of oxygen, iron is carried in the form of ferrous salts, most frequently as carbonate, chloride and sulphate. In well-aerated surface waters, however, these salts are liable to hydrolysis and oxidation, with the production of ferric hydroxides, part of which are in a colloidal form. It is probable that a large part of the iron transported by rivers is carried in the form of ferric oxide hydrosols stabilized by colloidal organic substances and adsorbed on clay minerals.

These colloids carry positive charges and can be transported long distances without suffering precipitation, provided that the concentration of electrolytes is low, and that negatively charged colloids are not present in sufficient quantities to cause co-precipitation.

On entering the sea, such colloidal suspensions are influenced by increased alkalinity so they flocculate and are deposited. The iron is still dispersed widely within the bottom sediment and in insufficient concentration to be an iron ore. To achieve that condition other factors must have some influence.

A marked reduction in clastic input from the land is needed and it is beneficial if a significant increase of iron is introduced into the depositional basin. The latter is probably best achieved by lateritic weathering in adjacent low-relief land areas, erosion supplying iron hydroxides, kaolinite and aluminium silicate and oxide weathering products. Some ferruginous duricrust (ferricrete) fragments may be physically transported and reworked in river systems before being finally washed into the sea.

Peloid and ooid structures, now present in many iron-formations, may have been inherited from old soil profiles. In some Phanerozoic ores a recycling of iron from older iron-formations or red-bed successions can be important. Volcanic and intra-basinal hydrothermal spring sources are also feasible.

The physiography of the sites of accumulation is the cause of much conjecture as true modern parallels are lacking. A shelf region bounding an inland or epeiric sea, with an irregular strand line, lagoons, shoals, offshore islands and basins of varying depths, seems to be the most appropriate setting. In the basins and on tidal flats berthierine, sideritic and pyritic muds would most likely accumulate, and in the shoals and shallow lagoon areas oolitic and peloid deposits. In marine successions, intensive bioturbation by *Cruziana facies* trace fossils, such as *Thalassinoides, Rhizocorallium* and *Chondrites,* characterizes all the sediments and reflects prolonged, relatively tranquil phases of deposition and geochemical stability within them.

Nonetheless, there is also much evidence for reworking of the soft sediments. Crossbedding, ripples, storm scour structures, graded bedding, intraclasts up to several centimetres long, and worn ooids are particularly evident in the coarser oolitic and granular grades. This remobilization of materials, including shell fragments, ensures that they are dispersed over wider areas and progressively displaced either further on shore or into quieter and deeper basins off shore.

The net result is to create ore deposits with particles of mixed origins. Some of these bodies show minor coarsening-upwards trends. In many Phanerozoic successions, this coarsening-upwards can be extended to include associated lithologies, so that a larger-scale cyclicity of mudrock, sandstone and iron-formation capping can be observed. This probably originates during repeated phases of marine transgression, represented by the mudrock, and marine regression, represented by the sandstone and ironformation.

In the late Devonian deposits of Libya most of the ten ooid ironstone beds overlie coarsening-upwards deltaic sequences, where they accumulated during the dying periods of outwards delta growth. The marine regression (or change in water level in lakes) aspect of ironformation genesis has several implications. The progressive shallowing of the water and sub-aerial exposure of shoal and mud-flat areas could account for the precipitation of goethite (limonite) as ooids and as coatings around and replacement of pre-existing berthierine ooids, peloids and intraclasts.

Conditions could become conducive to the formation of diagenetic haematite as cement or as coatings. Many iron-formations carry zoned chamosite-goethite, chamosite-haematite and goethite ooids, expressive of partial reworking within a highly oxygenated milieu. Sparry calcite cementation can be widespread in certain of these beds.

Under marine regressive circumstances it has been suggested that original calcareous ooids, forming part of a shoal, can become partly or completely converted into berthierine and goethite by persistently circulating iron-rich ground waters. In this event, which is an extension of an original Sorby hypothesis, the source of the iron, silica and alumina could be supra-adjacent lateritic soils formed during regressive phases, or iron-rich mudrocks.

In either instance the ferruginization process might be delayed for several thousand years. Ferruginized calcareous deposits have been recognized beneath degraded volcanic ashes in the Bahamas and Indonesia. Certain old iron-formations undoubtedly consist of a mixture of calcareous and ferruginous ooids, but the evidence is equivocal as to whether the berthierine (chamosite) has replaced the calcite or vice versa.

IRONSTONES AND IRON-FORMATIONS

These deposits include Phanerozoic 'minette-type' berthierine, chamosite, limonite and siderite ironstones, haematite ironstones and Precambrian siderite, haematite, magnetite, pyrite and greenalite banded iron-formations. In recent years it has been suggested that the name ironstone be substituted by iron-formation on the grounds that the twofold

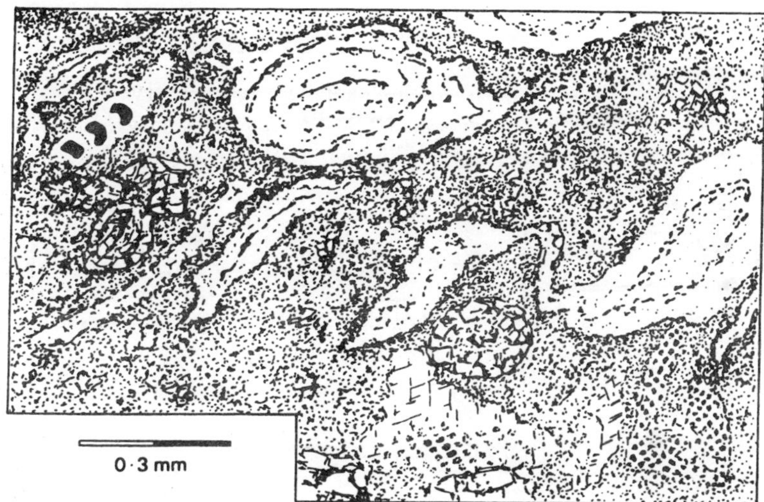

Figure 13.2 : Shelly berthierinitic sideritic berthierine oolite, Lias, northwest Scotland. Berthierine ooids, some spastolithic, and Shelly debris set in a fine-grained matrix of berthierine and rhombic siderite.

division is arbitrary, whether founded on mineralogy, lithology or age. However, ironstone is well established in the literature and will continue to be used, as appropriate, in this text. It has the advantage, normally, of referring to an individual bed or member, whereas iron-formation can be misleading in suggesting an association of beds, not all of which are necessarily iron-rich.

The descriptive nomenclature for iron-formations (and ironstones) is not well systematized. The Phanerozoic deposits are usually marine, oolitic and variably admixed with other allochems, matrix and cement. One method is to deal with the ooids and allochems independently of the matrix. For an oolitic textured rock with ooids of chamosite set in a matrix of limonite and siderite the name limonitic sideritic chamosite oolite is appropriate.

A shelly calcitic haematite oolite is an ironstone formed of shell debris and haematite ooids cemented by calcite, and so on. In the absence or near absence of ooids, the general term mudstone is used with qualifying nouns, such as siderite or chamosite, e.g. siderite mudstone. Mudstones are made up of silt- and claygrade particles, the former often being dominant. Limestones often show partial replacement of calcite by siderite, in which case the name sideritic limestone is used. The Liassic Marlstone ironstone of central England is a sideritic berthierine limestone in which calcareous ooids are peripherally replaced

IRON - FORMATION

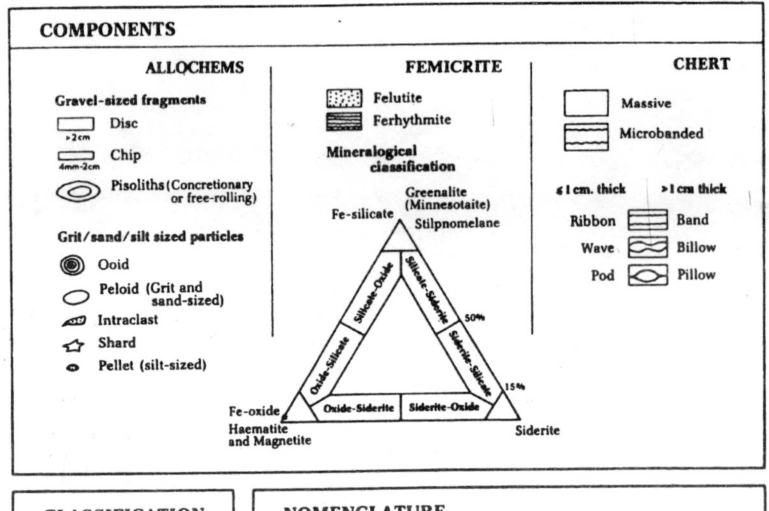

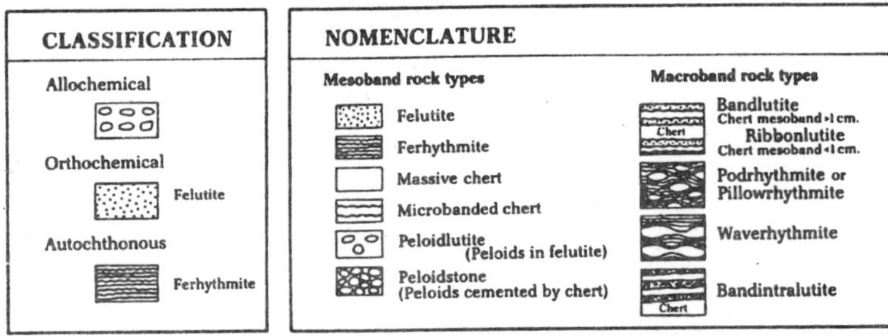

Figure 13.3 : Classification of iron-formations. Mainly suitable for Precambrian ironformations.

by berthierine. In the older types of iron-formation, in which chert bands and chert matrix are dominant, this nomenclature is more difficult to apply, even though some beds, as in the Lake Superior region, are clearly oolitic and peloidal.

Some authorities use a hybrid version of the Folk limestone terminology in conjunction with grain size. Ferruginous allochems (ooids, intraclasts, etc.) are distinguished from an orthochemical or felutite matrix. Felutite is a microcrystalline mixture of femicrite (iron minerals) and chert. Haematite-magnetite oo-felutite under this system would comprise haematite and magnetite ooids set in a fine-grained haematite-magnetite-chert matrix. As banding is common in the cherty ore bodies, a range of ancillary descriptive terms can be used, such as ferhythmite and ribbonlutite.

Minette-type Iron Formations

These are marine and fluviomarine successions, about 60m thick, and rarely extend over more than 20000sq. km. Although some of the best-known examples occur in the Mesozoic of Europe, there are many much older, as in the Upper Proterozoic of the Roper Bar and Constance Range regions of northern Australia.

The individual ore bodies are up to 15m thick, normally fossiliferous and bioturbated, and commonly accumulated at the top of coarsening- and shallowing-upwards sequences. They appear to reflect pauses in siliciclastic detrital input. Strong, probably storm-induced, reworking by powerful local currents is expressed by a range of primary shallow water structures. The thus remobilized and redistributed ferruginous materials are often cemented by sparry calcite.

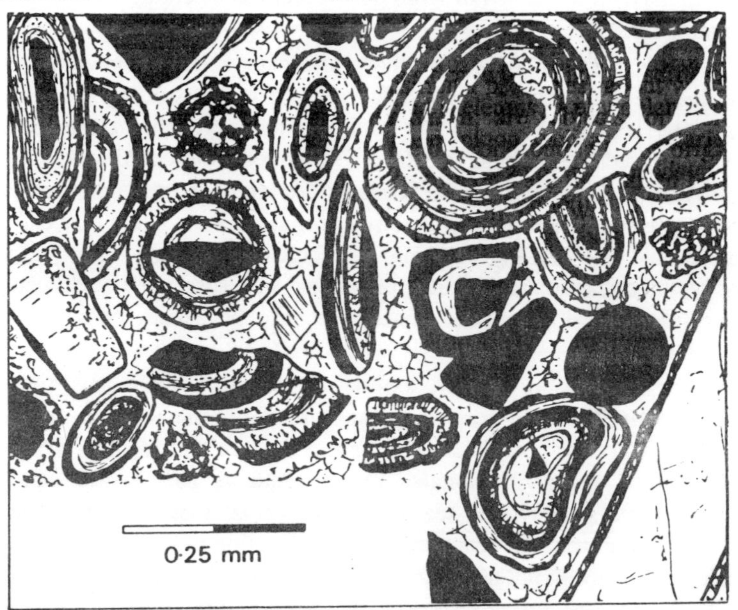

Figure 13.4: Sideritic berthierine oolite, Middle Jurassic, central England. Partially limonitized, berthierine ooids and calcareous shell debris set in a sideritic matrix. Some ooids are mildly spastolithic and others are much abraded.

The most prominent features of these iron ores are goethite, berthierine and chamosite ooids, and partly to completely ferruginized calcareous ooids, shell fragments and intraclasts. The ooids are variable in shape from spheroidal to ellipsoidal, and frequently concentric internally around a quartz or ferruginous mud nucleus. The matrix or cement is usually berthierine, chamosite and siderite in the higher quality ores and calcite in the lower quality. Siderite in microrhombic form is very prominent in some beds where it has replaced tote silicates, any calcareous matter and detritals. Secondary limonitization is widespread at weathered outcrop and gives an overall reddish-brown cast to what are, when fresh, pale green-grey rocks. In the Cleveland Ironstone Formation (Middle Lias) of northeastern England, original berthierine ooids sometimes have a white appearance due to decomposition into a mixture of finely divided opaline silica and kaolinite. In some instances the ooids have totally converted into kaolinite with loss in internal concentric structure.

Dense and opaque goethite ooids and peloids show little internal structure. In most cases, the goethite seems to have been directly precipitated from the sea or from pore waters as a coating around some suitable nucleus.

Ooids in fresh ores can show alternate zones of goethite and berthierine or chamosite, suggesting a periodicity in oxidizing and mildly reducing conditions within the soft sediment, probably close to the oxidation-reduction zonal interface. Under reducing conditions, in the presence of clay minerals, the ferric oxides are converted into a veneer of berthierine. When that situation is prolonged a complete conversion of the ooid into berthierine occurs. In the earliest stages of formation, the berthierine ooids are gel-like, plastic and very susceptible to distortion by gentle squeezing between adjacent grains. These distorted ooids are, spastolithic.

Not all minette ores are oolitic. Some mudstone seams consist almost entirely of fine-grained berthierine, chamosite and siderite. The siderite invariably post-dates and replaces any pre-existing iron silicates and oxides, and is precipitated in the reducing zone which can be several metres below the sea bottom (sideritization). The mudstones probably represent the progressive transformation of iron-bearing clays under quiet, though not necessarily deep, water conditions. A few workers believe that the ferruginization of micritic lime muds is another possibility.

Sedimentary Haematite-Chamosite Iron Formations

These can be considered as a special variety of minette ore in which haematite is a prominent additional member of the iron mineral association chamosite-siderite-pyrite. The beds usually carry a rich marine fauna and bear all the characteristics of shallow water deposition.

The best examples are found in Palaeozoic successions and it is possible that the age is significant, in that deep burial or mild metamorphism might have converted original berthierine, siderite and goethite into chamosite and haematite. On the other hand, the input of iron into shallow, highly oxygenated and warm waters may have been in such quantity that haematite especially could be readily precipitated. As with all minette ores there is ample evidence at some levels for the redistribution of the iron-rich components well away from the original sites of precipitation.

The Wabana iron-formation (Ordovician) of Newfoundland consist predominantly of sideritic chamositic haematite chamosite oolites. The ooids are built up of concentric layers of chamosite and haematite, suggestive of alternating reducing and oxidizing conditions. The ooids frequently have a clastic quartz nucleus and, indeed, the ore bodies as a whole belong to an arenaceous facies, limestones being absent. The interbedded sediments are principally sandstones with a chamosite cement.

In contrast, the Silurian Clinton ironstones of the eastern United States and certain Lower Carboniferous ores of Britain are essentially marine limestones in which various constituents have been replaced by diagenetic haematite. The Clinton deposits are mainly oolitic and shelly limestones in which the ooids and shell fragments are variably replaced by haematite.

In these, and similar deposits, there is usually abundant evidence for iron impregnation occurring at or before the time of final deposition. Altered particles are enclosed in clear sparry calcite and dolomite, or reworked haematized fragments are incorporated into unaffected sediment resting directly above. However, whether the iron was precipitated as haematite or some other iron mineral remains an unresolved problem.

Iron Formations and Banded Iron Formations

These are massive units, 30-1000 m thick, and traceable for many hundreds of kilometres along the strike. The bulk are early Proterozoic, such as the Biwabik Formation of the Lake Superior region, the Hamersley Group of the northwestern part of Western Australia and the Krivoi Rog series in the southern Ukraine, and formed about 2000 million years ago. All have undergone a degree of metamorphism and are low in alumina and phosphorus compared with Phanerozoic minette ores.

The origin of these rocks is unknown, but is likely to vary from region to region. Genuine stromatolites (benthic laminar mats of bacteria, cyanobacteria and algal protists) and degraded organic matter (kerogen) have been identified in a few iron-formations, such as the Gunflint of the Lake Superior region and the granular Frere Formation of Western

Australia, but they are uninformative on whether the deposits are marine or non-marine. Isotopic evidence, though confirming organic life, does not clearly distinguish marine from non-marine processes.

The stromatolites and richer organic deposits are usually present within granular, non-banded successions which carry peloids, ooids, oncoids and pisoliths, apart from exhibiting cross-bedding, ripples, cut-and-fills, intraformational breccias and sun-cracks. This association is consistent with deposition in shallow waters, probably little more than 5m in depth.

On the other hand, iron-formations are also characterized by banding with a total absence of fossils and shallow water structures. Many Archaean and Proterozoic ores are banded throughout their present outcrop and seem to have no 'granular facies' time-equivalent. The inference is that deposition was below the photic zone (about 120m in clear waters) and probably beneath wave base at a depth greater than 200m.

The banding in banded iron-formations (BIF) is on several scales. The Dale Gorge Member of the Brockman Iron-Formation in the Hamersley Group is 150m thick and is subdivided into 33 macrobands, 1-25m thick, each consisting of either an iron-formation (BIF) layer or a shale (S) layer. These layers alternate in a cyclical fashion. The BIF macrobands can be subdivided further into several mesobands, which are usually measurable in a few centimetre thicknesses.

At outcrop the banding causes a slaty appearance. Each mesoband consists of a layer of light coloured chert alternating with a darker coloured mixture of fine-grained iron minerals in a chert matrix (felutite). Many mesobands are, in turn, finely banded, the individual laminae being 0·1-1·5 mm thick and defined by varying proportions of iron minerals and chert.

These sub-mesoband laminae are often called aftbands and seem to be in the form of iron-rich and chert-rich couplets, suggesting a seasonal influence akin to that of pro-glacial varves. They do not have the basinwide continuity of the thicker bands and, in that respect, resemble the products of deposition in the quieter deeper parts of extensive bodies of water, such as large inland lakes and seas, or large partly barred basins on marine shelves. The macro- and mesobands reflect longer-term cyclical events in the depositional and source areas.

The iron minerals present in cherty iron-formations can be considered in terms of associations. The oxide association includes magnetite and haematite in which the haematite might be secondary after magnetite, or vice versa. Ooids showing complete or partial replacement by haematite may originally have been formed of goethite or berthierine. The silicate association includes greenalite, chamosite and other chlorites, and the carbonate association is predominantly siderite. The sulphide association is mainly pyrite.

The associations are commonly interbedded and changes from one to another also occur laterally. Evidence has been cited that haematite-bearing beds can be traced through directly into siderite-rich beds, with a gradual replacement of the siderite by haematite. In this respect the processes of diagenetic precipitation and replacement within banded iron-formations are no different from, and equally as complicated, as those in minette ores.

In the Hamersley Basin there is evidence for abrupt basinwide changes from oxide to silicate associations, which is taken to reflect major events affecting the geochemistry of the waters. Intermittent volcanicity has been cited as a sufficient cause for such changes.

Where Precambrian ores have been metamorphosed there are many mineralogic changes superimposed on to those effected during diagenesis. Greenalite converts into a talc-like ferrous silicate with sheaf and needle-like habits, minnesotaite. Stilpnomelane, a complex layered silicate, and pyrite replace greenalite and chamosite. If adequate sodium is available, fibrous riebeckite and its blue asbestos variety crocidolite form, probably at the expense of iron oxides.

At high grades of metamorphism the present mineralogy may only be a pale reflection of the sediment as laid down. Most of the pre-metamorphic oxides, silicates and carbonates are replaced by a range of iron-rich pyroxenes and amphiboles, such as grunerite $(Fe,Mg)_7.Si_8O_{22}(OH)_2$, and new magnetite can form. Chert and any original detrital quartz recrystallizes into coarser-grained quartz granules.

The origin of the iron and silica in cherty iron-formations probably involves a variable combination of material derived from land sources, including older red beds and laterites, and volcanic sources both on and off shore. Much of the iron was deposited as oxides either from solution or in some detrital form on the bottom of the basins, then converted into a range of oxides, silicates, carbonates and sulphides (via amorphous precursors), depending on the geochemical balance at and below the depositional interface.

The role of bacteria and algae, which were evolving and proliferating by early Proterozoic times, may have been a crucial factor in establishing that geochemical balance.

The silica problem is difficult to resolve. No siliceous organisms, as a potential source of chert, had evolved in Precambrian times. Leaching of hot volcanic rocks could have released silica into the waters, and it is feasible that a certain amount of chertification is secondary and formed by the redistribution of silica from interbedded shales during compaction and metamorphism.

Other workers advocate direct chemical precipitation of silica in some metastable gel-like colloidal form, carrying down with it $Fe(OH)_3$. This material, on coming to rest in layers resembling the present bands, dehydrates and segregates into chert and iron minerals. The small-scale and varve-like banding in the ores could reflect a critical chemical situation whereby the bottom waters were near to saturation with respect to silica.

During strong evaporative periods, climatically controlled and at air temperatures approaching 40°C, the balance could have tipped towards silica saturation and deposition, whereas during weak evaporative periods the conditions could have been more suitable for iron minerals to segregate and diagenetically aggregate from the silica gels already on the bottom. A certain amount of low-level current activity, concentrating the iron minerals further, is visible in aftbands.

NON-MARINE IRONSTONES

Sediments consisting principally of siderite occur as thin bands or as nodules in certain argillaceous rocks. Some of these sediments, such as the clay ironstones (sideritic mudstones) within the alluvial and deltaic Carboniferous and Mesozoic successions of northwest Europe, are fresh or brackish water in origin. The siderite appears to have been precipitated diagenetically as a relatively minor constituent of the mud, and later to have undergone segregation into nodules and layers.

The clay ironstones of the Coal Measures of northwest Europe are well known. They occur as bands of flattened nodules or as thin continuous beds, buff-brown to grey-brown in colour and compact in texture. Fossils, either plants or freshwater bivalves, are common in them, and often the nodule has clearly been segregated round a single organism.

The blackband ironstones of the Staffordshire and Scottish coalfields in Britain contain from 10% to 20% of drifted coaly matter and are practically free of argillaceous materials. Precipitation of the siderite occurred in shallow freshwater lagoons amid the coal swamps, screened off from all muddy sediment. The siderite is fine grained, the usual grain size being about 0.01 mm. Calcium and magnesium carbonates, and a little manganese carbonate, may form about 10% of these rocks,

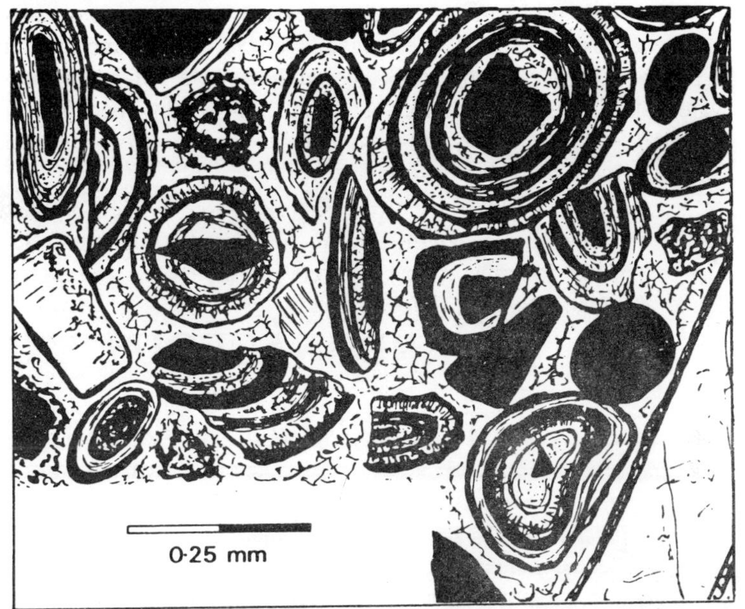

Figure 13.5 : Sideritic kaolinite mudstone, Coal Measures, northern England. Grossly spastolithic kaolinite ooids and globules (white) set in a dense matrix of partly limonitized rhombic siderite.

and probably exist in solid solution in the siderite, except in those cases where calcite occurs in the form of fossil shells. In some situations, however, diagenetic siderite is precipitated as spherical aggregates of radiating blade-like crystals.

These are known as sphaerosiderites and have been recognized immediately beneath old soil profiles, as in the Upper Carboniferous Coal Measures and Lower Cretaceous Wealden Beds of Britain. They evidently formed in stagnant, waterlogged and reducing conditions prevailing beneath the contemporary soils. Kaolinite ooids are known at some levels.

14

Carbonaceous and Bituminous Deposits

Carbonaceous and bituminous sediments include peats, coals, oil-shales and other deposits, which consist to a large extent of altered organic matter. Many contain recognisable. remains of plants, including algae. The distinction drawn between carbonaceous and bituminous products is often obscure in the literature and the terms are often used synonymously. Carbonaceous means like coal or containing carbon and refers to the products of diagenetic and low-temperature metamorphic coalification processes.

The changes leading to the formation of coal from plant material are initiated by bacterial and fungal decay at the surface and continue during deep burial and structural deformation. In the case of humic coals there is an early conversion of forest and swampy peat materials into humic substances with a relatively high oxygen content. Wood charcoal is produced under high oxidising circumstances.

Bituminous describes the generally soluble (in organic solvents) organic constituents of sediments, especially oil and fatty hydrocarbons. Bitumens are present in humic coals in relatively minor amounts in the form of waxes, resins and spores, but in some bituminous shales can form a high proportion of the total constituents. Oil-shales, which are frequently highly bituminous, derive most of their bitumens from the degradation of algal and animal matter.

Oil-shales and certain varieties of coal, such as boghead coal, are sapropelic in origin, having formed by the putrefaction of non-woody organic materials in non-marine and marine environments and under anaerobic conditions. These materials do not undergo peatification and humification changes.

The modern circumstances under which some sapropels accumulate are illustrated by periodic events in shallow lakes, such as The Coorong in coastal South Australia and in embayments at the southern end of Lake Balkash in Kazakhstan. Surface algal blooms of the genus *Botryococcus* (which is rich in hydrocarbons) become stranded as the lakes dry out and convert into a yellow-brown substance resembling rubber. In other lakes the algal concentrations are laid down on the bottom and consist principally of the remains of blue-green algae (cyanophytes) rather than the chlorophyte *Botryococcus*. All such sapropels are potential sources of hydrocarbons.

Both coals and bituminous sediments contain kerogen, a mixture of large organic hydrocarbon molecules (lipids, isoprenoids, terpenoids, steroids), which are insoluble in organic solvents. With burial and enhanced geothermal heating the kerogen progressively loses hydrogen and produces liquid hydrocarbons and wet gas (gas containing liquid hydrocarbons). This breakdown by heat accounts for many natural occurrences of liquid and gaseous hydrocarbons and is the main basis for the commercial extraction of oil from oil-shales.

Coals of commercial value are present in cyclic successions of all periods from the Devonian. Older coals or carbonised deposits are known, as in the Silurian of east China, but they are uneconomic. In Devonian and younger coals the humic matter is predominantly of land plant origin, whereas the older coals originate by the carbonisation of accumulations of algae and other organisms. In Mexico, anthracites and lignites range in age from Upper Triassic to Tertiary; in east China from Silurian through to Neogene. In Britain Carboniferous and Middle Jurassic bituminous coals outcrop, as well as Tertiary lignites in Northern Ireland, Devon and the Hampshire Basin.

The most widespread coal-bearing successions are deposited during major orogeneses when appropriate structures are created for burial and preservation. Major basins are formed along the edges of continental masses and extend laterally for thousands of kilometres. They usually have access to the open sea. A prime representative is the Upper Carboniferous succession laid down to the north of a mid-European continental mass. Coal seams, some several metres in thickness, can be traced for hundreds of kilometres in this paralic belt.

Other basins, designated limnic, are intracontinental, disconnected from the sea and much more limited in lateral extent. The area of coal accumulation, with individual seams occasionally 100 m or more in thickness, is generally small in relation to the total area of the basin. The Upper Carboniferous Lorraine, Brittany and Vosges coal basins of France are limnic in character, as are many late Cretaceous to Tertiary coal basins of Columbia, Venezuela and western North America.

The name coal-bearing province is given to the vast regions where conditions were appropriate for the synchronous deposition of coals. The Upper Carboniferous 'Euroamerican Province' is one such typified by bituminous to anthracitic coals. The Permian 'Gondwana Province' is another, although it is now dismembered into several coal regions and individual coalfields in Brazil, South Africa, peninsula India, Australia and Antarctica. In the Parana Basin in Brazil, coal beds are intercalated with deposits of glacial origin and the evidence here and elsewhere in the Province is now suggesting that the Gondwana coals need not have accumulated under tropical climates. Some probably developed in a humid, cool temperate climatic zone with seasonal rainfall.

ORGANIC CONSTITUTION OF COALS

Coals are either humic, formed of the degradation products of wood, or sapropelic, formed essentially of non-woody materials such as macerated plant debris, spores and algae. The identification of the constituents is very difficult when using a normal transmitted light petrological microscope, especially for the bituminous and anthracitic coals. Thin sections need to be about 5 pm thick before transparent particles can be reasonably categorised. Reflected 'white' light techniques using highly polished surfaces are better, for then many of the transparent and opaque fragments exhibit characteristic structures.

Fluorescent techniques, by which the organic matter is radiated by ultraviolet or blue light and variably emits luminous spectra, are a very useful adjunct to identification. The cause of fluorescence is unknown, but changes in fluorescence, often seen with increase in coal rank, may be due to changes in the amount of soluble organic matter and changes in the nature of constituent aromatic molecules.

The range of organic structures in coal particles is wide as the materials have been derived from the breakdown of the wood, bark and leaves of many species of trees and ferns. Pollen, spores, cuticles, resins and waxes are invariably present. All the primary biological fragments, clearly structured or not, are referred to as macerals, this being roughly equivalent to minerals in other sediments. Fusinite, for example, is a charcoal-like maceral with a wellpreserved cellular structure, sometimes mineral filled. Alginite is a maceral derived from algae, some of which are subspherical, 100-200μm in size, colonial and thick walled, with recognisable internal structure (alginite A or telalginite), and some of which are small, colonial, thin walled and lamellar in habit parallel to the bedding (alginite B or lamalginite). Other types of alginite appear almost devoid of structure and cannot be readily classified.

Macerals can be broadly grouped on the basis of their origin. The inertinite group is derived from structured lignitic debris, fibres and fungal matter which has oxidised under aerobic surface decay conditions, occasionally being converted into wood charcoal. The causes of charcoalification are ascribed to biochemical processes or spontaneous combustion leading to forest fires. In transmitted light these macerals are opaque and they do not fluoresce.

The vitrinite group consists of opaque macerals derived from wood, bark, leaves and roots under less oxygenated conditions than the inertinites. They are opaque in transmitted light but do show brown to red rims. Collinite is a structureless gel- or colloid-like maceral filling cell cavities. Two varieties are recognised. Telo-collinite (vitrinite A) has a higher lignitic content than gelocollinite (vitrinite B) which is richer in cellulose. Lignins andcelluloses are the main chemical constituents of humic coals. The maceral telinite is structured cell-wall material. The macerals in this group rarely fluoresce.

The exinite (liptinite) group is more mixed in character and encompasses pollen, spores, cuticles, fats, waxes and resins of plants and microscopic algae. Because of the heterogeneity of the group the macerals show a wide range of colours both in transmitted and fluorescent light. The lower-rank exinites usually have a very strong fluorescence. Resinite in transmitted light varies from yellow to red-brown.

Cutinite (surface layer of leaves) fluoresces from bright yellow to orange, and sporinite from green-yellow to brown-red. Sporinite comprises transparent yellow and brown mega- and microspores (with outer walls known as exines) which became opaque progressively with enhanced coalification. Fluorescence also changes steadily from green-yellow to orange, and finally to red in high-ranking coals.

The distribution of maceral groups within individual coal seams and coalbearing successions is not uniform. Bituminous coals often consist of alternate bright and dull laminae caused by defined assemblages of macerals. If the bright laminae are dominated by vitrinites (>95%) and the dull laminae by inertinites (>95%), the corresponding coal lithotypes are vitrain and fusain respectively. Durain and clarain are other lithotypes, and in refined classifications further subdivision is possible into coal micro-lithotypes.

In Australia Permian coals are rich in inertinites and poor in exinites compared with the Jurassic and Tertiary coals, which are rich in vitrinites and richer than the Permian in exinites. These differences are functions of the length of time available for diagenetic-metamorphic conversions and variations in the evolving floras and climates with the passage of time. Flowering plants (angiosperms) dominated the land flora from Cretaceous times onwards, displacing varieties of ferns, cycads and horsetails which were dominant during the Permo-Carboniferous.

The diagenetic-metamorphic changes affecting macerals commonly destroy any structure. This becomes increasingly so once the sub-bituminous and bituminous rank stages are reached at temperatures of 100-150°C. Alginite A arid organo-mineral complexes are typical diffuse, structureless organic materials found in sapropelic coals and oil-shales. Bituminite is another. The production of solid waxy and liquid bitumens during thermal transformation of macerals is commonplace. Exsudatinite is a form of oil found in humic coals, though not observed in sapropelic coals.

RANK IN COALS

The type of coal is determined from its maceral constitution. The rank of coal is more difficult to assess objectively. If humic coals are arranged in a series beginning with peat and continuing through lignites and bituminous coals to anthracites, it is observed that there are several progressive changes. In general, the moisture content decreases, as do the percentages of oxygen and volatile constituents. At the same time the percentages of fixed carbon and total carbon increase steadily. On these bases, the coals are divided into ranks as follows (increase in rank upwards):

Anthracite
Semi-anthracite
Semi-bituminous
coal Bituminous
coal Sub-bituminous
coal
Lignite
Peat

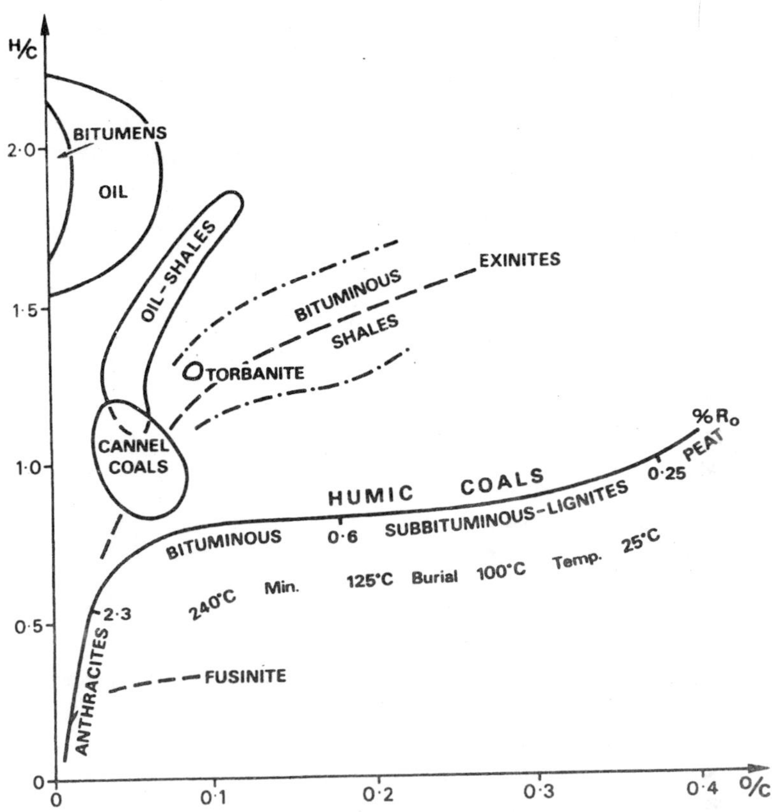

Figure 14.1: Compositional relationships of coals, macerals, sapropelic deposits and oils based on hydrogen/carbon and oxygen/carbon ratios.

Thus, coals of similar chemical composition and of approximately similar behaviour as fuels are grouped together in one rank, and coals at the top of the list belong to a higher rank than those below.

Unfortunately, rank cannot always be accurately assessed from the chemical factors listed above, so other techniques have to be adopted. In practice, the best of these, as it can be used over a wide rank range, is the optical technique in which the reflectance from the vitrinite group macerals is measured from polished surfaces using a photometer attached to a reflected light oil-immersion microscope. Polarised and non-polarised light is used on the rotating stage. Objectivity, accuracy and reproducibility of results are enhanced by automated image analysis. Vitrinite reflectance (R_{mean}, R_{max},/R_{min}) is less than 0.25 for peats, 0.25-0.6 for lignites and sub-bituminous coals, 0.6-2.3 for bituminous coals, and greater than 2.3

for anthracites. At values of *5* or over the rock is no longer a coal as it has been graphitised.

It has to be appreciated that the other maceral groups show reflectance. Inertinites are more reflective than vitrinites; exinites, whose reflectance verges on zero in the low-ranking coals, less. With increasing rank the reflectance of vitrinites increases and approaches that of inertinites. The reflectance of exinites of algal origin also increases to that of vitrinites in highrank coals.

Vitrinite reflectance of coals increases with depth of burial, burial diagenesis and low-grade metamorphism. Generally, the greater the depth of burial the greater the heat change, though if the geothermal gradient, e.g. 10-15 m/°C, and heat conductivity of the rocks are high at any locality then the higher coal ranks and higher reflectivities are reached at proportionately shallower depths.

Some Lower Carboniferous coals in the Moscow Basin are hard lignites and hard brown coals with low reflectivities, simply because the coals were never buried deeply enough and maximum temperatures probably only reached 20-25°C. Bituminous coals require minimum burial temperatures ranging from 125 to 240°C and anthracites reach even greater temperatures.

The intrusion of hot magma, especially in the form of thick extensive sills, can produce higher-ranking anthracitic coals adjacent to the heat source, as has been demonstrated in the Carboniferous coal-bearing successions of the Midland Valley of Scotland and the Alston Block of northern England. It is noticeable in these upgraded coals that the physical stresses in addition to heat have affected the vitrinites, so that they now show a greater optical anisotropy than hitherto.

The aromatic molecules and lamellae within the vitrinites have taken on a more pronounced preferred orientation, so that there is now a marked difference between the Rmax and R_m;.values. This condition, known as bireflectance, increases in vitrinites with increase in rank.

Significant features of rank variation in coals can be determined if seams are traced for a long distance. It is then found that the coals are of higher rank where they have been involved in crumpling, folding and faulting. The rank of coal increases with the intensity of orogenic forces and the natural heat treatment to which the coal has been subjected.

Regional variations in rank are well illustrated in the South Wales and Pembrokeshire Carboniferous coalfields, where the most highly disturbed seams are anthracites or comminuted anthracites, whereas the less disturbed are bituminous. The variation in rank is particularly noticeable as the anthracitic northwestern parts (Ammanford) of the Welsh coalfield is approached.

The Cretaceous and Tertiary coals of the Rocky Mountains field are of bituminous rank where they are involved in the folds of the mountain belt, but where they pass eastwards into the undisturbed region of the prairies, the seams change in character, and are represented by lignites and brown coals. A similar relationship between orogeny and regional changes in the rank of the coal seams affected can be traced in some of the eastern coalfields. In Pennsylvania, for example, there is a general increase in rank in an easterly direction as the Carboniferous seams become involved in the folds of the Appalachian Mountains.

HUMIC COALS

Lignites (or soft brown coals)

These are low-rank coals and are most commonly found in Tertiary and Mesozoic strata. They differ from ordinary bituminous coal in their brown or brownish-black colour, and in their greater content of moisture. Chemically the lignites are characterised by a high oxygen content, and by approximate equality in the proportions of volatile substances and fixed carbon. Because lignites contain from 25% to 75% of water they develop shrinkage cracks and crumble to small fragments when exposed to the atmosphere. Disintegrated or powdered lignite has a strong tendency to take fire by spontaneous combustion.

The porous texture of lignite is somewhat reminiscent of dried woody peat; the larger plant remains, such as wood, bark and leaves, are still recognisable and relatively uncrushed, and are embedded in a dark brown, structureless dense humic matrix. There are many varieties, some loose and fibrous, others compact and earthy. Lignites are rare in Britain, important examples being in Northern Ireland and at Bovey Tracey in Devon.

In Devon the beds, up to 4 m thick, are associated with kaolinitic ballclays of Neogene age. The bulk of the clay and woody materials, *e.g. Sequoia,* seem to have been carried into a lake from adjacent, much higher ground. True swamp vegetation is rare. Great deposits of lignite, partly allochthonous, partly au tochthonous, also occur in the Cretaceous and Tertiary strata of North America, and there are important beds in Germany, France and other parts of Europe referred to as *weichbraunkohlen* (soft brown coals).

Sub-Bituminous Coals (or hard brown coals)

These are intermediate in character between the lignites and bituminous coals. They are brown to black in colour, dull to bright, and, when freshly mined, have rather the appearance of ordinary bituminous household coals.

They are well bedded, though small-scale jointing (cleat) is not yet well developed. The high moisture content (over 20%) causes them to crumble on exposure to the air, and the disintegrating coal is liable to spontaneous combustion, especially if much pyrite is present. Volatile matter and fixed carbon are present in almost equal proportions, but the moisture content is lower than that of lignites.

In dull brown coals water is 25-35%, whereas in bright varieties it is 8-25%. An increase in compaction and reduction in porosity is matched by a progressive homogenization and gelification of humic materials. Lignins are converted into humic substances and celluloses are extensively altered.

In the bright brown coals the wood and bark have been converted into vitrinites and the Rmean approaches 0.6. As fuels, these coals ignite easily, burn with a bright smoky flame, and are capable of yielding much gas. Coals of this kind are usually of Mesozoic or Tertiary age, and are well developed in the Rocky Mountain Region of North America.

Bituminous Coals

These include the ordinary household and coking coals. They are generally well jointed or cleated and the joint faces are often covered with a thin layer of pyrite, ankerite or some other mineral. The coals consist of alternating bright and dull layers, the proportions of which vary in different samples. This laminated or banded structure results mainly from the interbedding of coal lithotypes, each with a different appearance on the fractured surface of a hand specimen and in reflected light inspection under the microscope.

The boundary between sub-bituminous and bituminous coals is transitional but, for convenience, is drawn at moisture contents of 8-10%, carbon 77% and oxygen 16%. Within the bituminous rank the coals show a considerable range in oxygen content, though the proportion of hydrogen is more or less constant, varying little from 5%.

The proportions of volatile constituents and fixed carbon vary reciprocally within a fairly wide range; coals with much oxygen have a low proportion of fixed carbon (low-rank bituminous coal), and those with least oxygen usually have a high proportion of fixed carbon (high-rank bituminous coals).

Macerals of all three maceral groups are well represented. The vitrinites increase in reflectivity through the rank and Rmeanvaries between 0.6 and 2.3 with bireflectance becoming more pronounced at higher values. Exinites lose their transparency and fluorescence gradually and eventually become indistinguishable optically and chemically from vitrinites. This is sometimes described as 'a coalification jump'.

The pore structure of bituminous coals is micron-sized and sieve-like, with very narrow interconnecting throats severely restricting the mobility of the larger hydrocarbon complexes. Only highly soluble organic matter and gases (carbon dioxide and methane mainly) are capable of seepage through these interstices.

Semi-Bituminous Coal (or high-rank steam coal)

This has a higher heating value than any other kind of coal, and burns with an almost smokeless flame if properly fired. It is a hard, brittle coal with a characteristic prismatic fracture, distinct from the rectangular cleat of bituminous coal. The content of volatile matter is low, being only one-third to one-sixth of the fixed carbon.

Anthracite

This differs from other coals in its extremely high content of fixed carbon, with a correspondingly low proportion of volatile matter, accompanied by low percentages of oxygen and hydrogen. It will ignite only at a high temperature, and burns with a short smokeless flame. Most varieties are clean to handle, owing to the absence of friable fusain, and broken surfaces show a submetallic lustre and conchoidal fracture. Most anthracites have a banded structure similar to that of bituminous household coals, but this is not an invariable characteristic, and massive varieties of anthracite are not uncommon.

In anthracites vitrinite becomes darker through the rank due to increasing aromatization and condensation of the molecular groups. Reflectance is high, 2.3-5.0, and approaches that of the inertinites. Bireflectance increases correspondingly as the organic macromolecules align parallel to the bedding. Microporosity is at a minimum compared with other humic coals because the long-chained macromolecules, distinctive of the middle-ranking coals, are degraded and reduced in size with the increased temperatures and pressures. The moisture content is about 0.5%.

SAPROPELIC COALS

Cannel Coal

This coal is unlaminated, and breaks with a glassy conchoidal fracture rather like that of pitch. Typical examples show little or no recognisable remains of wood, but consist of much-altered plant material (exinites or lipinites) containing variable, often high, quantities of miospores, and occasionally an appreciable amount of material derived from oil-bearing algae.

In their composition and structure they resemble the durain of bituminous coals, and there is no sharp distinction between dull bituminous coals and cannels, the name cannel being usually applied where woody layers are absent and the bulk of the coal consists of spore material, resin and cuticles.

Typical cannels have been water transported and deposited as organic sediments in stagnant ponds in the coal swamps. They often contain a considerable admixture of clastic sediment, shown in analyses by a high proportion of ash, and such deposits pass into coaly or carbonaceous shales. Cannel coals contain large proportions of volatile constituents, and burn with a bright, smoky flame like that of a candle.

Boghead Coal or Torbanite

This is a black to green-black sediment closely allied to some varieties of oilshale and consists of fresh-to-brackish water oil-bearing algae (telalginite up to 90% by volume) mixed with small quantities of detrital sediment. Thin sections reveal the presence of innumerable translucent yellow bodies which possibly represent the remains of colonies of algae, such as *Reinschia* and *Pila,* which resemble the modern *Botryococcus braunii;* these are set in a dark, almost opaque groundmass.

The environments in which boghead coals formed seem either to have been the central parts of large basins in which transport of organic material was restricted or small shallow basins with little input of surface waters. In each case the water was probably relatively free of colloidal matter and well oxygenated.

Jet

This is a compact, coaly substance found in isolated masses and lenticular seams in bituminous shales, such as the marine Upper Lias Jet Shales of northeast England. It always possesses a woody, vitrain-like structure, which may be revealed on etched surfaces or in suitably prepared thin sections.

Most masses of jet appear to have been formed from driftwood, which has suffered considerable compression since it was entombed in the sediment, so that the woody tissues, as seen under the microscope, are crushed and distorted. Occasionally a silicified core is found, preserving the woody tissues in their original, uncrushed form. Like vitrain, jet is believed to be derived from wood which at an early stage became waterlogged in an entirely anaerobic environment.

OIL-SHALES

There is no completely satisfactory definition of the name oil-shale. Perhaps the most suitable is 'a fine-textured rock, normally laminated, containing organic matter, especially exinites (liptinites), from which oil can be extracted by heating'. Oil-shales differ from marine black bituminous shales in having a higher exinite content, up to 95% of the whole rock, and being much richer in total organic content. Black shales have an organic content generally between 0.5% and 8%, and exinites rarely form more than 5%.

Nonetheless, the distinction between poor quality oil-shales and black shales, many of which contain alginite and other sapropelic matter, is arbitrary. Black shales whose organic matter is mainly sapropelic in origin are potential oil source rocks; those with a preponderance of humic materials are more likely to be a source of gas. Free-flowing oil is unusual in oil-shales though frequently there are small veins, blebs and pockets of solid and viscous bituminous matter, such as asphalt, gilsonite and ozokerite, which have become emplaced secondarily.

Carbon disulphide dissolves this bituminous matter but the bulk of the organic content in the shales, commonly referred to as kerogen, remains unaffected.

Oil yields ranging from 38 litres per tonne to 550 litres per tonne have been recorded from a range of rocks-limestones, marlstones, shales, siltstones and impure coals -all collectively referred to as oil-shales. The freshwater Scottish Lower Carboniferous oil-shales yield on average 90 litres per tonne; whereas the freshwater Eocene Green River Formation oil-shales of Colorado, Wyoming and Utah yield on average 125 litres per tonne. Torbanite shales in New South Wales and Queensland, Australia have yielded 230-800 litres per tonne.

In contrast, the Kimmeridgian oil-shales of Britain, Toarcian oil-shales of the Paris Basin and Cretaceous Toolebuc Formation oil-shales (marinites) of Queensland, all marine deposits, are rather poor in quality yielding on average 50 litres per tonne.

Individual seams of oil-shale vary in thickness from a few centimetres to several tens of metres and sometimes extend in uniformly bedded fashion for considerable distances.

Kerogen

The term kerogen ('oil-former') was first used with reference to the bituminous and carbonaceous matter in Scottish shales which gave rise to crude oil on distillation. In high quality oil-shales the evidence favouring animal matter as a significant contributor to the kerogen is slim, and most authorities claim plants as being the dominant source. In the majority of oilshales the principal constituent is exinite, with variable but lesser amounts of vitrinite and inertinite.

Alginite is usually the dominant maceral in the richer deposits, such as the marine Ordovician (kukersite) shales of Estonia, the marine Permo-Carboniferous (tasmanite) shales of Tasmania, and the freshwater lacustrine Scottish Carboniferous and Eocene Green River Formation deposits of North America.

The maceral represents either marine unicellular varieties of algae, such as *Tasmanites,* or mainly lacustrine colonial forms of the *Botryococcus* family. It fluoresces intensely, with colours changing from green to yellow and brown-red as the rank increases. Landderived resinite, sporinite and cutinite are commonly present in minor amounts and may be mixed in with varying amounts of other organic and inorganic detritus. Dinoflagellate and acritarch cysts are occasionally relatively abundant.

Environments of Deposition

Large Lakes

These are usually extensive land-enclosed basinal areas being particularly well developed where there has been block faulting and tectonic warping. The best-known examples are the Eocene Lakes Uinta and Gosiute of Colorado, Utah and Wyoming, the early Carboniferous Lake Cadell of the eastern end of the Midland Valley of Scotland, the lakes of New Brunswick, Canada, in which the Carboniferous Albert Shale was deposited, and Tertiary lakes of eastern Queensland. In the high yield oil-shales planktonic lamalginite is densely packed, whereas low yields are obtained where the lamalginite is dispersed as discrete small blobs.

The associated lacustrine sediments vary considerably. In the Scottish OilShale Group and the thick Tertiary sequences of Queensland the oil-shales are interbedded with sandstones, siltstones and mudrocks, in contrast to the thick Green River Formation which is rich in dolomitic limestones, marlstones and shales. Evaporite beds form during prolonged phases of aridity and desiccation, and volcanic rocks are interbedded in regions of marked crustal instability.

Small Lakes

Small depressions in swamps have plant and inorganic detritus swept into them which is admixed with indigenous organic materials. If conditions are appropriate, thin, lenticular canneloid and torbanite-like oilshales can form. Certain of these can be thick and of high grade, as in the Tertiary succession of Fushun, Manchuria. The dominant maceral is planktonic telalginite. Vitrinite and resinite, though small in amount, tend to be more abundant than in the shales laid down in larger lakes.

Shallow Marine Seas

These are of continental shelf and basin depths. The oilshales deposited within them are characteristically rich in telalginite derived from unicellular algae. Lamalginite from degraded dinoflagellates and acritarchs may be present. The rocks have a poor oil yield of about 60 litres per tonne. Individual seams vary between a few centimetres and 10m, and some persist over several thousands of square kilometres.

Scottish Oil-Shales

Relatively rich beds of oil-shale, worked until the 1950s, are present within the cyclic Carboniferous Oil-Shale Group at the eastern end of the Midland Valley of Scotland. The oil-shales constitute about 3% by volume of the predominantly siliciclastic succession, which is about 1 km thick. Thin limestones, marls and tuffs are present at some levels.

Typical specimens of oil-shale are fine-grained rocks, brown or black in colour, and showing a brown streak when scratched. The shale is finely laminated and sometimes varved, but this structure is not always visible in hand specimens. Beds having the appearance of being internally crumpled and slickensided are known by the miners as 'curly shale', as distinct from the normal undistorted 'plain shale'.

Under the microscope the rock is seen to contain alginite of planktonic origins, analagous to modern *Botryococcus,* a high proportion of fungal and bacterial remains related to this material, and a small proportion of animal remains. This autochthonous material is intermingled with varying quantities of detrital debris, including vitrinite, sporinite and resinite fragments derived from plants living on low-lying marsh areas bounding the lakes.

The presence of so much well-preserved cellular and structural organic matter indicates accumulation under stagnant bottom conditions which eventually arrested further decomposition. The anoxic conditions appear to have been established during short periods of reduced sediment input into a freshwater lake. At these times a meromict (permanently stratified) state existed in the lake with phytoplanktonic organisms,

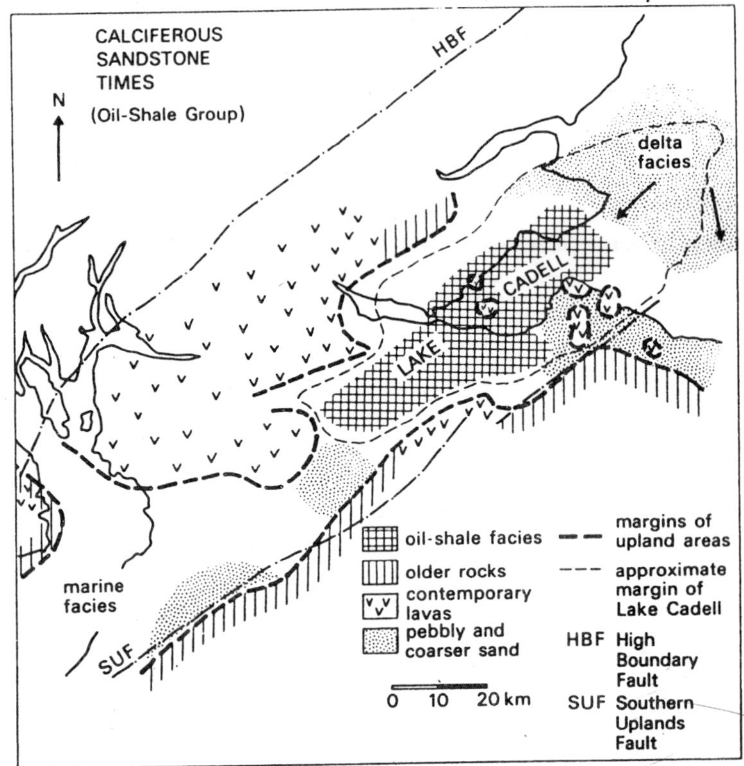

Figure 14.2 : Palaeogeography of Scottish Oil-Shale Group times. Although the bulk of the clastic material was fed into the lake via the major deltas to the northeast, there was a steady, but restricted, input from the uplands bounding the lake.

especially algae, proliferating in the near surface epilimnion, whereas in the hypolimnion beneath, conditions were oxygen-deficient and suitable for organic ooze accumulation.

The freshwater Lake Cadell was situated at a palaeolatitude of 4°S and close to sea-level. Hence, the climate was tropical and humid, such that the surrounding low-lying areas were swampy and hydrophytic land plants proliferated. Thin coals, alluvial and fluviodeltaic sediments characterise the bulk of the lake margin deposits but, at intervals, conditions changed so that lime deposits, often rich in algae, were laid down.

The latter sometimes covered large tracts at the lake margins and some of the gentle bottom slopes. The diminution of terrigene input and deposition of lime sediment often seem to have been precursors to the establishment in the deeper parts of the lake of widespread meromictic conditions suitable for oil-shale formation. At its maximum extent Lake Cadell occupied some 3500sq. km of the Edinburgh Basin, a graben-like subsiding area. The freshwater lake existed as a major physiographic unit on at least eleven occasions, and on each occasion anything up to five separate oil-shale seams, now 0.2-6m thick, were formed.

Contemporaneous vulcanicity frequently modified local depths and the shape of the lake. There is no evidence that maximum water depths ever exceeded 50m. There is considerable evidence for repeated partial desiccation of the lake. Mud-cracking and brecciation of some of the oil-shales, associated algal and peloidal micrites and siltstones are recognised. The influx of water from the surrounding land masses, however, effectively prevented complete desiccation to a playa-like condition and the mass precipitation of salts.

Green River Formation oil-shales

The thick sequence of shales, siltstones and dolomitic marlstones constituting the Eocene Green River Formation were laid down in fault-bounded basins occupied by several continental lakes, the most important being Lake Uinta (Colorado and Utah) and Lake Gosiute (Wyoming). At its maximum extent in middle Eocene times Lake Uinta covered an area of 54000sq. km (cf. Lake Gosiute at 32000sq, km), this interval being marked by a very distinctive zone of rich and thick oil-shales yielding oil up to 350 litres per tonne. The zone is known as the 'Mahogany ledge'.

There are two distinct types of oil-shale in the formation and their characteristics may have been determined by the depth of water in which they accumulated. Type I is least common, black weathering to bluish-white, has organic matter which is recognizably organised and is often mud-cracked and brecciated. It probably formed in very shallow water towards the edge of the lakes and became intermittently exposed and desiccated during falls in lake level. Inshore drifting of planktonic algae during 'bloom' phases plus a small input of humic material from the land probably accounts for the distribution. Type 2, the abundant variety, is light to dark brown weathering to buff, carbonate-rich, and laminated in which lamalgirlite (alginite B) is the dominant maceral. There are minor amounts of sporinite from higher plants.

The type 2 deposits are dolomitized marlstones in which the original finegrained aragonite and calcite have been replaced by fine-grained diagenetic dolomite. There is evidence for subsequent reworking of the dolomite grains. Varve-like laminae with sharp top and bottom contacts typify this facies, the varvites consisting of a graded layer, usually 0.01

to 2mm in thickness, which is coarser-grained, lighter coloured and alginite-poor at the base and finer-grained, thinner (0.005-0.05 mm), lensoid, darker coloured and alginite-rich at the top. Bitumens occur interstitially in the coarser layers. Albertite, a solid black bitumen, is also found in veins and fissures.

The varvites, which persist and can be correlated over large distances, are explained by differential gravity settlement of mineral and organic constituents from seasonally and chemically stratified lake waters (meromictic condition). The production peaks of both the heavier specific gravity clastic and chemical carbonates and the lower specific gravity plankton were in the summer season.

The total evidence indicates relatively long warm summers and cool moist winters, the mean annual temperature being estimated at 20°C. The input of water into the lakes seems to have been sluggish generally with most of the coarse detritals being deposited at the margins.

As with certain peak phases of the Scottish oil-shale deposition, it is possible to trace gross changes of facies from the deepest parts of the lakes, where the bulk of the oil-shales formed, towards the shore (algal, oolitic, peloid and ostracod limestones, shell marls) and eventually into the marginal swamps and deltas (clays, siltstones, sandstones, palaeosols and coals). At certain times, tuffs were deposited within the lakes and these materials have played some part in the precipitation of thin seams consisting almost entirely of diagenetic euhedral analcite. Magadi-type chert also occurs as continuous thin beds interlayered with the oil-shales.

The desiccation stages in the Green River lakes were characterised by extensive precipitation of a host of unusual salts at their depositional centres. At these time the lakes probably comprised several discrete playa-type areas. Nahcolite ($NaHCO_3$) occurs as individual small grains, masses more than a metre in size, and beds within the oil-shales. Dawsonite ($NaAI(OH)_3CO_3$) is usually disseminated as small grains in the shales. In Lake Gosiute, about 90000 million tonnes of bedded trona ($Na_2CO_3NaHCO_3.2H_zO$) or trona interlayed with halite, with almost an equivalent amount of shortite ($Na_2CO_3.2CaCO_3$), were deposited.

15

PHOSPHATIC DEPOSITS

Sedimentary phosphate rocks are of major economic importance and rank very high in terms of gross annual tonnage and volume of world trade. Between 80% and 90% of phosphate, which is mainly used as a fertiliser, is derived from ancient marine successions. The rocks encompass a variety of types with differing origins. The phosphate (P_20_5) content in some low-grade economic deposits can be as low as 4% in contrast to those of high grade which can approach 40%. A precise definition of what constitutes a sedimentary phosphatic rock is thus a continuing source of debate.

The most common marine variety, phosphorite, has been defined as one with a high enough content of phosphate minerals to be of economic interest. High-quality phosphorites contain at least 18% P_2O_5, roughly equivalent to 50% apatite which is the dominant mineral constituent. These high-grade apatite-rich phosphorites have also been called phosphatites.

SEDIMENTARY PHOSPHATE MINERALS

Phosphorus is an essential constituent of all living cells. It is widespread in phytoplankton and bacteria, and commonly occurs in bones, teeth and in some invertebrate shells as dahllite (carbonate hydroxyapatite-$Ca_{10}(P0_4,CO_3,OH)_6(OH,F)_2$). On the death of the organisms this relatively unstable chemical complex alters in marine waters into the more stable apatite francolite (carbonate fluorapatite -general formula $Ca_{10}Na.Mg(PO_4)_6.(CO_3)$ F).

Francolite is the dominant mineral in phosphorites and, in turn, is susceptible to alteration, but only when it is removed from the marine situation and subjected to weathering. Under those circumstances leaching of the carbonate, including that of any associated calcareous shells and cement, progressively brings about conversion towards fluorapatite ($Ca_{10}(PO_4)_6.F_2$). Both francolites and dahllites are optically anisotropic in contrast with collophanite which is not. Collophanite appears to be amorphous under the normal petrographic microscope, although it is now known to be a crypto- or micro-crystalline form of carbonate fluorapatite. It is a useful name when the actual composition of the phosphate material cannot be readily diagnosed, but some authorities would prefer to abandon it.

A characteristic feature of these apatite complexes is ionic substitution in which Ca ions are substituted by such as rare earths, uranium (U^4), manganese, magnesium, barium and sodium, and PO_4 ions are substituted by such as vanadium and arsenic oxide radicals. In Eocene deposits of Bakouma (Central African Republic) the U concentration reaches 5600ppm and in Wyoming and Utah, in the Green River Formation, up to 3000ppm. Moreover, humic acids can be incorporated giving a variable yellow, brown to black colouration to individual minerals, particles and ultimately to the rock as a whole. Hence the detailed composition of phosphate rocks is quite complex.

Not all phosphate minerals are apatite complexes. The lateritic weathering of pre-existing phosphate deposits resting on alumino-silicate foundation rocks can produce minerals like wavellite ($Al_6(PO_4)_4(OH).9H_2O$) and crandallite. Vivianite ($Fe_3(PO_4)_1.8H_2O$) is recognised in modern and ancient alluvial and freshwater lake deposits.

CONSTITUTION AND CLASSIFICATION OF PHOSPHORITES

Phosphorites commonly consist of grains which can be recognised as phosphatic bone and shell fragments, peloids, intraclasts and, less frequently, ooids set in a cement and/or matrix. The cement is often of a collophanitetype in which individual crypto-crystalline grains (< 1 um) cannot be easily recognised. The terms microsphorite and microsphatite are preferably used for this fine-grained material.

The matrix, in contrast, can be a mixture of contemporaneously deposited detrital siliciclastic and carbonate grains. These materials are exogangue constituents. Materials incorporated into phosphate grains, most often seen in peloids, are endogangue. Endogangue includes yellow-brown to opaque humic matter of marine planktonic origin, siliciclastic grains and authigenic pyrite framboids and secondary apatite crystals.

The grains which cause the most controversy are the peloids, which show a remarkable uniformity in size, about 0.5mm, and well-rounded shape throughout geological time. Excluding endogangue components, they are structureless internally and have a single veneer, a few microns thick, of glassy-looking apatite.

In some instances the veneer is duplicated or triplicated and the peloid then begins to take on the appearance of an ooid. The peloids almost certainly have multiple origins. Some may be phosphatised faecal coprolites, though their uniformity in size and the apparent sparseness of appropriate benthonic life in marine phosphate environments does not provide strong support for this concept.

The mechanical ripping up, rounding and sorting of microsphorite intraclasts could be an important mechanisms, as could the rounding and sorting of diagenetically replaced shell fragments. This style of replacement by finegrained phosphate is known as phosphomicritization. Other hypotheses include the diagenetic precipitation of phosphate within the soft bottom sediment around suitable organic nuclei, for example radiolaria or cysts of dinoflagellates. This is a kind of micro-concretionary effect in which the role of humic acid coatings in conjunction with bacterial activity probably creates the necessary geochemical conditions.

It is possible to classify marine phosphorites on the basis of grain size using a modified version of the classification used for limestones. Thus names like phosphorudite, phosarenite, phosphosiltite and phospholutite occur. Dunham's textural scheme for limestones, as adapted by Cook and Shergold, is also useful.

The clay or lutite grade component is usually microsphorite. Qualifying terms such as peloid, ooid, skeletal, oncolitic and intraclastic are used as a prefix if those constituents are especially prominent. Stromatolitic deposits, though rare, fall into the boundstone phosphorite

Table 15.1: Textural Classification of Phosphorites.

Depositional texture recognisable				
Contains mud			Lacks mud and is grain-supported	Original components bound together
Mud-supported		Grain-supported		
Less than 10% grains	More than 10% grains			
MUDSTONE PHOSPHORITE	HAACKESTONE PHOSPHORITE	PACKS TONE PHOSPHORITE	GRAINSTONE PHOSPHORITE	BOI NDSTONE PHOSPHORITE

category. Prime examples occur in the Aravalli Precambrian deposits of Jhamarkotra, Rajasthan in India and in the Middle Cambrian Georgina Basin of Australia. Only a few phosphatic sediments do not fall easily into the scheme and these include the subaerial guanos and phoscretes. The classification cannot cope with the complex textures (concretionary, botryoidal, encrusting, vuggy) of these secondary yellowish-white rocks.

CONTROLS ON MARINE PHOSPHORITE DEPOSITION

The original prime source of phosphorus entering sediments is fluorapatite Which is present in small amounts in most igneous rocks. The averagae P_2O_5 percentage in crustal rocks in 0.23% and oceanic water at present averages 70μg P per litre either in solution or in suspension. Although the quantity of phosphorus in oceans reaches maxima varying between 100 and 200μg per litre, as in parts of the Pacific and Black Sea, these concentrations plus direct contributions from crustal rocks are totally inadequate for phosphogenesis.

A 2-million fold enrichment is necessary to produce a high grade phosphorite (30-40% P_2O_5) and is partly achieved by living organisms which have an estimated concentration power of about X 150000. Certain planktonic microorganisms, diatoms in particular but also dinoflagellates, cyanophytes and hystrichospheres, contain up to 3% P_2O_5 locked up in their skeletal and softer parts, crustacea up to 8%, and other invertebrates and vertebrates as much as 40%.

The seasonal injection from moderate slope depths of nutrient-enriched (nitrates especially) waters on to warm, well-lit shelf margins, known as upwellings, stimulates high organic productivity by plankton and nekton. Planktonic 'blooms' are known to give a 5-6 times P_2O_5 concentration in these circumstances. Although upwellings occur around seamounts, they are most common on the western margins of continents at latitudes less than 40°.

They are caused generally by prevailing winds driving surface waters off shore or near-parallel to the shore. This displacement causes deeper waters to rise up the shelf slopes and intrude on to the shelf, the pattern of movement being controlled by local bathymetry and seasonably variable factors including salinity and surface temperature.

Phosphorite deposition requires the concentration of dead tissue and skeletal matter on the sea bottom because it is only then that further necessary enrichment, by a factor of 5-10, can occur. The mass accumulation is dependent on many factors. An adequate and persistent supply is needed over a prolonged period. This is occasionally enhanced by the mass mortality of organisms, especially fish, which appear to be poisoned by toxic substances produced by dinoflagellates 'in bloom'-the 'red-tide' phenomenon. The debris is required to be incorporated quickly into the bottom sediment to prevent phosphate dispersal and there, under alkaline (pH c. 7.5 or more) and mildly anoxic conditions and with Eh approximating zero, the interstitial waters become enriched to between 25 and 100 times that of the adjacent sea water.

The balanced oxidation-reduction state within the sediment is probably due to oxygen deficiency in the sea water. It is well known that in shelf areas of modern phosphogenesis, such as offshore from Namibia and Chile-Peru, there are oxygen-minimum zones or layers at depths of 30-300m where the bulk of phosphates accumulate. Within the zones the dissolved oxygen is low partly due to the oxygen-deficient nature of the dense upwelling currents. Once these overall conditions are established, the concentration of phosphate in the sediment can be such that authigenic precipitation occurs in the form of gel-like pods, peloids, lenses and laminae. These progressively lithify, at the same time increasing their phosphate content by diagenetic processes.

Another factor influencing the formation of phosphorites is the vexed one of Mg^{++} ions, which are common inhibitors of direct phosphate precipitation within the bottom muds, probably because the Mg^{++} competes with Ca^{++} for sites in the apatite lattice. In some circumstances the Mg^{++} ions are depleted by absorption into pre-existing clay minerals, such as smectites, or by the formation of authigenic Mg-silicate minerals, such as palygorskite and sepiolite. In others, dolomitization can give depletion, especially in bays, estuaries and insular atoll lagoons.

Problems involving Mg-inhibition of direct precipitation are diminished if phosphates originate by diagenetic replacement of pre-existing soft sediment, a process generally known as phosphatization. There is a large amount of supportive evidence for replacement. Calcareous and siliceous skeletons, siliciclastic grains and pre-existing phosphatic materials are all capable of being pseudomorphed by francolite precipitated from interstitial waters. Sometimes the francolite is in a microsphorite form and the replacive process is then known as phosphomicritization (cf. micritization in limestones). As a consequence of this evidence it is now recognised that marine phosphorites are the end product of both authigenic and diagenetic processes.

Sea-level variations are important in creating this end product. The Upper Proterozoic-Cambrian Phosphate Province of Australia, extending into eastern Asia, and the Cretaceous-Eocene Tethyan Phosphate Province reflect major episodes of marine transgression with periodic extensions of oxygenminimum zones across shelves. The continuance of phosphate deposition, albeit on a small scale, on some modern shelf areas is partly consequent on the Flandrian Transgression over the last 10000 years during which time global sea-level has risen about 100m.

Even small and more localised sea-level changes can have profound effects due to shelf irregularities altering the patterns and intensities of water movement. Depressions can be hydrodynamic traps in which phosphate muds can accumulate slowly and relatively quietly, and where gentle reworking can occur.

The winnowing away of siliciclastic exogangue in nearshore locations is capable of converting a commercially low-grade deposit into a more uniform high-grade one. In Morocco some Upper Cretaceous-Middle Eocene traps, tens of kilometres wide, were protected off shore by carbonate barriers and others delimited by contemporary shelf deformation accompanied by occasional salt diapir emplacement. Sedimentation rates of about 1 m per million years have been suggested for these high-grade deposits, a figure comparable with the estimated growth of phosphatic nodules off the Peruvian coast.

In summary, major marine phosphate deposits are laid down in low latitudes on shelves abutting land masses supplying little detritus. They require high organic productivity which is best achieved where upwelling nutrient-rich and cold (about 12-16°C) waters gain access to suitable traps created during marine transgressions across shelves.

Quiet and slow sedimentation from lowly oxygenated and warm (about 15-22°C) waters allows the buildup of phosphorus in the sediment and interstitial waters to the point at which authigenic and diagenetic processes become dominant. As many ancient phosphorites, such as the Proterozoic-Cambrian deposits of South Australia, are now located well away from low latitudes, it is clear that considerable crustal displacement has occurred since they were laid down.

BEDDED MARINE PHOSPHORITES

Many phosphatic deposits are of local character or form special phases within formations of a different nature, such as limestones or shales. In a few regions, however, phosphatic deposits are developed on a much larger scale and constitute thin but independent marine successions covering a considerable area. The most extensive of such bodies are of Cambrian, Permian, Upper Cretaceous-Eocene and Miocene age, but examples are found in all geological periods from the late Proterozoic onwards. The Middle Cambrian Georgina Basin of north-west Queensland and Northern Territories of Australia contains 18 phosphorite beds which were mostly laid down around its periphery, a distance of about 1000km.

The associated strata are cherts and phosphatic limestones, sandstones, siltstones and shales. The phosphorites accumulated near topographic highs on a marine shelf opening out to the south-east and in more restricted lagoonal and estuarine coastal environments flanking the basin. As a consequence they vary considerably in thickness. The more open shelf bodies are up to 38 m thick and can be traced for up to 23 km along the strike, whereas those laid down in shore are thin and lens-like, extending for 600m or less.

The bodies are well bedded to laminated and some layers show a fining-upwards from grainstone to mudstone-wackestone phosphorite rich in humic acids. It is clear that there has been some reworking and redeposition by the bottom currents. In fact, the maturity of some of the grainstone phosphorites is such that the materials have probably been moved a long distance away from their point of origin.

A feature typical of many major phosphorites also occurs in the Georgina Basin successions and that is medium-scale cyclicity with repeated alternations of phosphorites with siltstones, shales, chert and limestones. The individual cycles vary from a few to tens of metres in thickness and seem to have been caused by changes in sea-level allied to changes in rates of sedimentation.

The Middle Permian Phosphoria Formation of Montana, Idaho, Wyoming, Nevada and Utah is another major phosphatic deposit laid down in a basin at least 160000 sq. km in extent. This basin was located towards the edge of a continental shelf and isolated in part from the open sea to the west by a series of islands or submarine barriers. Within the deeper quieter parts highquality phosphorites accumulated and as many as 17 individual beds have been recognised in south-east Idaho.

Thicknesses are rarely more than 2 m and lamination is usual. Their P_2O_5 content is as much as 35%, mainly in the form of microcrystalline francolite, and the organic carbon content reaches 9%; pyrite is common. Many of the beds have peloids as their dominant component and these are variably intermingled with phosphate nodules and intraclasts, fish debris and phosphatised shells of brachiopods. Ooids are comparatively rare. The matrix is microsphorite present in very variable amounts. The phosphorites are interbedded with spicular cherts, phosphatic mudstones and black carbonaceous shales in the deeper water facies, and phosphatic mudstones, limestones and glauconitic sheet sandstones in the shallower water facies. Reworking of the last two brought about good sorting and cross-bedding locally.

Nearly 60% of the world's commercial phosphate resource occurs in North Africa, more especially in Morocco. The coastal basins of Morocco, however, are only part of a large Upper Cretaceous-Eocene Phosphate Province extending from Syria and Iraq in the Middle East to Senegal and Togo in West Africa. Within this province a whole array of shelf basins formed along the southern flanks of the Tethys, where they were partly protected by barriers of many types.

The beds are normally up to 2 m thick and 23 such are recognised in Morocco. In Jordan some exceed 10m in thickness. The beds are remarkably similar in texture and constitution across the whole province, a uniformity probably indicating that all the deposits were subject to periodic mechanical winnowing processes. Peloids are abundant at many levels, intraclasts and conglomeratic concretions at others, but ooids are again very rare except in Jordan. Cyclicity is common, the phosphorite levels (phosarenites with up to 30% P_2O_5) being interspersed repetitively with phosphatic chalks, limestones and marls (up to 15% P_2O_5), clays and cherts.

PHOSPHATIC NODULES

Phosphatic nodules or concretions, rounded to highly irregular in shape, up to 1 m long and weighing up to 70 kg, are present on shelves, ridges and basin slopes in parts of modern oceans. Nearly all the deposits occur at depths of 30-300m and are frequently associated with peloidal phosphorites and diatomaceous muds.

Not all nodules are modern, however, and it is believed that most are relict, having originated during the mid- to late Tertiary and Pleistocene and since then have been reworked and displaced many times. The only known areas of modern accumulation are off the coasts of southwest Africa, and Peru and Chile.

Initially, the nodules are light grey in colour and relatively soft with earthy surfaces, but as lithification proceeds they become light brown and commonly develop glazed smooth outer surfaces enriched in goethite and manganese oxides. Many nodules have a concretionary conglomeratic structure in which a number of discrete nodules are locked together by microsphoritic francolite cement.

A range of endo- and exogangue constituents occur within both the cement and the nodules. The origin of the nodules is controversial and the explanations are as varied, and closely resemble, those put forward for peloids. Some authorities suggest authigenic growth within soft pelagic phosphatic muds by nucleation processes around decaying organic debris, including coprolitic material. Others favour a more complex genesis, whereby the muds are phosphatised authigenically and diagenetically and then, after a degree of lithification, are subject to erosion and physical disintegration during rises and falls in sea-level.

The concept has arisen as a consequence of the discovery of 'phosphatic pavements', a type of hardground consisting mainly of peloidal phosphorite mudstones, at depths of about 33 m in present oceans. These appear to have broken up and then been re-cemented several times during Cenozoic sea-level changes, producing at various shallow water stages nodules, conglomeratic concretions, intraclasts and peloids.

Certain of the Miocene deposits of the Blake Plateau off South Carolina and the Chatham Rise, east of New Zealand, have such a history. The pavements during their reworking have produced lag-deposits of loose nodules and peloids several metres in thickness on the present sea bottom. On a lesser scale 'phosphatic pavements' up to 1 m thick are present in the Cretaceous chalks of northwest Europe. The chalks are well cemented, the calcite having been partly diagenetically replaced by phosphates during pauses in sedimentation. Channels cutting through the pavements contain lag-deposits (phosphatised intraclasts, coprolites and bones) derived from the submarine erosion of the pavements.

TERRESTRIAL PHOSPHATES

Pebble Phosphates and Phosphatic Pebble Beds

The Pliocene to Quaternary 'land-pebble' phosphates of Florida and North Carolina and their underlying marine source rocks, the Hawthorn Group of Miocene age, account for 30-40% of the world's total output. The phosphatic Hawthorn deposits, up to 200m thick in basinal areas, were laid down in a range of coastal, shallow platform and offshore environments away from the immediate influence of terrigenous detritus.

Sand-grade peloids are the dominant component of the phosphatic beds, with varying proportions of phosphatic intraclasts and skeletal debris, microsphorite and dolomite. In Pliocene times a phase of marine regression initiated extensive subaerial weathering and supergene alteration of the Hawthorn Group and these processes have continued intermittently through to the present day. The resultant deposits, laid down in fluvial channels, estuaries and immediate offshore areas, are phosphate gravels, sands and silts, the 'land-pebble' phosphates. They

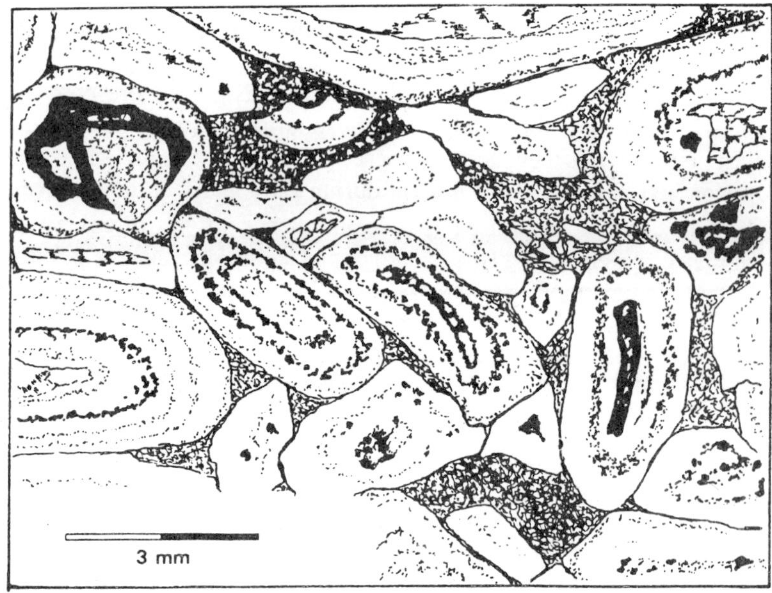

Figure 15.1 : Phosphate pebble bed, Lower Carboniferous, Scotland. Well-rounded phosphatic pebbles set in a dark granular cement of ferroan dolomite. Some pebbles show distortion due to early compaction.

are characterised by rapid facies and thickness changes, individual lenses reaching 7.5m in thickness. The highest-grade deposits consist of a loose aggregate of pebbles and peloids (about 33%) and quartz sand (about 33%) set in a fine matrix of phosphates, clay minerals and other siliciclastics.

On a much smaller scale than the pebble phosphates of Florida, and of less economic potential, averaging 15% P_2O_5, are phosphatic pebble beds which are familiar features of marine successions from the Proterozoic to the present day. Well-known examples occur in the Cambrian successions of the Baltic Platform and in the Mesozoic successions of northwest Europe and the Russian Platform.

In Britain the Cretaceous Cambridge Greensand is a type of phosphatic basal conglomerate, rich in glauconite, with pebbles derived from the erosion of the underlying Gault Clay. Another is the Rhaetic Bone Bed of western England, rich in phosphatic vertebrate debris which has been swept together into a layer only a few centimetres thick. All these deposits are thin and tend to have formed during periods of reduced sedimentation. Hence their association with condensed faunal sequences, non-sequences and unconformities.

Guano

Guano deposits are principally formed on rocky offshore islands in low latitudes, such as Nauru and Ocean Island in the western Pacific Ocean and Christmas Island in the eastern Indian Ocean. They are up to 10m thick and are composed of the excremental and skeletal remains of birds intermingled with non-digested fish and other animal debris.

The P_2O_5 content of fresh material is as low as 4%, but chemical changes soon after deposition, accelerated by appropriate climatic conditions (warm to hot temperatures; sporadic rainfall) dissolves soluble nitrates and oxalates, releases volatiles and generates bodies with up to 40% P_2O_5. These high concentrations are ultimately attained by magnesium and ammonium phosphates reacting with the underlying country rocks and producing a range of new, though metastable, minerals. Whitlockite ($Ca_3(PO_4)_2$) typically forms on limestone foundations, whereas on igneous rocks crandallite ($Ca_2Al_6(PO_4)_4(OH)_{10} \cdot 2H_2O$) is just one of several such new minerals. All these new products are deficient in fluorine.

The depositional history of any given guano is complicated as it reflects intermittent accumulation through late Tertiary to Quaternary times, during which sea-levels and climates have fluctuated. Some phosphorites on oceanic seamounts (the 'insular-type'), for example, may have a guano component in their make-up, generated when sea-levels were very low and the mounts exposed. On the Aldabra atoll in the western Indian Ocean, there are two recognisable styles of phosphate accumulation which can be related to climatic and sea-level variations.

A thinly bedded style comprises lenses up to 2 m thick carrying ooliths, peloids, intraclasts, bone and phosphatised shell fragments, particularly prominent in the phosarenite grade. These beds were probably laid down in shallow waters during higher sea-levels than at present. There is also evidence for contemporaneous bird activity and guano formation, suggesting that parts of the atoll were subaerially exposed. Moreover, there is a range of internal cavities within all the deposits from vugs to small caves.

These are variably filled with physically displaced fragments and chemical precipitates and indicate subaerial elevation and leaching on several occasions.

The second style of deposition is represented near to sea-level by phosphatecemented fanglomerates and beach sediments, which indicate longer-term wetter periods when it is unlikely that any marked quantities of avian guano could accumulate. None seems to be forming at present.

Phoscrete

Phoscretes are subaerial weathering surfaces, encrustations or duricrusts up to 1 m thick, which intermittently blanket underlying phosphorites and phosphatic sediments. They are comparable in origin to caliche profiles resting on limestones, and form in a similar fashion by subjecting the substrate to alternate periods of wetting and drying.

The consequent leaching, circulation of phosphatic pore waters, reprecipitation and recrystallization create a yellowish-white, sometimes porcellanous, rock of considerable structural and textural complexity. Some irregular, multi-coloured, laminated crusts and veins are well cemented by phosphomicrite, others not. The contact with the underlying strata is usually gradational and within the transitional zone there is often a vaguely peloid or clotted (grumous) texture. Phoscretes are well developed in the Georgina Basin in Australia and are equally well developed in other major Phosphate Provinces.

16

EVAPORITES

These are deposits whose origin can be referred directly to processes of precipitation and crystallisation from saturated solutions. The most important of these processes is undoubtedly evaporation and this takes place extensively in large inland salt lakes, around the margins of seas and within the more central parts of partly isolated marine basins. Some terrestrial deposits are of a playa-type.

Most of the hundred or so basins in the arid Atacama Desert and semi-arid Andean Highlands of Chile are occupied by halite- and gypsum-encrusted playas or salinas. The Wilkins Peak Member of the Eocene Green River Formation of Wyoming, typified by bedded halite and trona, is an older example.

The products thus formed are known as evaporites or salt deposits. Dolomitized lime deposits are commonly closely associated with evaporites, especially in coastal sabkhas.

CONTROLS ON EVAPORITE PRECIPITATION

High evaporation rates combined with comparatively low rainfall (less than 200 mm per year) and low influx of replenishing water favours the accumulation and preservation of most deposits. However, if the influx of salt in the waters is matched or exceeded by the outflow (reflux) of bottom brine, as happens in the present Mediterranean Sea, then saturation point is not reached and precipitation of the common evaporite minerals does not take place. Relatively high surface temperatures of the order of 16 °C and above are beneficial for salt precipitation, but need not be consistently high throughout the year. The bulk of the halite deposition in the Larnaca coastal sauna in Cyprus occurs in the summer months, when the air temperatures are above 30°C. In winter, with air temperatures as low as 10°C, saturation point is not reached and no halite is deposited.

The conditions which favour terrestrial and coastal evaporite precipitation also determine to a large measure the position of deserts. The Bocana de Virrila estuary in Peru, which is a modern site of halite and gypsum deposition with salinities as high as 355%, is located at 6°S latitude in an arid, persistently hot, desert situation. Air temperatures rarely fall below 24 °C and water temperatures below 25-27°C.

Lake Eyre in South Australia, at 8°S latitude and on the flanks of the Simpson Desert, is a vast salt-producing body where the mean annual precipitation of 200 mm is closely matched by the mean annual evaporation. In contrast, the evaporite deposits of the Lop Nor salina in the Sinkiang-Uighur region of China, which is at 40°N latitude and at the eastern end of the Takla Makan desert, are associated with a wide temperature range, varying between, on average, 24°C in summer and minus 12°C in winter.

In the geological column it is well established that many of the major marine and terrestrial salt deposits grade laterally into or are interbedded with strata showing evidence of arid to semi-arid desert-like conditions, including red beds. The Devonian, Permo-Triassic and Neogene successions in the Northern Hemisphere show these relationships well.

The most important salt deposits occurring on a large scale are the chlorides and sulphates of sodium, potassium, magnesium and calcium. These form simple salts, such as sodium chloride (halite) and calcium sulphate (anhydrite), but are also found in the form of hydrated and double salts (polyhalite) or even more complex combinations. The kinds and proportions of salts deposited depend on the composition of the water and the ionic concentration of the salts in solution to which all the normal laws of solution chemistry apply.

When the salts in solution become sufficiently concentrated then crystallisation begins. Present-day sea water will first precipitate carbonates and then, at salt concentrations beyond 4 times normal will form sulphates; beyond 12 times normal, halite; beyond 64 times normal, magnesium-potassium salts; and beyond 120 times, bischofite. Beyond halite precipitation concentrations the brines and salts are commonly referred to as bitterns. Sylvite, bischofite, kainite and borax are bitterns. Most of these minerals are capable of being precipitated in salt lakes (salinas) and playas, although the actual proportions can differ from those of the sea. There tends to be much more individuality about the chemistry of lake and inland sea waters, depending on a range of local factors, such as the nature of the immediate country rocks over which the feeder streams and river flow.

Table 16.1 : Principal Sedimentary Chlorides and Sulphates.

Class	*State*	*Species*	*Formula*
		halite	$NaCl$
	anhydrous	sylvite	KCI
CHLORIDES		bischofite	$MgCl_2.6H_2O$
	hydrous	carnallite	$KMgC1,.6H_20$
		glauberite	$Na_2SO_4.CaSO,$
	anhydrous	anhydrite	$CaSO_4$
		barytes	$BaSO_4$
		langbeinite	$K_2SO_4.2MgSO_4$
		mirabilite	$Na_2SO_4.10H_2O$
SULPHATES		gypsum	$CaSO_4.2H_2O$
	hydrous	polyhalite	$Ca_2K_2Mg(SO_4)_4.2H_2O$
		hexahydrite	$MgSO_4.6H_2O$
		epsomite	$MgSO_4.7H_2O$
		kainite	$KCI.MgSO_4.3H_2O$

Evaporite minerals are precipitated at the surface of highly saline bodies of water, from within the denser lowermost brine layers of water bodies, and authigenically at and immediately below the water-sediment interface. In the first two circumstances, the individual grains gravitate downwards and accumulate progressively on the bottom.

They are then, in effect, clastic and having a low specific gravity can be held in suspension for some time and even be remobilised by currents. In shelf situations of varying depths crossbedding, ripples, cut-and-fills, rip-up clasts and breccias are sometimes created. Materials can also be transported en masse into deeper waters to form graded layers.

As a consequence of this mobilisation of very soluble minerals some may partly or completely dissolve if moved into less saturated waters. Hopper structure is a curious internal feature of halite crystals which probably reflects alternate phases of growth in suspension (the darker zones) and authigenic growth (the cleaner zones) within the bottom sediment. Solubility is also expressed in other ways, especially in ancient successions.

At outcrop, evaporite-bearing successions may be devoid of salts and represented by residual sediments carrying salt pseudomorphs and by a range of localised collapse structures, usually of an intensely brecciated nature. Pockets of evaporite-solution breccias are commonplace in the Permian succession of northeastern England.

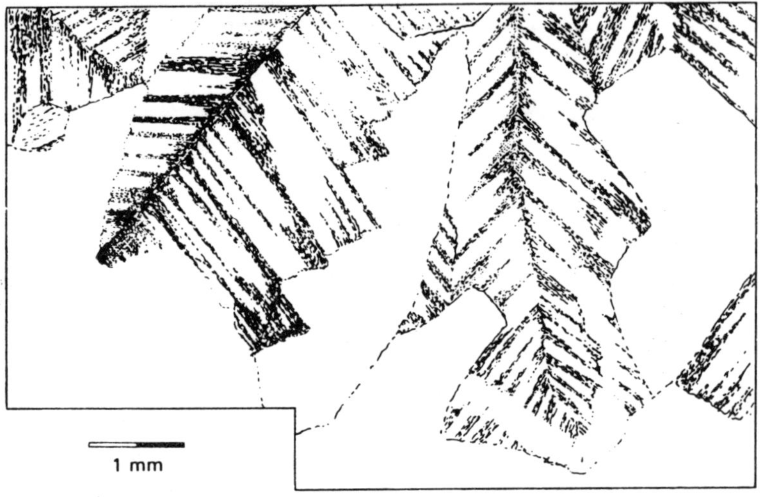

Figure 16.1 : Hopper or chevron structure in halite.

Metasomatic replacement and recrystallisation are widespread in all major evaporite bodies and are mainly a consequence of the solubility of the minerals, the progressive concentration of residual dense and mobile interstitial brines towards the end of prolonged periods of basin desiccation, and subsequent burial diagenesis. Whole beds may show mineral replacement. Some dissolution and replacement can occur during phases of prolonged subaerial exposure between high water levels.

Pressure ridge polygons, salt pinnacles, salt nodules and salt solution tubes several metres across can form under these conditions by movement of surface and subsurface waters. Moist winds can create linear grooves in salt crusts exposed on the flanks of inland salinas. Winds also have the capacity to dislodge silt and sand-sized salt fragments and concentrate them into rippled sheets, low mounds and even dune fields marginal to the salt bodies.

Under certain conditions salt, known as cyclic salt, can be carried by wind for long distances inland in the form of fine dust or as a spray.

In India great quantities of such salt, as much as 130000 tonnes per year, are carried northeastwards into Rajasthan Province from the Rann of Kutch and nearby coast during the hot period preceding the summer monsoon. This salt is dissolved by the later summer rainfall and carried down into lakes and reprecipitated. The hills and valleys surrounding inland salars in Chile are often mantled with cyclic salt which permeates the soil layers and forms a thin saline crust.

Modern hypersaline salinas and lagoons have a high organic productivity and content expressed by planktonic algae and algal mats, together with fragmented and drifted saltmarsh vegetation, such as *Salicornia* and *Zostera*. Drifted foraminifera, bivalves and gastropod shells can also be present in quantity. The salt deposits reflect this productivity by incorporating and preserving some of the organic matter.

In ancient evaporite beds, including those of deeper water origin, diagenesis and recrystallisation usually chemically and physically degrade organic matter into opaque pigmented bituminous and carbonaceous compounds, so that original organic structures are not now easily observable. However, carbonaceous inclusions in halite, and carbonaceous filaments resembling degraded algal mats, have been recognised in Upper Permian evaporites of Yorkshire, England.

CLASSIFICATION

No satisfactory textural classification of evaporites exist. This is because very few are unaffected by recrystallisation, metasomatic replacement or physical displacement. Recrystallisation and replacement textures are particularly common in beds which have been deeply buried or have been mobilised sufficiently to form veins, dykes, sills and diapiric salt domes. The Permian chloride zone minerals of North America and northwest Europe have recrystallised on a large scale in certain areas.

Gypsum and carbonates are commonly replaced by anhydrite, and halite and anhydrite by polyhalite. Conversely, anhydrite quickly hydrates to gypsum under the influence of circulating ground waters and there are few deposits where unaltered anhydrite beds actually outcrop.

The consequence of this inherent complexity is that normal sediment textural terminology is frequently supplemented by descriptive terms associated with metamorphic rocks. Thus we get names such as porphyroblastic schistose gypsum-anhydrite rock, in which the anhydrite flow lineation is disrupted by the growth of large secondary grains of gypsum.

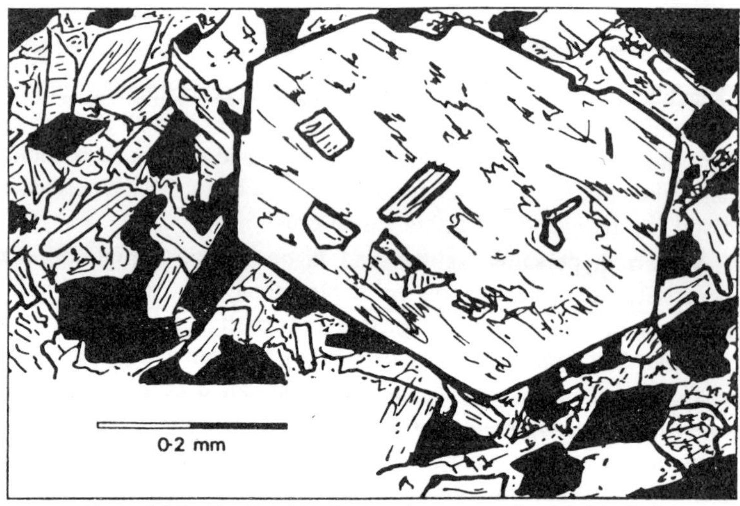

Figure 16.2 : Porphyroblastic secondary gypsum in anhydrite bed. Relict anhydrite grains enclosed within the euhedral gypsum.

TERRESTRIAL SALT DEPOSITS

Inland drainage systems are found at the present day in the drier parts of the world where the surface drainage is entirely balanced by evaporation, and there is no excess of water to overflow to the ocean. The salt of a few lakes, such as the Great Salt Lake of Utah, shows a curious similarity to that of the ocean, but the waters of most inland lakes are distinguished by individual pecularities of composition, and are often characterised by the predominance of salts or radicals which in sea water are subordinate or present only in small traces.

Alkaline lakes usually contain a considerable proportion of sodium carbonate. Lake Magadi in Kenya is a trona-precipitating lake. Other lakes may be dominated by sodium or magnesium sulphate, or may contain unusually large quantities of magnesium chloride. A few lakes are remarkable in containing notable quantities of borates, while others show abnormal proportions of bromine, potassium or dissolved silica.

The analyses in Table elsewhere in this chapter of the waters of some typical salt lakes illustrate only a part of the wide variations in composition that are possible; these are for the most part cases of extreme concentration, in which the water is saturated, at any rate for some of the salts, and deposition is actually taking place.

These figures indicate the existence of two fairly well-marked types: (1) salt lake proper, with dominant chlorides, of which the Dead Sea is a good example; (2) bitter lakes, with sulphates and alkaline

Table 16.2: Salt Content of Selected Lakes (in parts per 1000).

	Lake Urmia, Iran	*Great Salt Lake, Utah*	*Lake Dead Sea*	*Carson El'ton, Russia*	*Domo- Lake, Nevada*	*Lake shakovo, Siberia*
sodium chloride	190.47	118.63	63.86	38.30	64.94	3.55
magnesium chloride	5.22	14.91	163.67	197.50	-	6.08
calcium sulphate	1.81	0.86	0.78	-	-	2.84
magnesium sulphate	8.00	-	-	53.20	-	-
sodium sulphate	-	9.32	-	-	13.76	132.82
sodium carbonate	-	-	-	-	29.25	0.21
total salinity	205.50	143.72	228.31	289.00	107.95	145.50

carbonates, e.g. Carson Lake, Nevada and Lake Natron in Tanzania. Lake El'ton, near Volvograd, is intermediate between these two groups, being rich in both chlorides and sulphates. It is comparable with the Bitter Lake in Egypt.

The Dead Sea

The sea, which now has an area of about 1000 sq. km was initiated as a marine embayment in Pliocene times along the line of the Great Rift Valley, but soon lost its oceanic connection, so that by Pleistocene times it was totally landlocked. A post-Pliocene continental salt sequence, at least 4000 m thick, was laid down in the subsiding rift, and this style of deposition has persisted through to the present day.

Quaternary climatic changes have caused periodic diminished inflow and increased evaporation, which in turn have caused the level of the lake waters to fluctuate. During low inflow phases the salts became increasingly concentrated and, as now, were deposited in greater volumes at certain seasons of the year. About 70 m of halite, gypsum, aragonite and some muds have accumulated towards the southern end of the sea over the last 12000 years.

It has been shown that the waters of the River Jordan, which is the principal tributary of the Dead Sea, are unusually rich in dissolved mineral matter, because it flows through a region largely composed of salt-bearing rocks of Pleistocene, Tertiary and possibly Cambrian age. Besides this source of supply there are, in the neighbourhood, many brine springs, some of which are hot and connected with the volcanism that accompanied the subsidence of the rift zone in which the Dead Sea lies. It is also believed that there are subaqueous mineral springs, some of which yield bituminous products. The average salinity of the lake is 25·7%, which is the highest value of all the greater salt lakes in the world.

A remarkable feature of the water is the presence of a large proportion of bromine, and the Cl:Br ratio in the modern halite deposits at the southern end reaches as much as 2500:1. This figure contrasts with halite and allied chloride deposits of modern and ancient marine basin origin, which generally have a ratio falling between 3000 and 15000:1. As shown by analysis the waters of the Dead Sea are extraordinarily rich in magnesium chloride. Since this salt is much more soluble than sodium chloride, and the water is not yet saturated with it, compounds of magnesium are not being deposited. On the other hand, since the presence of magnesium chloride diminishes the solubility of sodium chloride, the water is saturated for sodium chloride.

Consequently crystals of common salt, together with gypsum, are abundantly present in the muddy deposits now being laid down on the floor of the lake, and the sodium and calcium salts brought in by the Jordan and other rivers are at once precipitated. The principle here enunciated is very important in the study of salt lakes; and it accounts for the fact that common salt is now being deposited in many lakes in the waters of which very different proportions of this salt are contained, since the saturation point for any one salt is controlled by the proportions of other salts present in the solution.

Playas (salinas)

The deposition of salts in playas, which often form in arid and semi-arid interior drainage basins, follows a comparatively simple pattern and frequently illustrates how the chemical evolution of brines in such an environment is mainly controlled by the bulk composition of the parent water and the degree of evaporation.

In the Basin and Range Province of Nevada and California, where daytime temperatures normally reach about 40°C and annual rainfall is 50 mm, there are several shallow lakes and salinas, such as Death Valley, where borax $(Na_2B_4O_5OH)_4.8H_2O)$ and other borates occur in association with halite, gypsum, anhydrite, trona and many other salt minerals.

Searles Lake, north of the Mojave Desert, and covering about 130 sq. km in area, is another such and comprises desiccated salt flats and marshes. The borates appear to have been leached out of adjacent bedded deposits of Pliocene age rich in hydrated borates by the action of invasive brine solutions rich in sodium chloride. The ultimate source of the boron compounds is believed to be volcanic or hydrothermal. Other examples of borate deposits are present in Tibet where they are known as tincal.

Important deposits of potassium nitrate (saltpetre) and sodium nitrate occur in South America. They are found in the desert region of the Pacific coast, especially in the provinces of Tarapaca and Antofagasta, in Chile. The nitrate area is several kilometres inland, and rises to a height of over 1 km above sea-level. It stretches from north to south for about 410 km. It seems probable that the nitrates have been formed in small quantities over a large area by the oxidation of organic matter present in the soil, and that the resulting surface efflorescences were dissolved during occasional tropical floods and redeposited in a concentrated form from temporary lakes into which the flood waters drained.

There can be little doubt that in their present form most of the South American nitrate deposits have crystallised from evaporating saline waters.

In the Saline Valley of California a zonal distribution of evaporites intermingled with the sands and muds can be detected (Fig. 13.3). The evaporite minerals also occur as surface efflorescences. Water flowing into the playa emanates from springs and mountain streams but, by the time it reaches the playa area, is subsurface. Maximum inundation by flash-flooding only amounts to a few centimetres.

Initially, the water is relatively rich in sodium, calcium, sulphate and bicarbonate ions but the response to a high rate of evaporation ensures progressive enrichment in sodium, chloride and sulphate ions towards the playa centre. The sequence of events is: first, the precipitation of calcite in alluvial fans adjacent to the playa which decreases the concentration of calcium and bicarbonate ions in the brines; secondly the precipitation of gypsum at the playa edge, which modifies the sulphate ion concentration; and finally, the concentration of sodium and chloride ions in the central parts which progressively gives rise to the deposition of glauberite and halite in variable proportions. Contemporary precipitates of analcite ($NaAlSi_1O_6.H_2O$) and sepiolite ($Mg_2Si_3O_8.nH_2O$) are distributed through the playa sediments.

The various salts are usually precipitated from capillary pore waters above the water table and form hard crusts, a metre or so in thickness. Around the margins of the playas even sand dunes may become effectively stabilised by this process. The surface crusts show extensive desiccation cracking and pressure ridging due to the crystallisation of new salts.

They also show corrosion into pseudo-ripples, cusps and channels, suggestive of spasmodic free flow of surface water. Salt wedges and tufa-like pinnacles, domes and terraces are associated features. Good examples of tufa pinnacles are present at Mono Lake, California.

During the final stages of desiccation, the concentration of salts in the interstitial brines of salinas may reach 1000 times the original concentration and the brines become 'chemically aggressive'. When volcanic ash is present in quantity, diagenetic zeolites form. Parts of the Eocene Green River Formation have playa-like characteristics and consist of beds containing up to 70% fine-grained analcite, believed to have resulted from vitric rhyolitic ash reacting vigorously with residual brines. In other situations, intensive salt crystallisation from the brines progressively destroys primary bedding and lamination.

Alkali Lakes

The character of alkali lakes differs little from that of many inland salinas, except that the body of water may be more persistent and remains highly alkaline throughout any repeated desiccation or high water level cycles. Lake Magadi in Kenya intermittently dries out except for brine pools along its margins. Lake Natron in Tanzania, also set in an arid climatic regime, varies in depth between 1 and 3 m and the waters are very rich in sodium and bicarbonate ions derived from hydrothermal springs near the lake margin and weathered carbonatite lavas.

The lake bottom sediments consist predominantly of alternating layers of trona and organic-rich clay carrying up to 25% authigenic analcite. There is some resemblance between these deposits and parts of the Triassic Lockatong Formation of New Jersey and Pennsylvania. The sodium carbonates in alkali lakes are invariably accompanied by sodium chloride and sodium sulphate, together with other carbonates and sulphates. Analyses of the water of small lakes near to Mount Ararat in northeast Turkey show extraordinary high salt contents, amounting in one case to 239 parts per thousand of sodium carbonate and 53 parts per thousand of sodium sulphate.

The Caspian Sea

Around the Caspian Sea the formation of salt deposits takes place in partially isolated areas having limited communication with the main body of water. Under such geographical conditions and where high temperature causes strong evaporation, while a small rainfall limits the supply of fresh water, the concentration of salts may be carried to the point of saturation and precipitation. A classic example is the

Karabogaz Gulf, some 18000sq. km in size, on the eastern side of the Caspian Sea. The waters of this Gulf are about 3 m deep and have a composition very different from that of the Sea.

Table 16.3 : Salts in the Caspian Sea and in the Karabogaz Gulf

	Caspian	*Karabogaz*
sodium chloride	8.116	83.284
potassium chloride	0.134	9.956
magnesium chloride	0.612	129.377
magnesium sulphate	3.086	61.935
salinity per mille	11.948	284.552

The Gulf is separated from the Caspian by a narrow strait some 10 km long, 100-150m wide and about 6 m deep. A continual current runs through the strait into the Gulf to supply water lost by evaporation, hence there is a continual addition of salts in solution. Chemical analyses furnish clear evidence that a great concentration of salts of sodium and magnesium has taken place in the waters of the Gulf over the last three or more decades.

This is due to falls in the Caspian sea-level, partly caused by water abstraction along the rivers Volga and Ural. As a consequence the flow of replenishing brine and sea water has been cut sharply. Sodium and magnesium salts, such as halite, epsomite and astrakhanite ($MgSO_4.Na_2SO_4.4H_2O$), are now being precipitated over about 75% of the water-covered area (about 10000sq. km). Prior to the 1930s the principal mineral deposited was glauberite but, since then, the proportion of other salts has increased due to increase in brine concentration. Halite was first precipitated in 1939.

During present summers halite, epsomite and astrakhanite are deposited, mainly in the northern and eastern parts of the Gulf where there are the highest brine concentrations. In winter, mirabilite with a little epsomite is precipitated. Mirabilite is also deposited in the southern and western parts during the winter months. Gypsum is forming fairly continuously at the margins of the Gulf, particularly on the western side adjacent to the strait.

However, the boundaries between the evaporite facies are variable in position, depending on the season, volume of inflow and the displacement of surface brines by the intensity of the prevailing winds. Beneath the modern salt layers are three older layers separated by carbonate beds. It appears that this triple succession reflects late-

Quaternary rises and falls in surface level of the Black Sea, the carbonate oozes being laid down during transgressive, high sea-level phases, and the salts being precipitated during regressive, lower sea-level phases.

MARINE SALT DEPOSITS

Marine Salinas

Crystallisation of sodium chloride is taking place at the present day in numerous marine salinas of all sizes. The salt ponds of the Bahamas are shallow pools lying behind the coast, and in most cases separated from the sea by a pervious barrier. One of the larger examples in the Mediterranian region is the salt lake at Larnaca in Cyprus.

Strong evaporation in summer leads to salt concentration and keeps the levels of the lake a metre or so below that of the open sea. Pervious sand and gravel barriers allow a constant slow inward seepage of sea water and the annual desiccation that ensues leads to the deposition of a layer of halite up to several centimetres thick. This layer is scraped off regularly for commercial use.

Persian Gulf

The extensive coastal plains on the arid western side of the Persian Gulf are up to 300 km long and 25 km or more wide, and are prograding areas dominated by lime sediment accumulation. Prime examples occur along the Trucial Coast at Abu Dhabi and Sabkhat Matti. The sediments are aragonitic and calcitic micritic muds, silts and sands of supratidal, intertidal and shallow water subtidal origins, commonly shelly and extensively bioturbated. Algal mats and stromatolites are a regular feature of the intertidal facies, and the supratidal facies also includes much windblown material.

The plains are flat, reach elevations of about 1-2m above high water mark, and are commonly salt-encrusted. The Arabic term for them is sabkha (or sebkha). Inland sabkhas are also recognisable. In coastal sabkhas the ground waters are of the nature of brines and essentially marine in origin. Replenishment or recharge occurs during the flooding of the peripheral marshes by high spring tides and storm surges, and by subsurface seepage from the open sea.

At times the flooding extends inland by as much as 12 km. The upper surface of the brine-saturated sediment is about 1-2 m below the sabkha surface and the salinity of the brines commonly reaches ten times that of the adjacent open sea, due to very high evaporation rates.

The reason why coastal sabkhas have been the focus of much attention over the last 30 years is that they have hosted a range of diagenetic evaporite minerals and secondary textures and structures which closely resemble those in several ancient successions once believed to be of deeper water lagoonal or marine basin origin. The sylvite-, carnallite- and halite-bearing Permian Middle and Upper Potash deposits of Yorkshire are now considered to be of sabkha origin. Lower Carboniferous and Purbeckian successions of central and southeastern England also carry thin sabkha-type evaporites.

The changes that take place are those of displacement and replacement within the host carbonate sediment. Early diagenetic gypsum is widespread in the high tidal flat and marsh muds and algal sediments, often as crystal mushes several centimetres thick. Open-textured stromatolitic layers are particularly susceptible to gypsum precipitation. From the concentrated capillary brines above to just below the water table further inland, anhydrite is precipitated in preference to gypsum.

The sulphate ions for the anhydrite are mainly derived from the gypsum. As the anhydrite crystals grow they progressively coalesce into rows of nodules and nodular layers, up to 10 cm in thickness. Initially, these are rather soft and plastic. The growth process appears to be one mainly of displacement of the host sediment rather than replacement of carbonate grains. At the contact between mutually interfering nodules, thin zones of original impurities and microcrystalline anhydrite are trapped and the whole sediment takes on the appearance of a coarse irregular net. This is 'chicken-wire' structure.

Recrystallisation of anhydrite to gypsum, or vice versa, plus compaction leads also to a style of nodular growth in which an elongation and convolution is developed, similar to that of animal intestines, a structure known as enterolithic folding.

Both the anhydrite and gypsum in the sabkha areas are susceptible to further diagenetic changes. Gypsum is known to replace anhydrite if the circulating pore waters become less concentrated with salts. Corroded anhydrite laths are enveloped by euhedral diagenetic gypsum where flashflooding dilutes the ground water. Calcite, dolomite, celestite and chalcedonic silica are known to replace crystals of anhydrite and gypsum. Bladed grains of secondary gypsum, up to 20cm long, and 'desert-rose' aggregates typify the surface of sabkhas. These grains commonly enclose quartz sand.

Halite is widespread as ephemeral crusts but does not occur as distinct layers. Instead it occurs as veneers around clastic particles or as cubes enclosing sand. It is very prone to solution during spasmodic flash-flooding of the sabkha surface and is then carried back down into

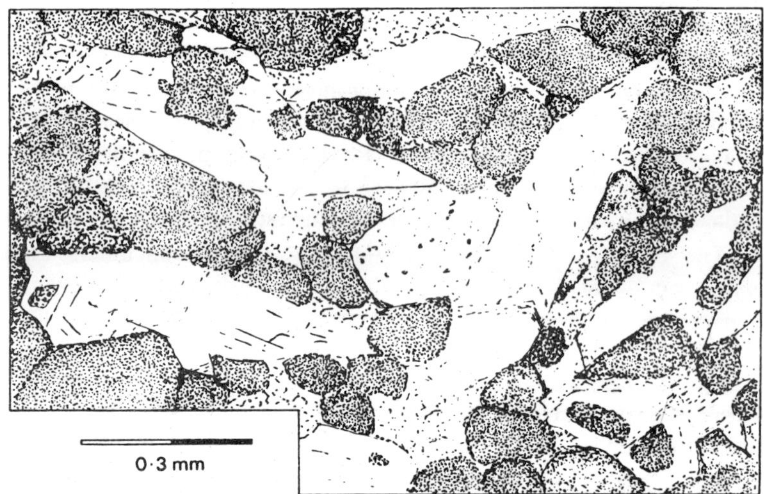

Figure 16.3: Replacement of authigenic gypsum by calcite. The original rock was a Jurassic peloid lime mud, probably laid down in an intertidal-lagoonal situation.

the subsurface brines. On dried-out surfaces the halite grains may be transported physically for some distance.

Celestite ($SrSO_4$), a minor diagenetic constituent of coastal sabkha successions, is most abundant in areas of intense dolomitisation and is a product of the release of Sr^{++} during the replacement of aragonite by dolomite. Dolomitization of fine-grained carbonate muds is very common, the appropriate high Mg^{++} : Ca^{++} ratios between 5 and 35 being attained in the areas about 2-10 km away from the open sea.

The high ratios in the brines are caused by the precipitation of calcium ions as sulphate and carbonate minerals. Diagenetic magnesite ($MgCO_3$) is usually present in the upper 20 cm of the host sediment and seems to be a precursor of the main phases of dolomitization.

ANCIENT EVAPORITES

Marine evaporites are present in strata of late Precambrian age and occur throughout the geological column up to the present day. Some of the oldest are in Pakistan, Iran and Siberia. The United States has evaporites in every geological system from the Ordovician onwards, and at certain times the accumulations were extensive and thick.

Over 260000sq. km of the eastern Great Lakes region is underlain by Upper Silurian gypsum, anhydrite and halite beds which, in places, reach thicknesses of 450 m. Thick anhydrite and gypsum beds are also present in the Carboniferous of Utah (1200m), the Permian of southeastern New Mexico (1350m) and the Cretaceous of southcentral Florida (1800 m). In the Whitby district of northeast England at least 529m of Permian evaporites have been proved in bore holes.

In many of these successions there is evidence that some of the deposits were laid down in current-swept, relatively shallow seas and under sabkha-type conditions. Ripple marks partly of corrosive origin, cross-bedded units up to 30 cm thick, minor unconformities, desiccation cracks and 'shale ball' pebbles, associated with marine fauna and flora, are found towards the edge of basins.

Table 16.4 : Permian evaporite succession of north-east England.

Succession	*Major cycles*	*Maximum thickness (m)*
Upper Halite and Upper Potash		45
Upper Anhydrite	4	9
Upgang Formation (dolomite)		1
Carnallitic Marl		20
Boulby Halite and Boulby Potash		90
Billingham Main Anhydrite	3	15
Upper Magnesian Limestone		65
Fordon Evaporites	2	325
Kirkham Abbey Formation (dolomite)		125
Hayton Anhydrite		145
Lower Magnesian Limestone		
Marl Slate	1	+100

In the Upper Permian succession of eastern Yorkshire, England, there are repeated developments of lenticular and nodular anhydrite, exhibiting displacement structures and interbedded with fine-grained, algal dolomites. The deposits were laid down at the western margin of the Zechstein Sea in what seem to have been extensive sabkhas and coastal lagoons. Over one period, represented by the diachronous Billingham Anhydrite formation, the sabkha zone alone had a width perhaps as much as 80 km.

A feature of ancient sabkha successions, as in parts of the Lower Carboniferous of the Midlands and Peak District, England, are rhythmic units, about 30 cm thick, each constituted of at least a basal sparsely fossiliferous, algal limestone passing upwards into a sulphate-rich nodular layer. Palaeosols, marls, mudstones and shales may be intercalated between the units. The limestones are usually dolomitized.

In most thick marine evaporite successions, however, the evidence

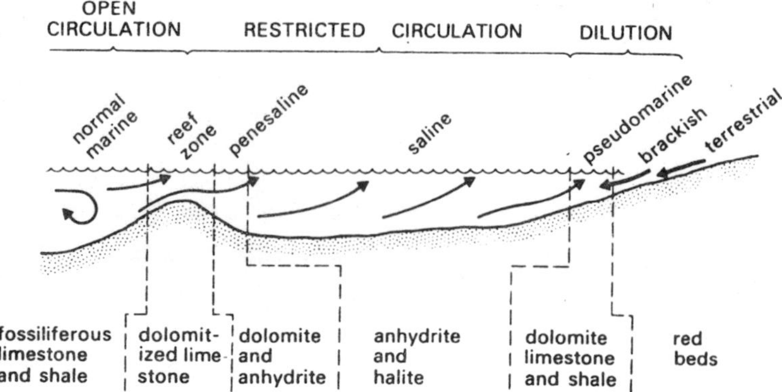

Figure 16.4: Back-reef evaporite deposition.

tends to favour a mixed genesis for the bulk of the salts. At the margins of the very large basins sabkha-like deposition commonly prevails, whereas in the deeper water sites the deposition appears to be mainly subtidal. Graded salt and carbonate beds with basal scour structures have been identified and seem to have been emplaced by turbidity currents.

Mass flow slump and slide deposits are represented by breccias with reworked evaporite and carbonate fragments. Individual beds of banded halite are commonly up to 25 m thick and collectively aggregate several hundreds of metres in thickness, which suggests deeper water conditions. Regularly laminated beds of halite, anhydrite and carbonates (laminites or varvites) can be traced over tens of kilometres.

The individual laminae are a few millimetres in thickness and indicate low energy conditions more readily found in deeper basins. In major evaporite basins characterised by large-scale cyclical successions it is assumed that marine incursions were restricted in some way, either by structural or physiographic barriers. Periodic breaching of these barriers, necessary to cause the cyclicity, might have been instigated by tectonic, climatic or eustatic events, though the relative importance of these events is conjectural.

The Mediterranean region in arid late Miocene (Messinian) times was partially isolated on at least two occasions from the Atlantic and Eurasian Tethyan Oceans. This was caused by plate convergance and eustatic lowerings of sea-level, estimated at about 60m maximum. Sills, probably closer to sea level than the present sill in the Strait of Gibralter (320 m depth), diminished the inflow of oceanic water and reduced reflux, such that the region became progressively more saline and, eventually, a major site of salt accumulation. In the first and most prolonged phase of isolation and desiccation four major subsiding basins, each rather complex in bathymetric detail, were established, Balearic in the west, Tyrrhenian and Ionian in the centre, and Levantine in the east.

Thicknesses of up to 2 km of sulphates and chlorides were precipitated. Each of the basins at an early stage had water depths, maybe as much as 1-2 km greater than the water depths over the coastal and intrabasinal platforms and ridges, on which shelf carbonates accumulated.

As desiccation progressed so water levels fell and sabkha and coastal salina conditions became established in the emergent shallows. Ultimately, lateral zonations on an irregular 'bull's-eye' pattern were generated with outer discontinuous carbonate-dominant zones grading inwards into sulphate dominant, then zones dominated by more soluble chlorides. Comparable patterns exist in the Permian Castile Formation of New Mexico and Texas and in the Permian Zechstein deposits of northwest Europe. They reflect an increase in brine concentration with reduction in size and volume of the residual bodies of water.

As the oldest Messinian salt beds (halite and polyhalite mainly) are so thick, it is unlikely that the residual bodies of water ever dried out completely, as in certain playas, though some workers think otherwise.

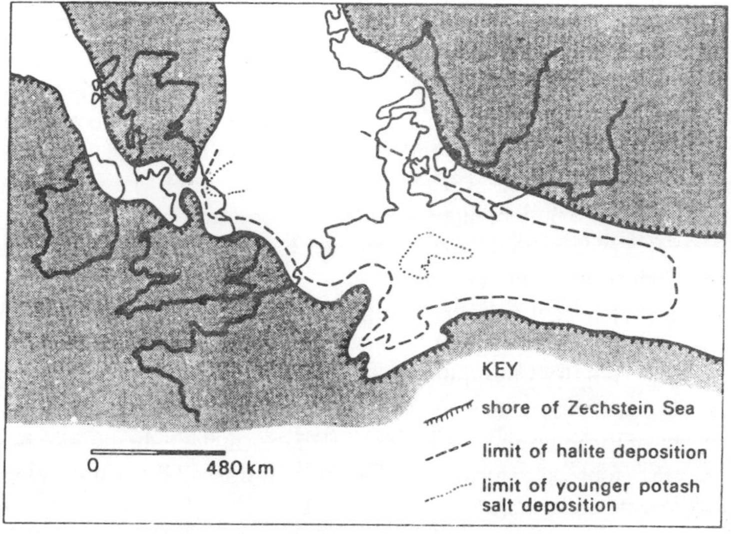

Figure 16.5: Permian Zechstein Sea, illustrating the 'bird's-eye' pattern of salt distribution.

It is more likely that replenishment was maintained from adjacent oceanic and land areas, though without any significant reflux. The waters may have remained quite shallow for most of the time during evaporite precipitation, despite basin subsidence. Density stratification probably played a significant role in concentrating the heavier chloride brines in the deepest parts.

The second phase of evaporite precipitation is represented by thinner and more widely distributed deposits, some now outcropping on land. They indicate a lowering of the sills to the west and possibly east, a major influx of oceanic water, followed by a renewed period of desiccation. In these younger deposits there is evidence again for sabkha and coastal salina deposition. Towards the end of Messinian times some of the basins appear to have been completely desiccated. The whole episode of evaporite formation ceased in early Pliocene times with the rapid flooding of the Mediterranean region, the re-establishment of high sea-levels and deep connections with open oceans, and the depositon of normal pelagic muds.

Minor physical barriers within basins are always likely to control evaporite distribution. Reefs flanking the shelf areas of the Permian Delaware Basin in Texas and New Mexico effectively reduced circulation in the back-reef lagoons and basins, allowing the precipitation of anhydrite and halite. Within this framework, lateral zoning resulted from the interaction of shoreward circulating salt water and diluting freshwater streams.

In some instances a minor lowering of sea-level can enhance the capacity for strong evaporation behind and between patch reef complexes. Such appears to be the case with the massive and laminated, granulose gypsum deposits of the Miocene age in Cyprus, as at Maroni, Polemi and Morphou. The irregularity of the sea bottom in these back- and inter-reef basins is sometimes expressed by turbidite-like grading and slump structures in the evaporites.

A partly barred large basin with relict sea characteristics was the Upper Permian Zechstein of Europe, which has an inferred extent of 250000 sq. km. The basin was located at about 15°N latitude towards the northern flanks of Pangea and had intermittent connection with the open ocean to the north. The barriers in the north were totally breached on at least four occasions, allowing mass influxes of oceanic water. These marine transgressions were related to worldwide sea-level changes consequent on the melting of ice sheets in southern Pangea (Gondwana) and the progressive breaking-up and drifting apart of Pangea.

In the later stages of the sequential desiccations, the floors of the depocentres, characterised by 'bull's-eye' patterns of deposition, probably lay a kilometre or more below that of the barriers. Sabkha and deep water styles of precipitation again appear to have been synchronous events. In northeastern England, beneath the North Sea and in Germany (Hanover and Thuringia districts) four to five major salt-bearing sequences are present, each representing a major episode of marine transgression followed by a prolonged period of desiccation and regression.

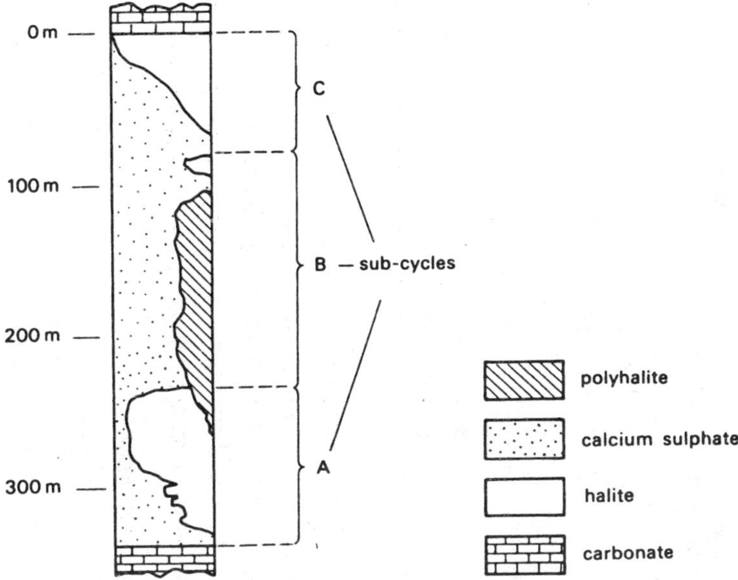

Figure 16.6: Subcycles in Permian Cycle 2 beds.

The initial deposits were shelly, oolitic, algal, reef and bituminous limestones, now dolomitized, marls and shales all of supratidal to relatively deep water origins. These grade vertically and laterally into sulphate and chloride beds. The thickest evaporite sequence in England is that of cycle 2 and it differs from the others in consisting of three subcycles of deposition, each indicating a minor transient marine transgressive episode.

The subcycles also show a gradual vertical change from anhydrite at the base through to halite at the top. The sequences are complicated, however, by the presence of polyhalite, a bittern about which there used to be some dispute regarding origin. Certain authorities suggested that the polyhalite was a direct precipitate from sea water, others remain convinced of a diagenetic replacement origin. The petrographic evidence favours the latter interpretation. The cycle 3 sequence is similar to the lower in being mainly composed of layers of anhydrite and halite with secondary polyhalite.

But cycle 4 differs in showing a complete vertical cycle of salt deposition with the least soluble carbonates at the base and most soluble chlorides, sylvite and halite at the top; polyhalite is absent. Smaller-scale cycles commonly involve the pairing of either halite with anhydrite or anhydrite with carbonate into varve-like laminae (laminites). These appear to be deep water in origin.

Where halite and anhydrite are paired the halite may represent a warmer water summer precipitate and the anhydrite the corresponding winter precipitate. This contrasts with the anhydritecarbonate pairs where the anhydrite may be the summer precipitate and carbonate the winter. But the assumption that these pairs are always genuine varves indicating one annual climatic cycle is open to considerable doubt, especially when the thickness exceeds 10 mm.

A more credible thickness for an annual evaporite varve is 1-2 mm. Hence, the thicker varves must originate in other ways. Some varve-like layers have almost certainly been formed by alternating precipitation and solution processes in which a mixture of salts is subject to partial solution, leaving behind a relatively insoluble residue. For example, if a halite-anhydrite-clay-water mixture which has just been deposited is subject to changes in the constitution of the brine, then the halite may be taken back into solution, leaving behind a thin anhydrite-clay seam.

The clay-grade particles in these seams are usually a mixture of quartz, chlorite and illite, some of which are undoubtedly authigenic. In the Middle Devonian Elk Point Group of Alberta, deposited in a basin extending over at least 750000sq. km, both sabkha-type and deep-water laminite evaporites are present.

The broad pattern of salt deposition is not 'bull's-eye' but markedly asymmetrical, with sulphate beds dominant adjacent to the northerly barrier to the basin and contemporaneous thick halite beds (up to 500m) confined to the innermost parts, some 100 km to the south. Reefs and associated shallow water carbonates acted as the barrier.

The general absence of anhydrite and gypsum precipitation in the halite areas suggests that the brines flowing into the innermost depressions were already deficient in calcium and sulphate ions and rich in sodium and chloride ions. In this respect, there is some resemblance to the late Middle Triassic Saliferous Beds of the Cheshire Basin in England, where the innermost depocentres were dominated by an alternate deposition of halite beds (15-105 m thick) and red mudstones.

17

VOLCANICLASTIC DEPOSITS

The term volcaniclastic encompasses a wide variety of rock types, all of which originate as a consequence of volcanic activity and consist essentially of volcanic ejectamenta. They are abundantly represented in converging, destructive plate and island arc successions.

CLASSIFICATION AND CONSTITUTION

The most important group of volcaniclastic sediments is pyroclastic in origin in which the discrete fragments are extruded from vents as a result of explosive volcanic eruptions. The mechanisms for such eruptions, which range from quiet to paroxysmal, are a combination of magmatic and phreatomagmatic.

With the former, the exsolution and expansion of hydrogen, carbon dioxide, sulphurous and other gases occurs as the magma approaches the ground surface, whereas with the latter the rising magma chills as it comes into contact with the ground and surface water. The autoclastic group forms by the physical fragmentation of lava within vents, pipes and lava flows.

The disruption generally occurs while the lava is moving, but can be brought about by gas explosions after it has come to rest. Hydroclastic rocks are those which fragment, often explosively, by rapid cooling in a body of water or beneath ice. The original rock is highly fragmented by this process and the fragments are very angular. As cooling is so rapid many of the rock fragments and matrix grains are vitreous, hence the group name hyaloclastites if glass is very prominent. Epiclastic rocks are those formed from the accumulation of weathered and eroded volcanic materials. Transport either through air or by water, or both, is implied. Volcaniclastic deposits are broadly subdivided on the grounds of their grain size and genetic characteristics.

The grain size boundaries correspond with those normally used for other varieties of fragmental rocks and the average size of the grains determines which category the deposit falls into. The degree of sorting in volcaniclastics is very variable, some agglomerates for example being almost devoid of tuff matrix, whereas others can carry as much as 25%. The larger clasts are then essentially matrixsupported as opposed to grain-supported.

With caution it is possible in pyroclastic rocks to distinguish between *essential* (juvenile) fragments derived directly from the erupting magma, *accessory* (cognate) fragments detached from co-magmatic earlier eruptives, and *accidental* fragments introduced from below or around the flanks of the volcanic centre. The last can be of any composition. In some Carboniferous pipes and vents of the Midland Valley of Scotland the tuffs carry nearcontemporaneous coal and limestone fragments and even Precambrian granulite fragments.

Constituents such as these are indirectly informative about the pre-volcanic history of a region. Volcanic bombs are the main constituents of agglomerates and are the product of the ejection through the air of solid and molten lava, usually of acid and intermediate composition. There is a well-documented example of an 8 tonne bomb being thrown for 1 km during an Hawaiian eruption, but most bombs are less than 1 m across _nd are dropped near the vents.

Spindle shapes are common, but others are spheroidal and even 'cowpat' shaped due to flattening on impact with the ground. Some spatter cones in Hawaii are formed of a solid mass of 'cowpats' with very little interstitial ash. Internally, bombs are frequently vesicular because of the continued expansion of gas during transit. Breadcrust bombs are deeply cracked. Chilled vitreous-looking crusts can often be seen.

Some small volcanic cones, rarely more than 200m high, consist predominantly of cinders (or scoria) of light, slaggy vesicular lava varying in size from coarse ash to boulders over a metre in diameter. Pumice is another variety of highly vesicular, but glassy, material which is light in weight, weakly permeable and highly porous. When deposited in water it floats on the surface until it becomes waterlogged

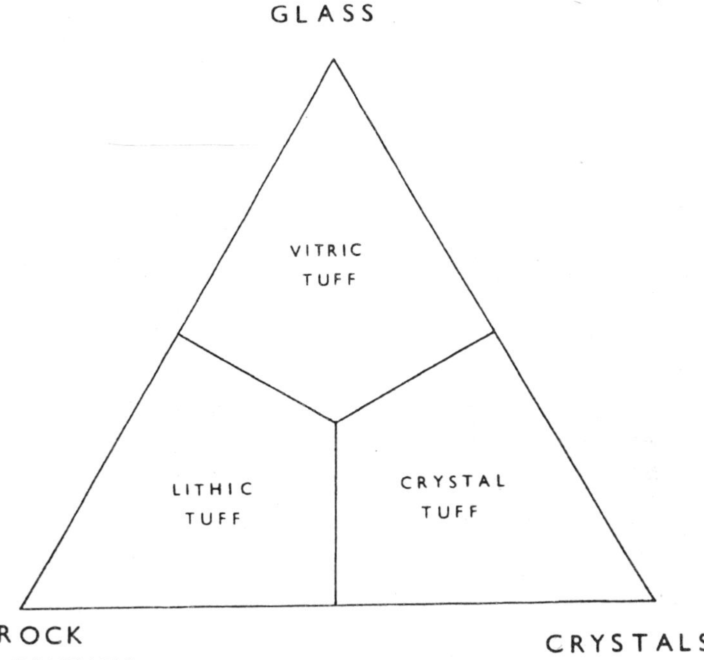

Figure 17.1 : Classification of tuffs (and ashes).

and slowly settles on the bottom. This gravitational settling and sorting out of the pumice from the other ash constituents does not occur to such a high degree when it is deposited directly on land and is a useful criterion, therefore, for distinguishing subaqueous from subaerial accumulation.

Lapilli are fragments which are very variable in shape, composition and internal texture. Some are angular and highly vesicular, others are rounded and dense. Certain spherical to ellipsoidal lapilli, with diameters reaching 1 cm, are of the accretionary type. They have a core of fine to coarse ash or tuff around which are a number of concentrically arranged rims. It has been suggested that accretion of the rims is by the rolling of ash grains across a freshly deposited ash surface or by the turbulent movement of ash grains through a water saturated ash cloud in the fashion of hailstones.

In the Ordovician Borrowdale Volcanic Series of the English Lake District accretionary lapilli tuffs have been called 'bird's-eye' tuffs because of their distinctive appearance. The finer-grained pyroclastic deposits, with grains less than 2 mm in size, are ashes or their consolidated equivalent tuffs. Further subdivision is made on the nature of the dominant particles (vitric, lithic, crystal), chemical and petrogr-aphic composition (e.g. acid tuff, andesitic tuff), manner of transport (e.g. ash-flow tuff) and environment of deposition (e.g. submarine tuff). The categories are not exclusive, hence a rock could be described as a subaerial acid vitric ash-flow tuff.

Tuff sequences and individual layers can vary a lot in composition and texture. Dacitic glassy material, for example, occurs as flattened bands and streaks, often with a flow-like fabric and interlaminated with vitroclastic andesitic tuff. This whole texture is eutaxitic. More discrete lenticular bodies of collapsed and altered pumice and glass, with a flame-like cross section, are called fiamme. These can be as much as a metre in length.

Vitric ashes and tuffs are made up principally of glass in the form of cuspate and rod-shaped shards and in the form of broken angular pieces of vesicular pumice. The material is ejected from the volcano in a liquid condition and with the sudden relief of pressure the dissolved steam and other volatile substances immediately explode into a gaseous condition. The expanding mass of foam-like lava solidifies to glass during the process and is shattered to dust by the pressure of liberated gases within each vesicle.

Glassy material produced in this way acquires a characteristic vitroclastic texture. In certain subaerial situations basalt glass fragments take on a smoother tear-shaped or spherical droplet form of small lapilli size known as 'Pele's tears'. They can be associated with very thin fibres of glass, more than 2 m in length, called 'Pele's hair', both being formed by the fountaining of frothy lava and wind displacement of the ejected fragments.

Vitric tuffs are commonly associated with the eruption of acid and intermediate lavas. This is also true of ignimbrites, a special variety of subaerial pyroclastic ash-flow deposit which is the product of a hot, turbulent, highly mobile and ground-hugging froth of disintegrated magma. Many major ash-flow deposits, such as the rhyolitic tuffs which cover 26000 sq. km in the North Island of New Zealand, are ignimbrites.

The 'volcanic sandflow' of the Valley of Ten Thousand Smokes, Mount Katmai, Alaska is a fairly recent (1912) example in which a whole major valley was plugged in three days by 300 m of ignimbrite for some 22 km downstream. At least 11 cu. km of material appear to have been expelled from vents in this short period. Gas escape structures typify the surface layers of this unstratified deposit, as they do for many older examples, and indicate steam emission from thousands of small fumarolic pipes for a few years after the flow came to rest. The fill of such pipes is autoclastic as the material is emplaced and redistributed by upwards gas-streaming. It includes such rock types as tuffisite.

Some levels within ignimbrite flows, both recent and ancient, are characterised by being hard and compact, and having a cherty conchoidal fracture and a subvitreous lustre. These are caused by the deformation, collapse, fusion and welding of the hot glass particles. Hence the name welded tuff for these rocks. The degree of welding is dependent on temperature, gas flow and viscosity factors, less so on loading stresses. Whether the welded tuff unit is 1 m or 30 m thick the quality of the welding is often similar.

Eutaxitic texture or structure, where bands and lenses of different colour, texture and composition are welded together, is commonplace. Where the welding has been intense the glass shards are stretched, densely packed and fused to such an extent that only ghostly traces of the original can be recognised. An anastomosing fine lamination, almost like a flow texture, is developed known as parataxitic.

In welded tuffs the welding is usually most pronounced at the base of the tuff unit and decreases progressively upwards. At the top there is a gradual or sharp passage into sillars, which are very weakly welded by escaping hot, gases and fluids if at all, and then only at point contacts of the shards. The shards are rarely flattened. Both sillars and welded tuffs frequently exhibit columnar jointing, somewhat akin to that found in lava flows proper.

The identification of vitric tuffs is sometimes difficult, as the glassy materials, especially in early Mesozoic and older examples, are usually devitrified. Axiolitic texture, a parallel intergrowth of fibrous looking sanidine feldspar and cristobalite (a cubic variety of silica) is typical of the alteration within shards. The new minerals grow inwards and eventually produce, where they meet, a distinctive narrow (axial) zone.

In Lower Palaeozoic tuffs of Snowdonia the glass is not only devitrified and recrystallised but often chloritized. Even so, the essential nature of such altered rocks can be deduced, providing that shard forms and vitroclastic textures are preserved. Many altered tuffs were originally recorded as lava flows because of these problems of recognition.

Lithic tuffs are dominated by fragments of previously formed volcanic rocks which may be fine-grained holocrystalline or partly glassy. Some of the fragments can reach boulder size and are described as xenoliths. The bulk of the materials are derived from the margins of the magma chambers, vents and fissures, but a certain amount is picked up during transit across adjacent surfaces.

All tuffs contain detached crystals and crystal fragments, but in crystal tuffs they are dominant to the extent of a $>$ 75% content. The crystals, often misleadingly referred to as phenocrysts, are formed by the disruption of magma which is in an advanced state of crystallisation at the time of eruption.

Sharp angular edges to many crystals imply breakage during the explosive phase and subsequent transportation, though further breakage occurs during compaction. The most common crystals found are quartz, sanidine feldspar and plagioclase with minor amounts of ferromagnesians and iron oxides. They are often set in a vitroclastic matrix.

PYROCLASTIC DEPOSITS

The fragmental air-transported material of pyroclastic origin is known collectively as tephra and comprises agglomerates, tuffs and ashes formed variously of large blocks, bombs, vesicular scoria, lapilli, pumice and mineral grains. The materials, especially those of a coarser nature, come to rest in the immediate neighbourhood of vents, but the finer particles are often spread as ash falls over a much larger area. Some fine dust may even achieve a worldwide distribution on land and sea, though the deposits in the sea are rarely greater than 30 cm in thickness.

As deposition from one eruptive phase can extend over a wide region, the product is of the same age and can be used for correlative and chronological purposes, areas of study known as tephrostratigraphy and tephrochronology respectively. All pyroclastic deposits are susceptible to reworking and epiclastic debris is thus a common, though usually minor component. If the materials are laid down in bodies of water they show a range of normal water-laid sedimentary structures, such as bedding, lamination, cross-bedding, slumps, grading and diagenetic concretions. Flame structures extending upwards into tuffs for 40 cm and large load casts are known to occur in sub-aqueous tuffs, suggesting rapid deposition on uneven surfaces.

Bioturbation may be so intensive as to destroy original structures. Dilution by non-volcanic organic and inorganic debris is also pronounced in lakes, seas and oceans, so that the volcanic component may be obscured and can only be inferred from the presence of diagenetic derivatives. Analcite ($Na(Al, Si_2) O_6.H_2O$), a zeolite created from the breakdown of volcanic glass (zeolitization), has been identified in quantity in Recent sediments of the Gulf of Naples and central Tyrrhenian Sea. It is also found as distinct layers in the Quaternary lake successions of the East African Rifts and the Eocene lacustrine Green River Formation of Colorado, Utah and Wyoming. Phillipsite ((K, Na, Ca) (Al_2, Si_4) O_{12}.4.5H_2O) is another common zeolitic derivative, and in the more volcanic parts of the Pacific Ocean forms up to 50% of the bottom sediment. These early diagenetic sedimentary occurrences should, however, be distinguished from those zeolites distributed along fractures and within cavities of thick old volcanic sequences.

The latter are formed secondarily by reaction between sea water pervading hot igneous materials and include analcite, heulandite ($Ca_3(K_2Al_8Si_{28})O_{72}.24H_2O$) and laumontite ($Ca_4Al_8Si_{16}O_{48}.16H_2O$). Studies of these thick sequences show a downwards succession of zeolite facies, each consisting of a specific assemblage of zeolites. In the Cretaceous ophiolite of Cyprus the uppermost levels are lavas and volcaniclastics of an analcite facies passing down into heulandite and laumontite facies.

Pyroclastic debris can be ejected into the atmosphere for a considerable height. For example, ash was blown into the air in the form of an ash cloud to a height of about 20 km when Mount St Helens in the State of Washington, USA erupted in 1980. Ashfall from the cloud was heavy for tens of kilometres downwind and 40 mm accumulated over a limited area 300 km eastwards after one paroxismal phase. The volcano Quizapu in Chile deposited 2 cm of ash on Buenos Aires, 1150 km away, in 1932.

Not all such debris is carried so far and in any given surge of activity, whether it be subaerial or submarine, much is moved down slope, closely hugging the surface as ash-flows or pyroclastic flows. These are hot and turbulent density flows of volcanic rock debris and gases that travel at speeds of up to 150 kph on land and are known to extend 100 km or more down valleys. They are usually accompanied by clouds of airborne dust full of lithic and pumice fragments.

On land this type of flow sometimes appears to be generated within the dense basal zone of rapidly moving *nuee ardente* ('glowing cloud') eruptions. Whether subaerial or submarine the resulting deposits, called ash-flow tuffs, form layers as much as 120 m thick, though usually they are 30 m or less. Submarine beds up to 100 m thick are present in the Ordovician of North Wales.

The progressive build up of successive ash-flows emanating from a vent of the central Plinean type or from a fissure system often produces successions of 300m thickness or more. In more extreme circumstances multiple and prolonged eruptions create vast ash-flow fields. Some of the largest on land occur in the North Island of New Zealand and in the Lake Toba region of Indonesia.

In southern Nevada and southwest Utah during late Tertiary times some 125 000 sq. km of ground were covered by up to 2400 m of such ash. Many of these deposits are ignimbrites. A threefold sequence has been observed in some individual flow deposits, especially near to the source of the eruption. The bottom thin part, moulded to the ground surface, consists of lenticular, trough cross-bedded and laminated units suggestive of turbulence and avalanching at the base of the fast-moving flow, and these are known as basal or ground surge deposits.

They are sharply separated from the overlying main body of the flow, which is usually massively bedded and poorly sorted. Flame and load cast structures occur at the base of this main body in some submarine flows. Reverse grading, with the clasts increasing in size upwards, is commonly present in the main body.

The uppermost part of the sequence is best developed on land and consists of ash cloud deposits. Again the bottom contact is sharp. Normal graded bedding, shown by a decrease upwards of lithic fragments, some of epiclastic origin, is often present. The uppermost material is a fine tuff, though there can be enclosures of pumice and lithic blocks indicating the early onset of another eruption.

HYDROCLASTIC (HYALOCLASTIC) DEPOSITS

Also known as hyaloclastites when glassy, these rocks are formed by the granulation of lavas beneath a water or ice cover. Occasionally, large blocks of lava with chilled margins are incorporated to such an extent that pillow breccia is a more accurate description. The typical constituents of hyaloclastites and the finer-grained hyalotuffs are chips of obsidian, pumice and sideromelane, the last being an isotropic variety of basaltic glass which readily alters into the hydrous product palagonite.

The shards appear to originate by the spalling of glassy crusts from the rapidly chilled lavas. Palagonite is an isotropic waxy or earthy substance, light yellow to brown in colour and, in turn, alters to microcrystalline micas, such as smectite, with a fibrous and spherulitic habit. Palagonitization is not confined to hyaloclastic rocks but can form in any deposit containing sideromelane.

Lithified palagonite-bearing rocks are found among the sediments of the Pacific islands. In thin section the fragments of altered basic glass are seen to be mixed with shells of foraminifera and other calcareous organic debris, the whole deposit being cemented by calcite. The cement in hydroclastics and hyaloclastites often consists of calcite, but zeolites can also be present. Both types of mineral probably derive in part from the diagenetic alteration of the glass.

ASSOCIATED SEDIMENTS

Lahars

During and subsequent to subaerial paroxysms the loose debris mantling the vents and flanks of volcanoes is in an unstable state and is very prone to downslope movement. Slumps, small-scale folds, erosional surfaces and channelling typify this movement. A wide range of brecciated and tuffaceous deposits collectively known as lahars are the product. These were originally defined as volcanic breccias transported down the slopes of a volcano by water, some of the boulders being as much as a metre in size.

More recent usage would extend the meaning to include volcanic mudflows. The water in subaerial situations originates from caldera lake spillover, from the rapid melting of ice and snow cappings, or can be due to torrential and persistent contemporary rainfall. Whatever the cause, the net result is a chaotic unsorted jumble of volcanic debris deposited very quickly adjacent to the volcano. Speeds of 335 kph have been estimated for certain Peruvian lahars, which are known to have extended down valley for 150 km. The displacement of material can be enormous and can exceed 50 million cu. m. Invariably much damage is caused and, in recent cases, the loss of human life is measurable in hundreds.

Umbers

Umbers are unusual, very fine-grained pelagic sediments composed predominantly of microspherulitic aggregates (framboids) of iron and manganese oxides and hydroxides with small amounts of quartz and clay minerals. They vary in colour from light yellow to deep brown and because of this have been used for centuries as a natural colouring pigment in paints, mortar and other materials. There are well described occurrences in the Middle East, including Cyprus and Turkey.

The Fe_2O_3 content usually varies between 35% and 60% and the MnO_2 between 11% and 23%. Generally they are not regarded as being volcaniclastic, though they are often interbedded with tuffs and can contain epiclastic detritus.

Umber bodies occur as discrete pockets and lenses, up to 40 m thick, though typically less, and several hundreds of metres long resting in hollows on subaqueous pillow lavas within ophiotite complexes. An oceanic ridge-spreading situation is the most likely setting, with water depths probably exceeding 2.5 km, as they are associated with radiolarian cherts. Lamination and occasional slump and scour structures indicate a certain amount of transport for the materials.

The origin of umbers is uncertain but there are suggestions that the metals are leached by sea water penetrating and reacting with hydrothermal fluids in the hot lavas and then being forcefully emitted via submarine springs at temperatures of 7-17°C. The outpouring metalliferous fluids are acid when they emerge from the springs and react with the alkaline sulphate-rich sea water. Within thin clouds and plumes rising to 180m above the bottom sulphides are precipitated which are then gently wafted, laid down and oxidised in suitable adjacent hollows. Whether bacterial agencies play some direct role in the precipitation remains unresolved at present.

DIAGENETIC DERIVATIVES

Many consolidated volcaniclastic rocks, and especially those which are mixtures of volcanic and detrital or calcareous material, have suffered little alteration other than that involved in cementation. There are, however, certain diagenetic derivatives which have not merely been subjected to lithification or recrystallization, but have undergone more or less fundamental chemical changes since deposition. Secondary silicification affects many volcanic ashes, producing extremely hard rocks such as halleflinta. Some of the more remarkable modifications of volcaniclastic deposits result from the devitrification or hydration of pumice tuffs, and a few of the resulting rocks are discussed below.

Bentonite

Bentonite is an argillaceous rock of peculiar character, which when fresh is light green or pale greenish-yellow in colour, and has a fracture rather like that of hard wax. When a small piece is moistened, it swells to many times its original volume before breaking down to a soapy paste. Bentonite from different localities adsorbs water in quantities varying up to about eight times its own original volume; many other substances, such as organic dyes, are also strongly adsorbed. Another striking property of this material is its marked capacity for base-exchange.

Bentonite consists essentially of montmorillonite, or closely allied clay minerals. Most samples also contain small particles of orthoclase, plagioclase and biotite, and some of the accessory minerals of igneous rocks. Thin sections show relics of the characteristic structure of pulverised pumice which suggests that they formed by the devitrification of volcanic ashes in which glass is the principal constituent. The Cretaceous bentonites of Arkansas are interbedded with volcanic tuffs made up of angular fragments of trachyte, and in this case there is no doubt that we are dealing with an altered trachytic glass, which was originally deposited as a pumice-ash. Other bentonites, such as those south of the Black Hills of South Dakota, appear to be derived from andesitic ashes.

In the alteration of volcanic glass to bentonite, devitrification is accompanied by certain chemical changes. The alkalis, which are important constituents in glasses of feldspathic composition, are almost completely removed, and a considerable proportion of silica is also lost. The principal additions are water of hydration, magnesium and iron. Diagenesis is believed to have taken place in contact with sea water.

Although bentonite deposits are usually unfossiliferous, they are almost invariably interstratified with marine sediments, and there can be little doubt that they were laid down in the sea. Individual beds are usually quite thin, varying from about 3 cm to 1 m in thickness, but in spite of this they remain surprisingly uniform in character over enormous areas.

The principal deposits are found in the Cretaceous of western North America, and in the Ordovician of the Appalachian region. Some of the rocks described as fuller's earths in the Silurian, Jurassic and Cretaceous of England and Wales should really be regarded as bentonites; the Lower Cretaceous fuller's earth of Bedfordshire and Surrey, for example, consists principally of montmorillonite and allied clay minerals.

Fuller's Earth

Fuller's earth is an argillaceous rock or clay, which, like bentonite, has remarkably strong powers of adsorption for water, colouring matters, and especially grease and certain oils. For this reason it was formerly employed for cleaning and whitening wool. At present fuller's earths and bentonites are used as thickeners in drilling muds, fillers in rubber and synthetic materials, absorbents and in the pharmaceutical trade.

Typical fuller's earth owes these properties to the presence of montmorillonite, and the ultimate origin of this mineral is probably to be sought in the devitrification of volcanic glass, as in the bentonites. A few deposits of fuller's earth resemble bentonite so closely in their microscopic structure and mineral constitution that they may safely be regarded as water-laid pumiceous tuffs which were devitrified *in situ;* in other cases, this direct origin as autochthonous diagenetic deposits cannot be so confidently assumed, and there is reason for supposing that the material may have been deposited as a montmorillonite-clay, derived from the erosion of bentonitic rocks originally laid down as ashes elsewhere.

Fuller's earth occurs in considerable quantity in the Cretaceous Lower Greensand of Surrey and Bedfordshire and in the Middle Jurassic of western England. Important deposits are also found in the United States, principally amongst the Tertiary and Quaternary sediments of the eastern coastal plain from Florida to New York, in the Mesozoic of South Dakota and Arkansas, and amongst strata of various ages in California.

Tonsteins

Tonsteins are brown, kaolinite-rich compact mudstones with a conchoidal fracture and are mainly found in Upper Devonian to Tertiary coal-bearing successions.

The beds are usually 3-10 cm thick with sharp contacts and are frequently incorporated as layers within coal seams. They can persist laterally over very large areas, much more so than the associated beds such as coals and seat earths. Thus they are useful as marker horizons for correlation purposes.

In thin section the dominant kaolinite can be seen to exist in the form of clear, twisted tabular vermicules, rolled-up masses (boules) and discrete grains, set in a light brown cryptocrystalline kaolinite matrix which is almost isotropic. Degraded plant matter, angular detrital quartz, illite, montmorillonite and a wide range of accessory constituents, including apatite, zircon and opaques, are variably present, though in small amounts. The latter commonly differ in shape and abundance from those in the associated strata.

After many decades of discussion there is general agreement that tonsteins are formed by the in-place alteration of volcanic ash-falls deposited in low salinity acid waters, such as would be expected in low energy terrestrial swamp environments. The ashes appear to have been mostly acid vitric types, probably more susceptible to breakdown into kaolinite than those of a basic nature. The fragmented, poorly sorted texture partly reflects the fabric of the original deposit, though there has been some redistribution and occasional flowage of the kaolinite. Shards are invariably destroyed and this loss may have been biochemically accelerated by the activities of fungi, bacteria and algae, providing that burial by subsequent deposits was very slow. With care the nature of the parent ash can be partly deduced from the distribution of trace elements on the basis of similarity with known ashes.

INDEX

C

D

E

F

G